Technological Oceanography

Technological Oceanography

Editors

Mikhail Emelianov
Georgy I. Shapiro

MDPI

Basel • Beijing • Wuhan • Barcelona • Belgrade • Novi Sad • Cluj • Manchester

Editors
Mikhail Emelianov
Instituto de Ciencias del Mar, CSIC
Barcelona
Spain

Georgy I. Shapiro
University of Plymouth
Plymouth
UK

Editorial Office
MDPI
St. Alban-Anlage 66
4052 Basel, Switzerland

This is a reprint of articles from the Special Issue published online in the open access journal *Journal of Marine Science and Engineering* (ISSN 2077-1312) (available at: https://www.mdpi.com/journal/jmse/special_issues/Technological_Oceanography).

For citation purposes, cite each article independently as indicated on the article page online and as indicated below:

Lastname, A.A.; Lastname, B.B. Article Title. *Journal Name* **Year**, *Volume Number*, Page Range.

ISBN 978-3-7258-0223-4 (Hbk)
ISBN 978-3-7258-0224-1 (PDF)
doi.org/10.3390/books978-3-7258-0224-1

Contents

Journal of
Marine Science and Engineering

MDPI

Editorial

Technological Oceanography

Mikhail Emelianov [1,*] and Georgy I. Shapiro [2]

[1] Instituto de Ciencias del Mar, CSIC, Passeig Marítim de la Barceloneta 37-49, 08003 Barcelona, Spain
[2] School of Biological and Marine Sciences, University of Plymouth, Drake Circus, Plymouth PL4 8AA, UK; g.shapiro@plymouth.ac.uk
* Correspondence: mikhail@icm.csic.es

Advances in our understanding of phenomena in the ocean would not be possible without innovation. Identifying new phenomena, assessing environmental risks, monitoring and forecasting the ocean state, and delivering observational data to modelers are all challenges requiring technological solutions.

This Special Issue of *JMSE*, entitled "Technological Oceanography", aims to present an oceanographic perspective on modern technology, allowing for a better understanding of the ocean in all its diversity and taking into account both societal and scientific needs. This Issue has collected original papers based on applications of novel approaches in the development of new sensors, observational programs, sampling strategies, new ocean modeling techniques and data processing. It is impossible to describe the current state of such a complex and constantly evolving science as technological oceanography in one volume. Thus, the articles presented in this volume are just some examples of modern trends in the development of new technologies in some areas of oceanography.

One of the important aspects of oceanography concerns the long-term measurement of depth profiles of hydrological, biological, chemical, and optical parameters at fixed locations. These in situ measurements are performed with automated tethered profilers, known as Yo-Yo, connected by a flexible cable with bottom anchors or other underwater supports [1]. Autonomous tethered profiling systems that operate in shallow waters often use winches [2]. Alexander G. Ostrovskii et al. (Contribution 1) discusses automated tethered profilers for hydrophysical and bio-optical measurements, advancements which allow researchers to monitor water column parameters from a fixed observational point with high spatial and temporal resolution. These breakthroughs permit the transmission of telematic data in real time, including the results of studies of biogeochemical cycles of carbon transformation throughout the euphotic zone of the sea. The original results of in situ measurements were obtained for the inner continental shelf part of the site using this new tethered profiler. Another innovation is related to marine plankton sampling. Different types and systems have been developed in recent decades to discriminate plankton samples at different strata in water columns [3,4]. Arturo Castellón and Maria Pilar Olivar (Contribution 2) presented a cod-end multisampler design, VERDA. This uses a carrousel-like system, similar to some sediment traps, that works like a revolver with six or eight compartments whose turning mechanism is triggered when the net arrives to a programmed depth level. This device allows research to discriminate plankton samples at different strata in the water column and is useful for all types of ships due to the relatively easy deployment operations.

It is well known that the coastline is susceptible to anthropogenic impact and climate change, particularly the rising sea level and more frequent extreme meteorological events [5]. However, the coastal waters are temporally and spatially undersampled [6] due to the difficulty of establishing oceanographic platforms capable of performing continuous sampling away from the coastline. Another concern is the limited accuracy and spatiotemporal resolution of satellites in the shelf waters. Inmaculada Ortigosa et al. (Contribution 3) described scientific adaptation of a vessel and initial measurements from a "patí a vela",

Citation: Emelianov, M.; Shapiro, G.I. Technological Oceanography. *J. Mar. Sci. Eng.* **2024**, *12*, 175. https://doi.org/10.3390/jmse12010175

Received: 28 November 2023
Accepted: 15 December 2023
Published: 17 January 2024

which is a very popular unipersonal catamaran in Barcelona. This versatile sailing vessel has been adapted to contain several low-cost sensors and instruments to measure water properties. Citizen sampling from fully sustainable sailing boats may turn into an effective strategy to monitor the urban coastal waters.

The effectiveness of laboratory experiments continues to improve, allowing researchers to study the physical mechanisms that produce ocean phenomena in detail [7–9]. Andrey G. Zatsepin et al. (Contribution 4) presented the results of a laboratory experiment conducted to quantitatively investigate turbulent exchange between two quasi-homogeneous layers of equal thickness and different density (salinity), as well as the fine structure of the density transition zone (interface) between the layers. It was shown that maximum mixing efficiency was achieved at a critical Richardson number, when the density interface was in a transitional state between the sharpening and diffusive modes.

One of the important tasks of physical oceanography is the assessment of vertical turbulent exchange in the stratified waters of seas and oceans. In particular, the vertical flux of nutrients from deep waters into the photic layer, where most biological production takes place, depends on the intensity of turbulent exchange [10–12]. Oleg I. Podymov et al. (Contribution 5) presented the analysis of the long-term regular time series of current velocity and conductivity, temperature, and depth (CTD) profiles, measured using the moored autonomous profiler Aqualog over the upper part of the continental slope at a fixed geographical location in the Northeastern Black Sea. This study focused on the fine structure of the density profiles to show that the fine-structured Cox number (C) is a power function of the Richardson number (Ri). The original analysis of the obtained data suggests that the estimations of the vertical turbulent mass exchange could be performed using CTD data only.

Since 2009, a new generation of satellites launched by ESA and NASA have provided radiometric measurements, allowing the retrieval of sea surface salinity [13]. In situ data are crucial for validating and enhancing the quality of satellite products obtained in far oceanic regions, where few such measurements have been taken [14,15]. Marta Umbert et al. (Contribution 6) showed the results of temperature and salinity measurements in the first 60 cm of a water column in the data-scarce areas of the World Ocean. Measurements were obtained during the Vendée Globe, which is the world's most famous solo, non-stop, unassisted sailing race with the CTD MicroCat, installed on the One Ocean One Planet boat. The measurements were performed every 30 s during navigation and this allowed us to obtain data in the sub-Antarctic zone, between the tropical and polar fronts, and it passed through areas of oceanographic interest such as Southern Patagonia (affected by glacier melting), the Brazil–Malvinas confluence, the Southern Pacific Ocean, and the entire Southern Indian Ocean.

Fine-resolution ocean modelling is becoming a ubiquitous practice to resolve important mesoscale and submesoscale features such as eddies, fronts, boundary currents, and localized upwellings, which play important roles in ocean dynamics (e.g., [16,17]). These localized models can be run by relatively small groups due to the availability of excellent ocean models such as ROMS or NEMO to the wider oceanographic community [18,19]. Georgy I. Shapiro and Jose M. Gonzalez-Ondina (Contribution 7) developed a simple and computationally efficient method for creating a high-resolution regional (child) model nested within a coarse-resolution, high-quality data assimilation (parent) model. This method, named Nesting with Downscaling and data Assimilation (NDA), reduces the bias and root-mean-square errors (RMSE) of the child model and does not allow the child model to deviate from reality. The NDA method utilizes data assimilation process in the coarse model as a way to spread information from observation over the whole domain, avoiding the use of simplified and less accurate methods such as spreading via parabolic equation. The NDA method reduces the RMSE, typically by a factor of two to five, but by occasionally more. The method is particularly efficient in areas with sparse observational data.

Due to intrinsic inaccuracies in the model equations and numerical schemes, and due to limitations in the quality of input data streams, even the best ocean models gradually deviate from reality and can only be considered estimates of the true ocean state [20]. The introduction of data assimilation (DA) techniques allows researchers to reduce the deviation of models from the true state, vastly improving the accuracy of ocean forecasting [21]. José M. González-Ondina et al. (Contribution 8) developed a new mathematical method for computing the background and observation error covariance functions and matrices. As data assimilation methods are invaluable tools for operational ocean models, the covariance matrices of background errors (differences between the numerical model and the true values) and the observation errors (differences between true and measured values) are required for correcting ocean model outputs by assimilating observational data using a variational approach. The authors demonstrate that in many cases the new method allows the use of the separable convolution mathematical algorithm to increase the computational speed significantly up to a given order of magnitude.

In recent years, underwater acoustic technology has been widely used to detect underwater targets, describe underwater overviews [22], measure hydrological information, and perform other tasks. Due to the influence of the water surface, water body, water bottom, and other features, sound wave propagation in the ocean presents a complicated distribution state [23]. Xian Ma et al. (Contribution 9) developed a direct solution method for the two-dimensional Helmholtz equation of ocean acoustic propagation without using simplified models. In their study, the authors applied two spectral methods, Chebyshev–Galerkin and Chebyshev–collocation approaches, to correctly solve the two-dimensional Helmholtz model equation. Due to a lack of model constraints, the Chebyshev–collocation method has a wide range of applications and provides results with high accuracy, which is of great significance in the calculation of realistic ocean sound fields.

Ocean acoustic waves propagate over long distances because their energy attenuates to a smaller degree in water than that of electromagnetic waves; consequently, acoustic waves have been widely used in target detection, environmental monitoring, and underwater communication [24]. Wei Liu et al. (Contribution 10) established a vector wavenumber integration (VWI) model to provide benchmark solutions for ocean current study. The depth-separated wave equation was solved using finite difference (FD) methods with second- and fourth-order accuracy, and the sound source singularity in this equation was treated using the matched interface and boundary method. Time-averaged sound intensity (TASI) was calculated using the pressure and particle velocity, and the TASI streamlines were traced to visualize the time-independent energy flow in the acoustic field and better understand the distribution of acoustic transmission loss.

In recent years, dictionary learning [25] has been widely used in the field of sound velocity profile data processing as a feature extraction method. Kaizhuang Yan et al. (Contribution 11) used the dictionary learning method to achieve sparse coding of the high-resolution sound speed profile and used a compressed sparse row method to compress and store the sparse characteristics of the data matrix. Their research revealed how this method significantly reduces the storage capacity of high-resolution sound speed profile data and ensures the accuracy of the data, providing technical support for efficient and convenient access to high-resolution sound speed profiles.

As Guest Editors for this Special Issue, we would like to sincerely thank the authors who have submitted their work to this.

Working with such a bright group of scholars has been an honour and a joy, and we sincerely appreciate their time, energy, and effort. The articles include distinct viewpoints and ideas on the subject matter, encompassing an extensive array of techniques and strategies concerning technological oceanography.

The calibre of the entries as well as the diligence and focus on detail exhibited by every author have impressed us. Because of their contributions, this Special Issue is a special compilation, one that we believe will be of great interest and value to readers in the field.

Author Contributions: Conceptualization, M.E. and G.I.S.; writing—original draft preparation, M.E. and G.I.S.; writing—review and editing, M.E. and G.I.S.; visualization, M.E. and G.I.S.; supervision, M.E. and G.I.S.; project administration, M.E. All authors have read and agreed to the published version of the manuscript.

Funding: This research received no external funding.

Acknowledgments: We would like to thank the reviewers of this Special Issue who help to keep a high academic standard of publications.

Conflicts of Interest: The authors declare no conflict of interest.

List of Contributions

1. Ostrovskii, A.G.; Emelianov, M.V.; Kochetov, O.Y.; Kremenetskiy, V.V.; Shvoev, D.A.; Volkov, S.V.; Zatsepin, A.G.; Korovchinsky, N.M.; Olshanskiy, V.M.; Olchev, A.V. Automated Tethered Profiler for Hydrophysical and Bio-Optical Measurements in the Black Sea Carbon Observational Site. *J. Mar. Sci. Eng.* **2022**, *10*, 322.
2. Castellón, A.; Olivar, M.P. VERDA: A Multisampler Tool for Mesopelagic Nets. *J. Mar. Sci. Eng.* **2023**, *11*, 72.
3. Ortigosa, I.; Bardaji, R.; Carbonell, A.; Carrasco, O.; Castells-Sanabra, M.; Figuerola, R.; Hoareau, N.; Mateu, J.; Piera, J.; Puigdefabregas, J. Barcelona Coastal Monitoring with the "Patí a Vela", a Traditional Sailboat Turned into an Oceanographic Platform. *J. Mar. Sci. Eng.* **2022**, *10*, 591.
4. Zatsepin, A.G.; Valerii V. Gerasimov and Alexander G. Ostrovskii. Laboratory Study of Turbulent Mass Exchange in a Stratified Fluid. *J. Mar. Sci. Eng.* **2022**, *10*, 756.
5. Podymov, O.I.; Zatsepin, A.G.; Ostrovskii, A.G. Fine Structure of Vertical Density Distribution in the Black Sea and Its Relationship with Vertical Turbulent Exchange. *J. Mar. Sci. Eng.* **2023**, *11*, 170.
6. Umbert, M.; Hoareau, N.; Salat, J.; Salvador, J.; Guimbard, S.; Olmedo, E.; Gabarró, C. The Contribution of the Vendée Globe Race to Improved Ocean Surface Information: A Validation of the Remotely Sensed Salinity in the Sub-Antarctic Zone. *J. Mar. Sci. Eng.* **2022**, *10*, 1078.
7. Shapiro, G.I.; Gonzalez-Ondina, J.M. An Efficient Method for Nested High-Resolution Ocean Modelling Incorporating a Data Assimilation Technique. *J. Mar. Sci. Eng.* **2022**, *10*, 432.
8. Gonzalez-Ondina, J.M.; Sampson, L., II; Shapiro, G.I. A Projection Method for the Estimation of Error Covariance Matrices for Variational Data Assimilation in Ocean Modelling. *J. Mar. Sci. Eng.* **2021**, *9*, 1461.
9. Ma, X.; YongxianWang; Zhu, X.; Liu, W.; Lan, Q.; Xiao, W. A Spectral Method for Two-Dimensional Ocean Acoustic Propagation. *J. Mar. Sci. Eng.* **2021**, *9*, 892.
10. Liu, W.; Zhang, L.; Wang, Y.; Cheng, X.; Xiao, W. A Vector Wave number Integration Model of Underwater Acoustic Propagation Based on the Matched Interface and Boundary Method. *J. Mar. Sci. Eng.* **2021**, *9*, 1134.
11. Yan, K.; Wang, Y.; Xiao, W. A New Compression and Storage Method for High-Resolution SSP Data Based-on Dictionary Learning. *J. Mar. Sci. Eng.* **2022**, *10*, 1095.

References

1. Prien, R.D.; Schulz-Bull, D.E. Technical note: GODESS—A profiling mooring in the Gotland Basin. *Ocean. Sci.* **2016**, *12*, 899–907. [CrossRef]
2. Aracri, S.; Borghini, M.; Canesso, D.; Chiggiato, J.; Durante, S.; Schroeder, K.; Sparnocchia, S.; Vetrano, A.; Honda, T.; Kitawaza, Y.; et al. Trials of an autonomous profiling buoy system. *J. Oper. Oceanogr.* **2016**, *9*, s176–s184. [CrossRef]
3. Weikert, H.; John, H.C. Experiences with a modified Bé multiple opening-closing plankton net. *J. Plankton Res.* **1981**, *3*, 167–176. [CrossRef]
4. McNeely, R.L. Development of the John M. Cobb pelagic trawl—A progress report. *Commer. Fish. Rev.* **1963**, *25*, 17–27.
5. Harley, C.D.G.; Hughes, A.R.; Hultgreen, K.M.; Miner, B.G.; Sorte, C.J.B.; Thornber, C.S.; Rodriguez, L.F.; Tomanek, L.; Williams, S.L. The impacts of climate change in coastal marine systems. *Ecol. Lett.* **2006**, *9*, 228–241. [CrossRef] [PubMed]
6. Alverson, K.; Baker, D.J. Taking the pulse of the oceans. *Science* **2006**, *314*, 1657. [CrossRef] [PubMed]

7. Zatsepin, A.G. On peculiarities and similarities of the coherent structure formation in stratified and rotating fluid. In *Turbulent Mixing in Geophysical Flows*; Linden, P.F., Redondo, J.M., Eds.; CIMNE: Barcelona, Spain, 2001; pp. 211–299.
8. Linden, P.F. Mixing across a density interface produced by grid turbulence. *J. Fluid Mech.* **1980**, *100*, 691–703. [CrossRef]
9. Krylov, A.D.; Zatsepin, A.G. Frazil ice formation due to difference in heat and salt exchange across a density interface. *J. Mar. Syst.* **1992**, *3*, 497–506. [CrossRef]
10. Johnson, K.S.; Riser, S.C.; Karl, D.M. Nitrate supply from deep to near-surface waters of the North Pacific subtropical gyre. *Nature* **2010**, *465*, 1062–1065. [CrossRef] [PubMed]
11. Mikaelyan, A.S.; Mosharov, S.A.; Kubryakov, A.A.; Pautova, L.A.; Fedorov, A.; Chasovnikov, V.K. The impact of physical processes on taxonomic composition, distribution and growth of phytoplankton in the open Black Sea. *J. Mar. Sys.* **2020**, *208*, 103368. [CrossRef]
12. Monin, A.S.; Ozmidov, R.V. *Oceanic Turbulence*; Gidrometeoizdat: Saint Petersburg, Russia, 1981; p. 320. (In Russian)
13. Tang, W.; Yueh, S.; Yang, D.; Fore, A.; Hayashi, A.; Lee, T.; Fournier, S.; Holt, B. The potential and challenges of using Soil Moisture Active Passive (SMAP) sea surface salinity to monitor Arctic Ocean freshwater changes. *Remote Sens.* **2018**, *10*, 869. [CrossRef]
14. Fournier, S.; Lee, T.; Wang, X.; Armitage, T.W.; Wang, O.; Fukumori, I.; Kwok, R. Sea surface salinity as a proxy for Arctic Ocean freshwater changes. *J. Geophys. Res. Ocean.* **2020**, *125*, e2020JC016110. [CrossRef]
15. Tang, W.; Yueh, S.H.; Fore, A.G.; Hayashi, A. Validation of A quarius sea surface salinity with in situ measurements from Argo floats and moored buoys. *J. Geophys. Res. Ocean.* **2014**, *119*, 6171–6189. [CrossRef]
16. GFDL. High-Resolution Modeling. 2020. Available online: https://www.gfdl.noaa.gov/high-resolution-modeling (accessed on 6 July 2020).
17. Dufour, C.O.; Griffies, S.; de Souza, G.; Frenger, I.; Morrison, A.; Palter, J.; Sarmiento, J.L.; Galbraith, E.; Dunne, J.P.; Anderson, W.G.; et al. Role of mesoscale eddies in cross-frontal transport of heat and biogeochemical tracers in the Southern Ocean. *J. Phys. Oceanogr.* **2015**, *45*, 3057–3081. [CrossRef]
18. NEMO. Annual Mean of Sea Surface Salinity in 1/12° (NEMO-WRF Coupling). Available online: https://www.nemo-ocean.eu (accessed on 16 June 2021).
19. ROMS. Regional Ocean Modeling System (ROMS). Available online: https://www.myroms.org (accessed on 16 June 2021).
20. Carrassi, A.; Bocquet, M.; Bertino, L.; Evensen, G. Data assimilation in the geosciences: An overview of methods, issues, and perspectives. *Rev. Clim. Chang.* **2018**, *9*, e535. [CrossRef]
21. Lahoz, W.A.; Schneider, P. Data assimilation: Making sense of Earth Observation. *Front. Environ. Sci.* **2014**, *2*, 16. [CrossRef]
22. Erchang, S. Progress of geoacoustic inversion in underwater acoustics. *J. Appl. Acoust.* **2019**, *38*, 468–476.
23. Bhowal, A.; Kshetrimayum, R. Optical Improved Quadrature Spatial Modulation for Cooperative Underwater Wireless Communication under Weak Oceanic Turbulence Conditions. *IET Optoelectron.* **2020**, *14*, 434–439. [CrossRef]
24. Etter, P.C. *Underwater Acoustic Modeling and Simulation*, 5th ed.; CRC Press: Boca Raton, FL, USA, 2018.
25. Tosic, I.; Frossard, P. Dictionary learning. *IEEE Signal Process. Mag.* **2011**, *28*, 27–38. [CrossRef]

Journal of
Marine Science and Engineering

MDPI

Article

Automated Tethered Profiler for Hydrophysical and Bio-Optical Measurements in the Black Sea Carbon Observational Site

Alexander G. Ostrovskii [1,*], Mikhail V. Emelianov [2], Oleg Y. Kochetov [1], Vyacheslav V. Kremenetskiy [1], Dmitry A. Shvoev [1], Sergey V. Volkov [1,3], Andrey G. Zatsepin [1], Nikolai M. Korovchinsky [3], Vladimir M. Olshanskiy [3] and Alexander V. Olchev [1,4]

1 Shirshov Institute of Oceanology, Russian Academy of Sciences, Nakhimovskiy Prospect 36, 117997 Moscow, Russia; realspinner@gmail.com (O.Y.K.); sk@ocean.ru (V.V.K.); dshv@ocean.ru (D.A.S.); mendur@outlook.com (S.V.V.); zatsepin@ocean.ru (A.G.Z.); aoltche@gmail.com (A.V.O.)
2 Institut de Ciències del Mar, CSIC, Passeig Marítim de la Barceloneta, 08003 Barcelona, Spain; mikhailivanolga@gmail.com
3 Severtsov Institute of Ecology and Evolution, Russian Academy of Sciences, Leninskiy Prospect 33, 119071 Moscow, Russia; nmkor@yandex.ru (N.M.K.); vmolsh@yandex.ru (V.M.O.)
4 Department of Meteorology and Climatology, Faculty of Geography, Lomonosov Moscow State University, 119991 Moscow, Russia
* Correspondence: osasha@ocean.ru

Abstract: Biogeochemical cycles of carbon transformation throughout the euphotic zone of the sea are controlled by physical processes, e.g., daily thermocline, variation in solar irradiance, thermohaline convection, and intermittent mixing. These processes should be regularly observed with sufficient time resolution at fixed geographical locations. This study provides a brief overview of the carbon observational site in the Northeastern Black Sea. The focus is on the design of a new tethered profiler Winchi for the inner continental shelf part of the site. The profiler hull and two outriggers comprise an open trimaran platform that is positively buoyant and tends to maintain a horizontal position in the water. The lower end of the winch wire is secured to the bottom anchor. By unwinding/winding the wire, the profiler ascends/descends while measuring the depth profiles of marine environment parameters ranging from the seafloor to air–sea interface. After surfacing, the profiler determines its location using the Global Positioning System (GPS) and transmits data to (and from) a server on land through the Global System for Mobile Communications (GSM). Initial field tests with the Winchi profiler at the Northeastern Black Sea shelf exhibited promising results. We report these early tests to demonstrate the use of Winchi.

Keywords: greenhouse gases; carbon flux; tethered profiler; winch; real-time data transmission; Black Sea

Citation: Ostrovskii, A.G.; Emelianov, M.V.; Kochetov, O.Y.; Kremenetskiy, V.V.; Shvoev, D.A.; Volkov, S.V.; Zatsepin, A.G.; Korovchinsky, N.M.; Olshanskiy, V.M.; Olchev, A.V. Automated Tethered Profiler for Hydrophysical and Bio-Optical Measurements in the Black Sea Carbon Observational Site. *J. Mar. Sci. Eng.* **2022**, *10*, 322. https://doi.org/10.3390/jmse10030322

Academic Editor: Weidong Zhai

Received: 12 January 2022
Accepted: 22 February 2022
Published: 25 February 2022

Publisher's Note: MDPI stays neutral with regard to jurisdictional claims in published maps and institutional affiliations.

1. Introduction

This study helps establish the monitoring network based on the observation of the processes of formation, accumulation, emission, and deposit of primary greenhouse gases (GHGs) in various ecosystems over Russia [1]. The research activity at the carbon observational site in the Gelendzhik area (Figure 1), located at the northeastern coast of the Black Sea, focuses on the development of infrastructure and technology for measurements of GHG fluxes over marine and coastal areas. The sea shelf part of the carbon observational site adjacent to Blue Bay of the Black Sea extends a bit beyond the shelf edge (the isobath of ~250 m), 4 to 5 nautical miles from the sea shore. The area near Blue Bay houses the Southern Branch of the Shirshov Institute of Oceanology of Russian Academy of Sciences (SIO RAS; https://ocean.ru/en/index.php/otdeleniya-i-filialy-io-ran/yuzhnoe-otdelenie, accessed on 22 November 2021), where several oceanographic instruments were designed and tested during 1950–1990. Our team has used this base while working on R&D projects for the moored automatic mobile profilers since 2005 [2–6]. For an overview of the Black Sea physical and biogeochemical characteristics, see References [7,8].

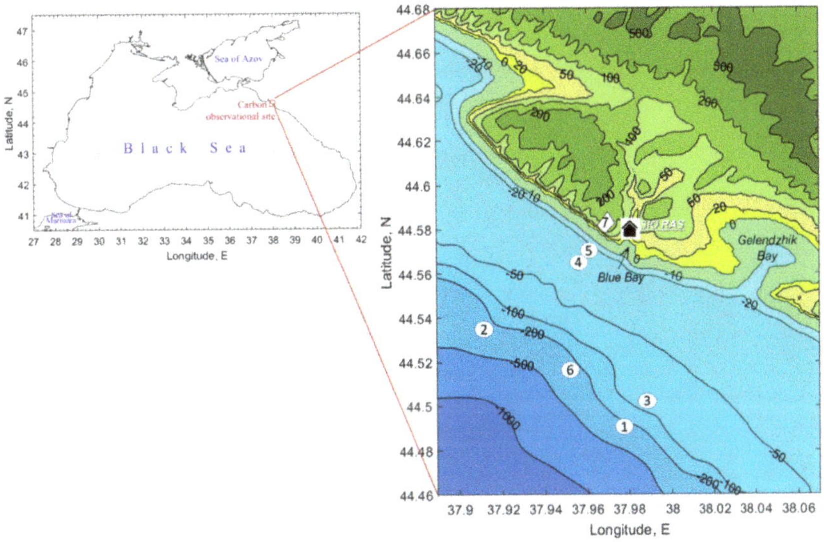

Figure 1. Left—The Black Sea coastline and location of the carbon observational site. **Right**—Bathymetry of the Northeastern Black Sea area near the carbon observational site. Bathymetry is based on the ETOPO1 Global Relief Model (shading (m)); (https://www.ngdc.noaa.gov/mgg/global/, last access: 24 September 2021). Numbers in the circles indicate observation instruments deployed or planned for deployment as follows: (1)—moored automatic mobile profiler for monitoring of the dissolved oxygen; (2)—moored automatic mobile profiler for environmental monitoring; (3)—mooring buoy for the monitoring of GNGs; (4)—shallow-water tethered profiler Winchi (presented in this study) to observe the environment; (5)—bottom mounted wave and current meter connected by an optical cable to the coast; (6)—subsurface mooring for suspended matter and carbon flux monitoring; and (7)—the terrestrial part of the observational site to observe the GHGs, carbon flux, wind, and weather.

An automated tethered profiler is required to frequently measure the hydrophysical control of the carbon system in the shelf water from the bottom part to the air–sea interface at a geographically fixed location. Although this study focuses on the research and development of the profiler, the organization of the carbon observational site in the Black Sea is beyond the scope of the study. While profiler mooring should provide information on time variations of water column properties with a sufficiently high temporal resolution, water sampling will be carried out one or two times per month over a wide area to provide a spatial perspective on the processes under investigation. Other activities include the chemical analysis of seawater samples for dissolved and particulate organic and inorganic carbon and organic and inorganic nitrogen and observations of GHG fluxes over the sea–land interface zone using the eddy covariance and chamber (mobile and stationary) methods based on the data obtained from surveys on the research vessel and the buoy mooring data. Available satellite observations are used for data interpretation. Hybrid process-oriented and statistical (machine learning) models have been developed to retrieve carbon stocks and fluxes from remote sensing data. This helps better understand the carbon cycle in the sea–atmosphere system, which is necessary to evaluate the GHG balance using numerical models. The development of technologies to increase the GHG uptake based on terrestrial and marine ecosystems is one of the key activities being conducted at the carbon experimental site. This research program should provide an opportunity for high-level training with regard to the observation and analysis of experimental data based on GHG fluxes.

In oceanography, automated tethered profilers, known as Yo-Yo, connected by a flexible cable with bottom anchors or other underwater supports (see, for review, Reference [9]) have been used for long-term measurements of depth profiles of hydrological, biological, chemical, and optical parameters at fixed locations, e.g., in the Baltic Sea [10] and the Arctic [11]. These profilers are often equipped with several underwater sensors to assess aquatic ecosystems. Tethered profilers are usually parked at the bottom and can ascend at a speed of 0.1 m s^{-1} and take measurements at a high-vertical resolution of 0.01–0.1 m, which is sufficient to resolve issues of fine-scale structures, e.g., thermohaline intrusions, sharp vertical gradients, boundary layers, and others.

Tethered profilers may obtain data on stratification at depths shallower than 1 m, which is useful to estimate the transfer across the air–water interface and verify satellite remote sensing data. Surfacing tethered shuttle buoys, known as coastal surface piercing profilers (CSPPs), are vulnerable to sea currents and internal and surface waves. For instance, at the inshore sites of the Ocean Observatories Initiative, CSPPs can be deployed between April and September, as the wave climate precludes deployment between October and March [12]. However, surfacing allows the profiler to transmit measurement data via radio and satellite links, which are essential for operational oceanography.

Autonomous tethered profiling systems that operate in shallow waters often use winches [13]. Few winch-type profiling systems include at least two electronic control devices: one at the winch and the second one at the movable floating carrier with an environmental sensor suite, which includes a pressure sensor. Both controllers interact via a cable, because information regarding the submersion depth of the sensor carrier is crucial for the winch operation. In such cases, the underwater winch should be equipped with a waterproof slip ring or rotary joint. The disadvantages of this design are complicated maintenance and high cost.

Unlike the systems that include two units connected by an underwater electrical cable, there were tethered profilers with all the modules (buoyancy, winch, battery pack, electronic control module, environmental sensor suit, and communication module) mounted on a single open chassis. A synthetic wire was wound around the profiler winch spool, and the second end of the rope was fixed to the bottom anchor. Such a profiler exhibited positive buoyancy; therefore, the winding/unwinding of the synthetic wire allowed it to descend/ascend. The most enhanced profilers of this type were custom-made underwater winches or automatic elevator system type 3 by Nichiyu Giken Kogyo Co., Ltd., Tokyo, Japan (https://www.nichigi.co.jp/en/en_products/en_nigkocean.html, accessed on 22 November 2021) and Thetis by WetLabs, Inc., Philomath, USA (production discontinued). The chassis structure of the profiler was designed to mount modules of various sizes; a lot of space within the chassis remained unused owing to large dimensions of the devices, e.g., the Thetis profiler had a diameter of 0.8 m, height of 2.1 m, and weight of 100–160 kg. The primary disadvantage of this design is a high hydrodynamic drag. Hence, to operate in a flow, it was necessary to increase the buoyancy, which led to an increase in the energy consumption to submerge the apparatus. These profilers could be deployed and recovered using A-frames and cranes from ships, which were not readily available for low-budget studies, especially in lakes and reservoirs. Hence, there is a need for a compact and lightweight profiler that can be deployed from a small boat. A prototype of such an autonomous profiling measurement system for the ocean surface boundary layer (APMOS) was presented by Reference [14]. Unfortunately, there have been no reports on the further development of this device.

This study presents the development of a small, tethered winch-type profiler suitable for deployment in a simple rowboat. This development is based on the invention of an underwater winch probe [15] in which a spool is aligned with the motor drive shaft. A synthetic wire is wound around the spool, which does not rotate. The roller supporting the synthetic wire rotates around the spool. The motor drive helps rotate the roller and reciprocates the motion of the spool, which ensures the uniform placement of the synthetic wire when it is wound around the spool. Based on this invention, a new device, Winchi, was developed and tested in lake water and sea shelf water, as described below.

2. Development of the Profiler

2.1. Design

The development focused on the profiler layout of the self-sufficient winch-type tethered device, where the motor drive, controller, and battery pack are placed in pressure-proof cylinder housing. The other end of the winch wire was secured to the bottom anchor or subsurface buoyancy. Originally, the positive buoyant profiler was designed to operate in an upright position. Hence, environmental sensors and positioning communication antennas were mounted on the upper-end cap of the profiler housing, as in an Argo ocean float, whereas the winch spool was located in the lower part of the profiler hull (Figure 2). Upon moving to a predefined parking depth controlled by a pressure sensor, the motor drive rotation was reversed, changing the direction of winding/unwinding. By winding and unwinding the spool wire, the profiler moved up and down in the water to measure the depth of the aquatic parameters using environmental sensors. For another application with regard to under-ice operations, the profiler was assumed to be negative in terms of the buoyancy and had to be turned upside down to hang down from the surface ice with the sensors facing downward.

Figure 2. Prototype of the automated tethered profiler (**left**). The prototype profiler pierces the sea surface (**right**). The orange collar is the stabilizing disk. RBRconcerto C.T.D++ is mounted at the upper-end cup.

The near-surface horizontal flow generates a hydrodynamic drag on the profiler. The hydrodynamic drag force is proportional to the square of the flow speed and the cross-sectional area of the profiler. In the absence of surface waves, a tethered profiler connected to the bottom anchor using a synthetic low-stretch wire drifts with the flow, while the wire is tensioned. At a high flow velocity, when the flow around the profiler becomes turbulent, the positive buoyant profiler may not reach the surface or can quickly be pulled back under the water after surfacing owing to the hydrodynamic drag and wire tension force. Field trials exhibited that the profiler in the upright position becomes unstable at the sea surface and submerges owing to the current, which has a high rate of 0.3 to 0.4 m/s.

The hydrodynamic drag force can be reduced by maintaining the horizontal position of the profiler in the water flow, such that the longitudinal axis of the profiler cylindrical hull is parallel to the oncoming flow. Therefore, we have redesigned the profiler to comply with the aforementioned requirements. This profiler is known as Winchi (Figure 3). A suspension bar is designed to maintain the horizontal position of the profiler. Additionally, the profiler is balanced. A polyethylene suspension bar is attached to the profiler hull using two claps. There are two outrigger floats, which are fixed at the structural portions of the clamps to generate buoyancy and lift the GPS and GNSS communication antennas out of the water, thereby mitigating the effect of the waves while the profiler is surfaced. The main hull and two outriggers include the open trimaran platform of the profiler.

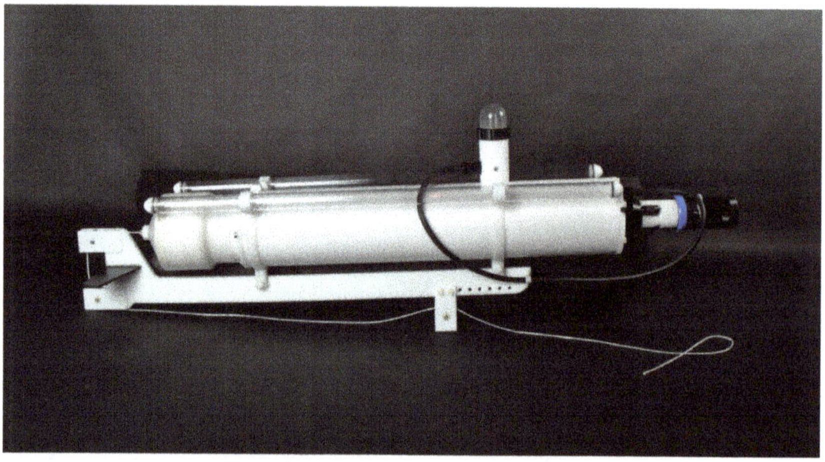

Figure 3. Automated tethered profiler—Winchi—with the suspension bar; GPS/GSM module; and conductivity, temperature, and depth probe RBRconcerto C.T.D++. The housings are made of polyacetal POM-C. The suspension bar is made of ultra-high molecular weight polyethylene. Photo from 2019.

For the profiler cylindrical hull in the horizontal position, the winch is located at the rear part of the profiler. The Dyneema® synthetic low-stretch wire (Figure 4) connects the spool around the bypass rollers at the tail of the suspension bar and the roller below the cylindrical hull located at a distance of approximately 1/3 of the length of the hull (Figure 5). There are vertical keels and horizontal stabilizers at the tail that contribute to the longitudinal and directional stabilization and control of the profiler in the flow. The hydrodynamic lift force acting on the horizontal stabilizer helps maintain the horizontal position of the profiler.

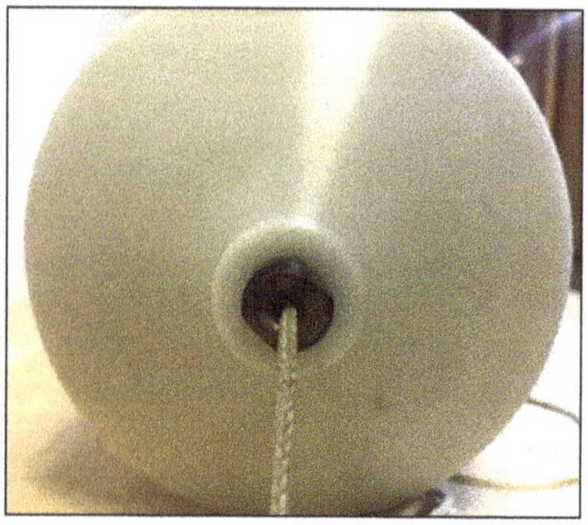

Figure 4. Back view of the profiler wet compartment that houses the winch. The winch wire passes through hawsehole of the wet compartment.

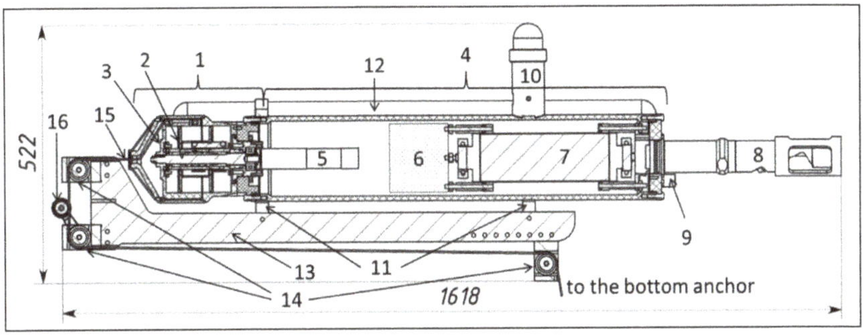

Figure 5. Layout of the tethered profiler Winchi: (1)—wet compartment for the winch spool (2) and line rollers mounted at the rotor (3), (4)—pressure-proof housing to accommodate the motor drive (5), microprocessor control unit (6) and battery (7), (8)—environmental sensor suite (see Section 2.2 below), (9)—connector, (10)—GPS/GSM module (see Section 2.3 below), (11)—clamps, (12)—outrigger, (13)—suspension bar, (14)—bypass rollers, (15)—wire, and (16)—wire tension sensor.

The Winchi profiler includes a wet compartment and pressure-proof housing (Figure 5). In the wet compartment, stationary oscillating spool and line rollers rotate around the spool. A line roller is mounted on the rotor, which, in turn, is mounted on the shaft of the motor drive. The line roller revolves around the stationary spool, which reciprocates back and forth, winding a line on the spool. The reciprocation is controlled by a mechanism that works with the rotation of the motor shaft. On the winch spool, a wire of 60 m in length and a diameter of 2.5–3 mm can be placed. The Dyneema® core of the wire has a working stretch of <1%, while the UV stable core cover protects against abrasions and generates the lowest friction in the rollers.

The pressure-proof housing includes a 24-V DC geared motor (Figure 6), a rechargeable battery pack 930 W h assembled with Nickel Manganese Cobalt cells (Figure 7), and a microprocessor control unit (MCU). The motor driver is configured to power up to 300 W. The profiler hardware is based on a 32-bit ARM microcontroller.

The profiler enters sleep mode between profiling cycles to conserve energy. For accurate time synchronization in sleep mode, the MCU uses a ±5 ppm clock generator with a consumption of up to 1 uA.

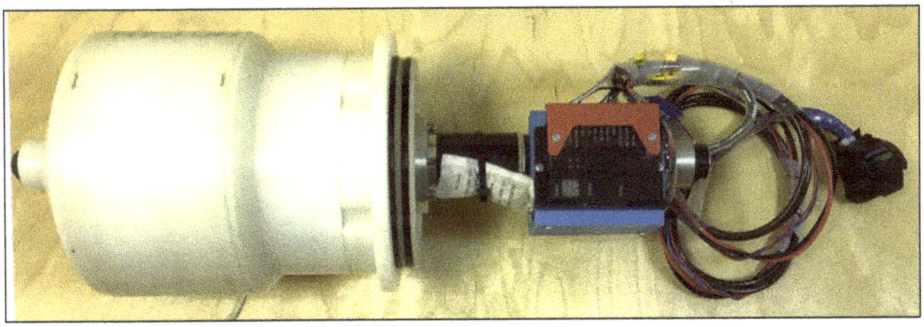

Figure 6. The winch mounted on the DC geared motor with a motor speed controller.

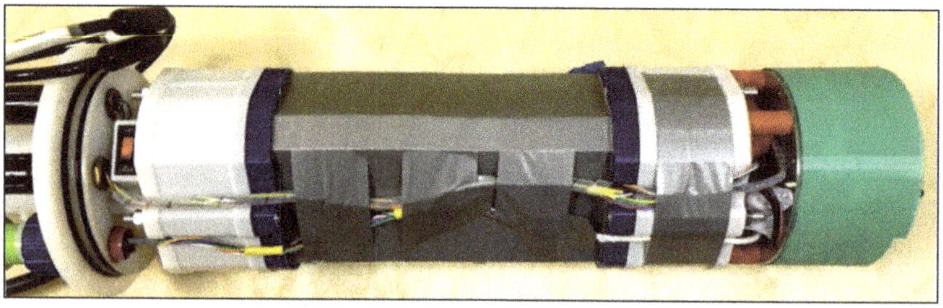

Figure 7. Battery and MCU that are mounted on the housing end cap. The MCU is protected by a plastic shield (green colored). The rechargeable battery with a protection circuit board is covered by thermal adhesive tape.

External devices can be connected via four serial ports of UART equipped with converters to the standard RS232, allowing data exchange at 90 kbps (maximum). A 24-bit ADC is used at each of the four analog channels to digitize signals of 0–5 V at 100 Hz (maximum). Two I2C buses with power amplifiers are used to connect low-power digital sensors. The telemetry and measurement data are stored in the SD card supporting the SD input/output (SDIO) specifications. At a short distance, the exchange of the command and data is carried out via Bluetooth. The environmental sensor suite is mounted on the profiler side, opposite to the wet compartment. Connectors are installed at the housing end cap. One connector can be used for an additional sensor. The other one is allocated to connect to the communication module.

The embedded software is an RTOS-based modular framework. Its functionality can be expanded by adding a custom code to accommodate specific functions of the profiler. The diagnostic and programming software is implemented as a cross-platform desktop GUI application, which can be expanded to match additional features.

The overall design of the profiler is compact. One technician of an inflatable boat can deploy and recover the profiler owing to its light weight (~30 kg) and small dimensions (length of 1620 mm and diameter of 170 mm).

2.2. Sensor Suit

The Global Ocean Observing System (GOOS) defines essential ocean variables, which are ancillary to ecosystem observations, to resolve the relevant stratifications as follows: temperature, salinity, dissolved oxygen, chlorophyll, backscatter, and pressure or depth of the measurements [16]. This set of parameters defines the backbone of the core biogeochemical and physical measurements. Hence, the Winchi profiler sensor suite featured the following: RBRconcerto C.T.D++ logger by RBR Ltd. (Ottawa, Canada) with a Seapoint Chlorophyll Fluorometer (SCF) and Turbidity Meter and Oxygen Optode 4831F of Aanderaa Data Instruments AS (Bergen, Norway) (Figure 8). The time constants of the temperature and pressure sensors are less than 0.01 and 1 s, respectively. For profiling, fast sensors are used; the dissolved oxygen (O_2) sensor response time (63%) with fast response foil is less than 8 s. The time constants of fluorometer and turbidimeter are about 0.1 s. The conductivity, temperature, and depth (CTD) data sets are obtained at an interval of 0.5 s. Considering 0.25 m s^{-1} for the profiler vertical movement speed, the reliable vertical resolution for the temperature profile is ~0.13 m. Dissolved oxygen, chlorophyll, and turbidity data are obtained at an interval of 1 s.

Figure 8. Sensor suite of Winchi: (1)—pressure-proof housing with end cap, (2)—CTD probe of logger RBRconcerto C.T.D++, (3)—Seapoint Chlorophyll Fluorometer (SCF), (4)—Seapoint Turbidity Meter, (5)— Aanderaa 4831F dissolved oxygen sensor, and (6)—pressure sensor (this port can be used for a custom sensor). It is assumed that this backbone of core biogeochemical and physical measurement instruments will be supplemented by pH, CO_2, and CH_4 sensors in the future.

The control electronics provides four serial ports with independent 12-V power supply lines and four 24-bit ADCs for analog sensors with independent 5-V and 12-V power supply lines. There are also two I2C bus ports with 3.3-V power boosters for digital sensors. The current configuration of the sensors can be expanded by 2 sensors with serial ports, 3 analog sensors, and a reasonably sized chains of digital sensors at I2C buses. The power demand of these extra sensors ultimately depends on their specifications. However, taking into account the power draw of the winch motor in motion, especially in the bottom-bound direction, it would be fair to estimate the decrease of the operation time as 10–15%.

In the future, O_2 sensors should be complemented with pH and pCO_2 sensors. However, implementation depends on the availability of sensors to operate stably in sea or land waters during long-term missions. As emphasized by Reference [17], although O_2 optodes are widely used in oceanography, pH and CO_2 optodes are still at the nascent level.

2.3. Communication and Positioning

In addition to GSM communication and GPS positioning devices, the external communication module includes a sensor to detect the air–sea interface and a flashing LED beacon to search for the device (Figure 9). The communication module is placed in a small, sealed housing with a transparent cap at the top. In this case, printed circuit boards are installed. An air–sea interface sensor and cable connector are installed in the lower part of the housing.

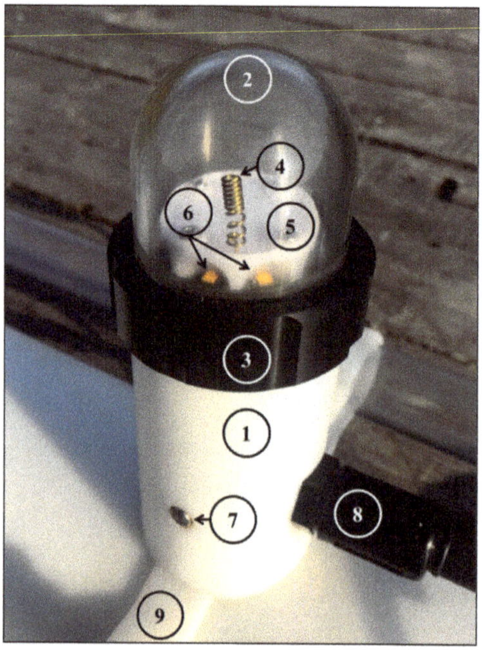

Figure 9. GPS/GSM module of the profiler—Winchi: (1)—pressure-proof housing, (2)—end cap made of acrylic, (3)—cap nut, (4)—GSM antenna (spiral), (5)—light reflector (white plastic caved cover), (6)—LED flashlights (only 2 pieces of the total 8 pieces are shown), (7)—air–sea interface sensor, (8)—data I/O and power connector, and (9)—mounting clamp. The antenna of the GNSS navigation receiver is located behind the light reflector (not shown).

The data transfer speed highly depends on the type of cellular network connection (LTE/3G/GSM) and signal strength. In the best-case scenario, the speed is about 38,400 baud. The limiting factor is the baud rate of the serial connection between the MCU and GSM communication modules.

In order to operate in real time, Winchi attempts to transfer data each time it reaches the surface. If the communication conditions are not favorable, Winchi puts the current data set transfer request into the queue for later transmission on the next surfacing. As the data files are stored on a SD card, the transfer queue holds only file names and counts of successfully transferred bytes, so its memory footprint is rather small. The most recent data sets are always transferred first. Furthermore, the user can send a remote operation command to transfer a specific file at any time during the mission. Such a command generates a corresponding data transfer request and puts it at the end of the queue. In normal communication conditions, the transfer of one data set takes about 2 to 3 min.

2.4. *Cyclogram of the Winchi Profiler's Operation*

The profiling schedule and control parameters, e.g., motor power rate, can be configured using Bluetooth before deployment, e.g., boarding a boat. In water, the device is operated using the MCU, and it repeats all operations multiple times, including sinking and rising. However, certain parameters can be configured remotely using a GSM link. The parameters are GSM communication timeout, LED flashing, duration of the pressure measurement in parking mode, and threshold height of the sea surface wave, which would be unsafe for profiling. A high-level description of the instructions to complete the profiler mission is defined step by step.

1. Program the device to operate: set the mission parameters. Upload the mission schedule. Set a timeout prior to the initiation of the mission (up to 30 min (by default)).
2. Start mission: Initiate automatic operation modes. Switch on the LED beacon. The wire is released to ascend to consider the starting position at the sea surface. Then, the external sensors are turned off.
3. Profiling cycle
   (a) Keep the profiler floating on the sea surface—turn on the SurfHold function. When the device submerges for more than 15 s owing to wave action or current, SurfHold switches on the motor to spool more wiring to ensure that the air–sea interface sensor remains above the waterline. As soon as the air–sea interface sensor rises above the waterline or the wire tension drops below a predefined threshold, SurfHold cuts off the motor power.
   (b) Navigate—Acquire a GPS position.
   (c) Communicate—Transmit the telemetry and measurement data and receive control commands. Turn off SurfHold after the communication is accomplished or timed out.
   (d) Count the profiling cycles—If the predefined maximum number of cycles is reached, the device switches to the user direct control mode with the transmission of telemetry every 15 min. Otherwise, a new cycle is initiated.
   (e) Descend—Switch on the external sensors and motor to wind the wire.
   (f) Save energy—Turn off the LED beacon while descending below a depth of 2 m.
   (g) Park—Turn off the sensors after descending to the predefined parking depth.
   (h) Sleep—Set the MCU to sleep mode. The external pressure is measured every 5 min. If the device depth is less than 2 m, cancel the automatic operation mode, set the error code to WireBreak, switch to the user direct control mode, acquire the GPS position, and transmit the alarm data every 15 min.
   (i) Wake up—Begin the measurement of the external pressure (the sampling is done every 0.1 s (by default)). Collect data on the pressure fluctuations (for 1 min (by default)) and estimate the surface wave height. If the wave height exceeds the threshold height of the sea surface wave (0.8 m by default), then the MCU program moves to step 3 h of this cyclogram, i.e., it turns to another sleep period.
   (j) Ascend—Switch on the sensors. The system conditions are checked. Switch on the motor to spool the wire
   (k) Display—Turn on the LED beacon above a depth of 2 m (the activation of the LED beacon can be canceled remotely using a user command during the next communication session).
   (l) Surfacing—Slow down the motor at a depth of 1.5 m. Check the air–sea interface sensor every 0.2 s. The MCU program has initiated at the beginning of the profiling cycle.
4. End of the mission—The device stays on the sea surface or can submerge to the specified depth for a user-defined period of time based on a user command received via the GSM link.

Figure 10 shows a chart of the profiling cycle track. As the profiler was positively buoyant, its vertical speed was 0.33 m/s while ascending and 0.25 m/s while descending.

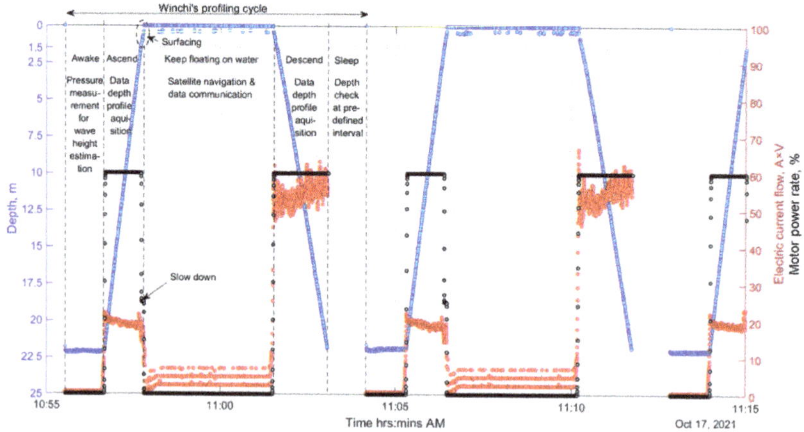

Figure 10. Typical profiling track (profiler depth versus time) and energy consumption under the conditions of a weak sea current less than 0.3 m/s. Blue circles indicate the depth of the profiler during deployment. Energy consumption is denoted by red dots. The black circles indicate regimes of the winch motor operation, i.e., the motor power rates are 60% when the profiler ascends and descends, 25% when the ascend speed slows down in the sea surface layer, and 0% when the profiler is on the sea surface. The microelectronic control unit of the system is switched to sleep mode when it is parked on the near-bottom layer.

If the profiler's battery is drained to the critical level (which can be programmed prior to the mission start or set via remote operation command), it aborts the mission and tries to ascend immediately. At the surface, Winchi sends its telemetry with a low battery warning, location coordinates, and Mission Aborted status. Then, the device continues to send its telemetry every 15 min while waiting for a remote operation command. The user has the option to issue a command to dive to a certain depth until a specified date and time, which would be proper if an immediate retrieval of the device is not possible. After receiving this command, the profiler will sleep safely underwater until the support boat arrives.

If the profiler unexpectedly rises above the sea surface, then it determines its GPS position, sends notification, including its geographical coordinates via the communication module, and starts to flash. Such S.O.S. messages are transmitted every 15 min.

3. Trials

The trials of the profiler Winchi were carried out at Lake Glubokoe (Moscow region) and at the SIO RAS carbon observational site on the Northeastern Black Sea shelf. For sea trials, the profiler mooring system was deployed at 25–30 m (location depicted as circle mark No. 4, as shown in Figure 1). The overall design of the mooring is shown in Figure 11. Anchor, acoustic release, and recovery flotation were used in the lower part of the profiler mooring system to recover the mooring. A profiler winch wire was attached to the top of the recovery float. If the data very close to the bottom layer were wanted to be obtained, the U-style mooring could be used.. In a U-style mooring system, a taut line runs from the tethered profiler anchor to the second anchor (located at a distance approximately equal to the depth of the deployment site) which is joined by another line to a surface buoy.

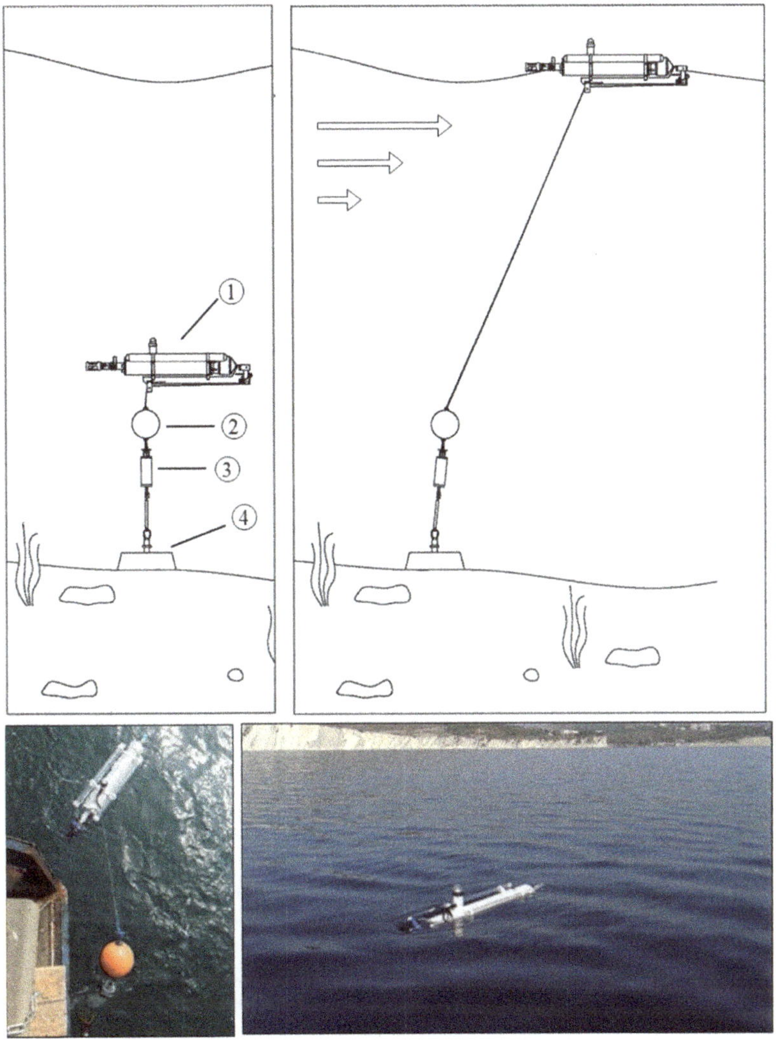

Figure 11. **Upper part**—Schematic of the profiler mooring system. **Top left**—The profiler at the parking position: (1)—the profiler Winchi, (2)—flotation, (3)—acoustic release, and (4)—anchor. **Top right**—The profiler after ascent to the sea surface. **Lower part**—Sea trials near Blue Bay in the Northeastern Black Sea. **Bottom left**—Deployment of the profiler Winchi mooring system. **Bottom right**—The profiler on the sea surface.

The profiler was configured in advance for automatic operation. The measurement depths, intervals, and other settings could be modified based on GSM communication every time the profiler surfaced. A sample plot of the profiler track is shown in Figure 12. To check the operability of the profiler, the device was scheduled to generate a few fast cycles at the beginning of the mission; therefore, descent and ascent were carried out without long stops at the parking depth. Subsequently, the device automatically switched to a profiling interval of 2 h. After 4 days, the profiler received a command from the operator to change the profiling interval to 1 h. The next day, this setting was changed (Figure 12).

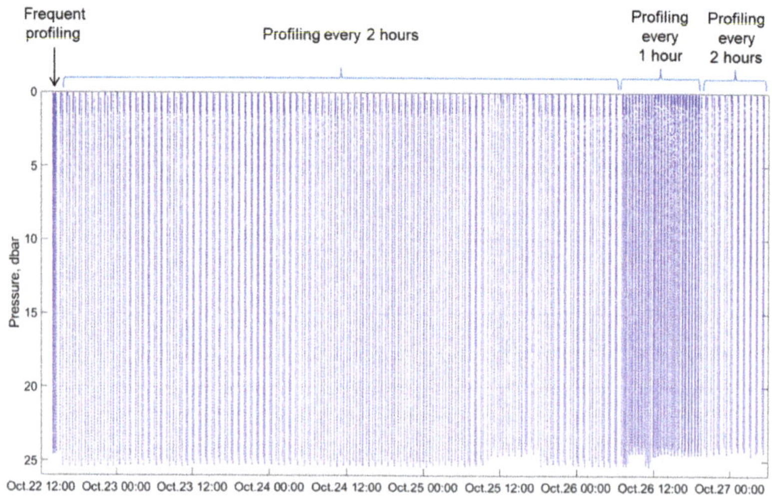

Figure 12. Time–depth plot of a typical track of the profiler. Data were acquired during the trials in the shelf area near Blue Bay in the Northeastern Black Sea in October 2019. Tick marks along the horizontal axis are placed at the Coordinated Universal Time (UTC).

As an example of data that can be obtained using Winchi, Figure 13 shows the distributions of the temperature and chlorophyll-a for the same period as illustrated in Figure 12. The profiling allowed us to observe temporal variations in the vertical stratification. The temperature measurements (upper panel of Figure 13) exhibited a water mass up to 19.5–20.5 °C on the upper layer (15–20 m) of the shelf water. The narrow thermocline separated the warm water mass from the near-bottom layer, which was colder than that at 13 °C. In October, strong stratification can be supported by the along-shelf advection of light, warm water from the southeastern part of the sea and upwelling events, which led to the propagation of dense, cold water from the deep part of the sea above the continental slope and over the shelf [18].

The warmer upper layer was rather poor in chlorophyll-a (Figure 13, lower panel). The chlorophyll-a concentration increased to 2–2.5 μg/L near the sea bottom, where the water was thought to be nutrient-rich. The observed maximum of chlorophyll-a below 15 m in depth is a characteristic feature of the Black Sea ecosystem [19–21]. The clear-cut diurnal cycle of the chlorophyll-a measurement data obtained by the SCF sensor in the upper 10-m layer (lower panel of Figure 13) should be treated with caution. A decrease in daytime indicates a period of nonphotochemical quenching under continuous light, particularly at high intensity, and the maximum fluorescence emission settles down to a lower value compared with the value measured shortly after a long dark period [22]. This period occurred when the solar irradiance increased and a diurnal warm layer developed. The timing of the diurnal minimum coincided with that of the maximum temperature in the surface layer.

In the short-period band, vertical displacements of the thermocline at a depth range of 15–25 m were irregular. However, few of these displacements could be associated with the internal waves of periods of 10 min, which were observed using temperature sensor chains [23]. Intermittent high-frequency waves are aliased in the Winchi profiler observations, which help sampling low-frequency dynamics, e.g., near-inertial waves.

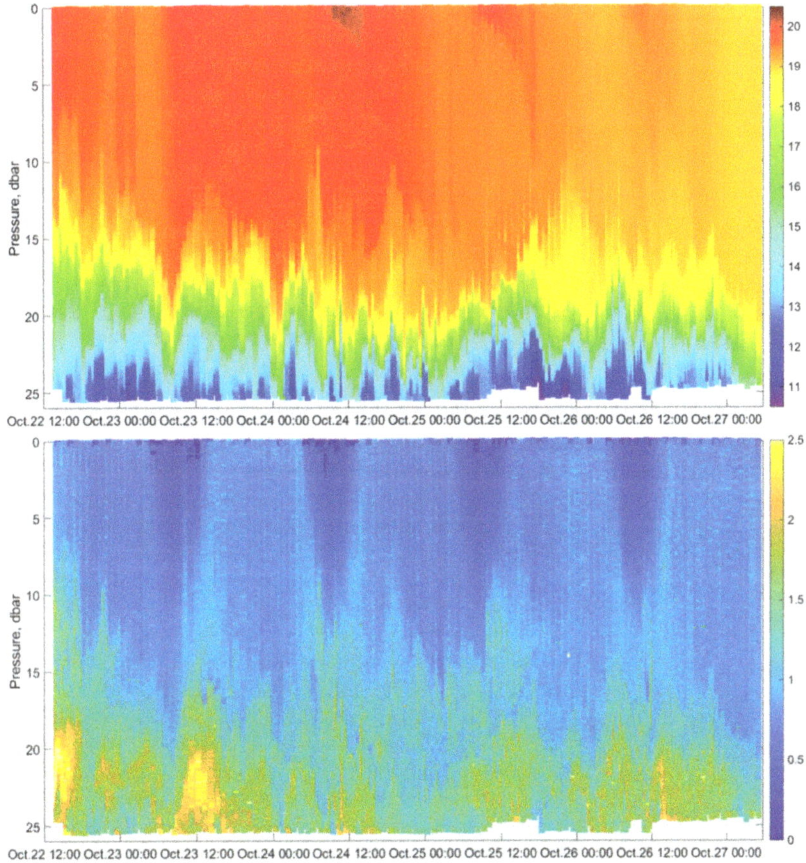

Figure 13. Profiles of the temperature (**upper panel**) and fluorescence-based chlorophyll-a concentrations (**lower panel**) depicted in color for the same period as in Figure 11.

The profiler sensor can be configured to operate when it ascends and/or descends. The sampling frequency of the CTD probe is set to 0.5 Hz, and the sampling frequencies of the dissolved oxygen, fluorescence, and turbidity sensors are set to 1 Hz. Owing to a sensor response time longer than the sampling period, there can be differences between the data obtained from the ascending leg and those from the descending leg. Such differences should be addressed during post-deployment data analyses.

4. Discussion and Conclusions

The SIO RAS carbon observation site near Blue Bay in the Black Sea includes devices to observe the marine environment. To obtain a long time series of data in the deep part of the site, two subsurface moorings have been installed. The subsurface moored profiler Aqualog stores the observation data in the internal memory for a retrospective study after the recovery of the mooring. Examples of such analyses can be found in References [6,24]. In future, these systems should be complemented by sediment traps installed at fixed depths below the profiler's operational depth to assess the composition of the sinking material in the deep layers of the Black Sea. However, the Aqualog profiler cannot measure through the sea surface layer, because its profiling range is limited to the top by subsurface floatation, which supports the mooring wire in an upright position [3]. Sampling through the top layer (0–30 m) in the coastal zone of the Black Sea cannot be achieved using Argo floats, because they do not operate in shallow areas [25]. Sensor chains (strings), which are often used in shallow seas [18], are not sufficient for carbon sites, because they are usually limited to temperature and salinity sensors such as those developed by RBR Ltd. (Ottawa, Canada) (https://rbr-global.com/products/standard-loggers/thermistor-strings (accessed on 21 February 2022)), GeoPrecision GmbH (Ettlingen, Germany) (https://www.thermistor-string.com/ (accessed on 21 February 2022)), Aanderaa Data Instruments AS (Bergen, Norway) (https://www.aanderaa.com/productsdetail.php?Aanderaa-SeaGuard-String-31 (accessed on 21 February 2022)), Star Oddi Ltd. (Gardabaer, Iceland) (https://www.star-oddi.com/products/submersible-water-sensors/salinity-logger-probe-CTD (accessed on 21 February 2022)), and others. Therefore, specific instruments are required to study the carbon system. If multidisciplinary profiles are required, the cost of several sensor suits, along with the mooring line, makes a difference. In addition, the sensor chains need to be deployed at subsurface moorings to avoid the impact of waves; the data are not available for the upper ocean boundary layer.

This study describes the new tethered profiler Winchi, which is surface-pierced to obtain full-water column profiles. However, it is capable of sampling through the sea surface layer regularly and frequently during extended deployment. It uses a single sensor suite, thus reducing the cost of the initial investment and maintenance, including post-survey calibration of the sensors. In the future, the profiler will accommodate additional sensors to obtain data on dissolved gases, e.g., carbon dioxide and methane, photosynthetically available radiation (PAR), abundance of chlorophyll, colored dissolved organic matter (CDOM), and pH value, which are required to comprehensively assess the carbon fluxes.

We focused on the control of the movement of Winchi to maintain sufficient tension of the wire. It is risky for the profiler mission to ascend in the high seas. During ascent, loosening and looping of the wire (when it is unwound from the drum) may occur owing to the surface waves, thereby posing a problem for underwater winches. Therefore, before ascent, the profiler obtains data on surface waves to avoid operation in stormy weather conditions. If the wire tension weakens, then overlapping of its turns on the winch drum may occur. Hence, it is of utmost importance to control the wire tension and reverse the direction of the winch drum rotation to generate the required tension for the wire. Another problem is that strong currents can sink the device. To reduce this effect, the winch was programmed to spool more wiring to allow the profiler to rise to the sea surface.

When the profiler surfaced, it transmitted the observational data to a server on the land. This makes the profiler useful for operational oceanography applications, particularly for providing the observational data for assimilation in mathematical models for diagnosis and forecasting. Data transfer to and from the profiler Winchi has been designed with regard to the mobile phone link. In the future, satellite communication options will be included.

The effort to build the carbon observational site in the Gelendzhik area is in accordance with two projects: Advancing Black Sea Research and Innovation to Co-Develop Blue Growth within Resilient Ecosystems (BRIDGE-BS) (https://cordis.europa.eu/project/id/101000240, last accessed: 12 December 2021) and Developing Optimal and Open Research Support for the Black Sea (DOORS) (https://cordis.europa.eu/project/id/101000518, last accessed: 12 December 2021), both for 2021–2025 under the umbrella of the EU Horizon 2020 program. In particular, the BRIDGE-BS project targets the development of capabilities necessary to understand and predict the impact of climate-driven and anthropogenic multi-stressors on services stemming from Black Sea ecosystems. The key ecosystem services at the Northeastern Black Sea pilot site, which shares the same infrastructure as the carbon observational site in the Gelendzhik area, will include carbon sequestration in the sea. In a broad oceanographic context, by developing the new observational profiling platform—Winchi, we are taking a step forward to create and apply innovative technologies that can help in long-term multidisciplinary oceanographic observations at fixed locations in the sea coastal zone.

Author Contributions: Conceptualization, A.G.O. and A.V.O.; methodology, D.A.S. and S.V.V.; software, O.Y.K.; validation, N.M.K. and V.M.O.; formal analysis, A.G.O.; investigation, O.Y.K. and D.A.S.; resources, A.G.Z.; data curation, O.Y.K.; writing—original draft preparation, A.G.O.; writing—review and editing, M.V.E.; visualization, A.G.O. and O.Y.K.; supervision, A.G.O.; project administration, A.G.Z.; and funding acquisition, V.V.K. All authors have read and agreed to the published version of the manuscript.

Funding: The research and development of the new tethered profiler Winchi was carried out within the framework of the assignment of the Ministry of Science and Higher Education of Russia No 0128-2021-0018 and supported in part by the Russian Fund for Basic Research via grant No 19-05-00459.

Acknowledgments: The collaboration with colleagues from the southern branch of SIO RAS is appreciated. We sincerely thank the two anonymous reviewers for their careful reading of our manuscript and their many insightful comments and suggestions.

Conflicts of Interest: The authors declare no conflict of interest.

References

1. Abakumov, E.; Polyakov, V. Carbon polygons and carbon offsets: Current state, key challenges and pedological aspects. *Agronomy* **2021**, *11*, 2013. [CrossRef]
2. Ostrovskii, A.G.; Zatsepin, A.G. Short-term hydrophysical and biological variability over the northeastern Black Sea continental slope as inferred from multiparametric tethered profiler surveys. *Ocean Dyn.* **2011**, *61*, 797–806. [CrossRef]
3. Ostrovskii, A.G.; Zatsepin, A.G.; Soloviev, V.A.; Tsibulsky, A.L.; Shvoev, D.A. Autonomous system for vertical profiling of the marine environment at a moored station. *Oceanology* **2013**, *53*, 233–242. [CrossRef]
4. Ostrovskii, A.G.; Zatsepin, A.G. Intense ventilation of the Black Sea pycnocline due to vertical turbulent exchange in the Rim Current area. *Deep Sea Res. Part I Oceanogr. Res. Pap.* **2016**, *116*, 1–13. [CrossRef]
5. Ostrovskii, A.G.; Zatsepin, A.G.; Solovyev, V.A.; Soloviev, D.M. The short timescale variability of the oxygen inventory in the NE Black Sea slope water. *Ocean Sci.* **2018**, *14*, 1567–1579. [CrossRef]
6. Ostrovskii, A.G.; Arashkevich, E.G.; Solovyev, V.A.; Shvoev, D.A. Seasonal variation of the sound-scattering zooplankton vertical distribution in the oxygen-deficient waters of the NE Black Sea. *Ocean Sci.* **2021**, *17*, 953–974. [CrossRef]
7. Oguz, T.; Tuğrul, S.; Kideys, A.; Ediger, V.; Kubilay, N. Physical and biogeochemical characteristics of the Black Sea. In *The Sea*; Robinson, A.R., Brink, K.H., Eds.; The President and Fellows of Harvard College: Massachusetts, MA, USA, 2005; Volume 14, pp. 1331–1369.
8. Poulos, S.E. The Mediterranean and Black Sea Marine System: An overview of its physico-geographic and oceanographic characteristics. *Earth-Sci. Rev.* **2020**, *200*, 103004. [CrossRef]
9. Carlson, D.F.; Ostrovskii, A.; Kebkal, K.; Gildor, H. Moored automatic mobile profilers and their applications. In *Advances in Marine Robotics*; Oren, G., Ed.; Lambert Academic Publishing: Chisinau, Republic of Moldova, 2013; pp. 169–206. ISBN 978-3-659-41689-7.
10. Prien, R.D.; Schulz-Bull, D.E. Technical note: GODESS—A profiling mooring in the Gotland Basin. *Ocean Sci.* **2016**, *12*, 899–907. [CrossRef]
11. Matsumoto, H.; Haxel, J.; Kahn, B.; Roche, L.; Dziak, R.P.; Turpin, A.; Childress, J.; Sexton, K.; Klinck, H.; Nakamura, T. Field testing and performance evaluation of the Long-term Acoustic Real-Time Sensor for Polar Areas (LARA). In Proceedings of the OCEANS 2019 MTS/IEEE SEATTLE, Seattle, WA, USA, 27–31 October 2019; pp. 1–7. [CrossRef]
12. Trowbridge, J.; Weller, R.; Kelley, D.; Dever, E.; Plueddemann, A.; Barth, J.A.; Kawka, O. The Ocean Observatories Initiative. *Front. Mar. Sci.* **2019**, *6*, 74. [CrossRef]
13. Aracri, S.; Borghini, M.; Canesso, D.; Chiggiato, J.; Durante, S.; Schroeder, K.; Sparnocchia, S.; Vetrano, A.; Honda, T.; Kitawaza, Y.; et al. Trials of an autonomous profiling buoy system. *J. Oper. Oceanogr.* **2016**, *9*, s176–s184. [CrossRef]
14. Gerrlets, M.; Badewlen, T.H.; Braun, A.; Behrens, G.; Nicolai, H.; Zielinski, O. APMOS: An Autonomous Profiling Measurement System for the ocean surface boundary layer. In Proceedings of the OCEANS 2017, Aberdeen, UK, 19–22 June 2017; pp. 1–6. [CrossRef]
15. Ostrovskij, A.G.; Shvoev, D.A. Underwater Winch Probe. International Patent Application. No. RU2642677C1, 25 January 2018.
16. Bailey, K.; Steinberg, C.; Davies, C.; Galibert, G.; Hidas, M.; McManus, M.A.; Murphy, T.; Newton, J.; Roughan, M.; Schaeffer, A. Coastal Mooring Observing Networks and Their Data Products: Recommendations for the Next Decade. *Front. Mar. Sci.* **2019**, *6*, 180. [CrossRef]
17. Staudinger, C.; Strobl, M.; Fischer, J.P.; Thar, R.; Mayr, T.; Aigner, D.; Müller, B.J.; Müller, B.; Lehner, P.; Mistlberger, G.; et al. A versatile optode system for oxygen, carbon dioxide, and pH measurements in seawater with integrated battery and logger. *Limnol. Oceanogr. Methods* **2018**, *16*, 459–473. [CrossRef]
18. Silvestrova, K.; Myslenkov, S.; Zatsepin, A. Variability of wind-driven coastal upwelling in the north-Eastern Black Sea in 1979–2016 according to NCEP/CFSR data. *Pure Appl. Geophys.* **2018**, *175*, 4007–4015. [CrossRef]
19. Finenko, Z.; Churilova, T.; Lee, R. Dynamics of the vertical distributions of chlorophyll and phytoplankton biomass in the Black Sea. *Oceanology* **2005**, *4*, 112–126.

20. Churilova, T.; Suslin, V.; Krivenko, O.; Efimova, T.; Moiseeva, N.; Mukhanov, V.; Smirnova, L. Light Absorption by phytoplankton in the upper mixed layer of the Black Sea: Seasonality and parametrization. *Front. Mar. Sci.* **2017**, *4*, 90. [CrossRef]
21. Churilova, T.; Suslin, V.; Sosik, H.M.; Efimova, T.; Moiseeva, N.; Moncheva, S.; Mukhanov, V.; Rylkova, O.; Krivenko, O. Phytoplankton light absorption in the deep chlorophyll maximum layer of the Black Sea. *Eur. J. Remote Sens.* **2018**, *52*, 123–136. [CrossRef]
22. Kirk, J.T.O. *Light and Photosynthesis in Aquatic Ecosystems*, 3rd ed.; Cambridge University Press: Cambridge, UK, 2010; 649p. [CrossRef]
23. Ocherednik, V.V.; Zatsepin, A.G.; Kuklev, S.B.; Baranov, V.I.; Mashura, V.V. Examples of approaches to studying the temperature variability of Black Sea shelf waters with a cluster of temperature sensor chains. *Oceanology* **2020**, *60*, 149–160. [CrossRef]
24. Kolyuchkina, G.A.; Syomin, V.L.; Simakova, U.V.; Sergeeva, N.G.; Ananiev, R.A.; Dmitrevsky, N.N.; Lyubimov, I.V.; Zenina, M.A.; Podymov, O.I.; Basin, A.B.; et al. Benthic community structure near the margin of the oxic zone: A case study on the Black Sea. *J. Mar. Syst.* **2022**, *227*, 103691. [CrossRef]
25. Palazov, A.; Ciliberti, S.; Peneva, E.; Gregoire, M.; Staneva, J.; Lemieux-Dudon, B.; Masina, S.; Pinardi, N.; Vandenbulcke, L.; Behrens, A.; et al. Black Sea Observing System. *Front. Mar. Sci.* **2019**, *6*, 315. [CrossRef]

Journal of
Marine Science and Engineering

MDPI

Article

VERDA: A Multisampler Tool for Mesopelagic Nets

Arturo Castellón [1,*] and María Pilar Olivar [2]

[1] Unidad de Tecnología Marina (UTM), Consejo Superior de Investigaciones Científicas (CSIC), Passeig Marítim 37–49, 08003 Barcelona, Spain
[2] Institut de Ciències del Mar (ICM), Consejo Superior de Investigaciones Científicas (CSIC), Passeig Marítim 37–49, 08003 Barcelona, Spain
* Correspondence: arturoc@utm.csic.es

Abstract: Different types and systems to discriminate plankton samples at different strata in the water column have been developed in recent decades. For sampling at sufficient depth, opening and closing zooplankton multinets are ideal because there is no contamination of one sample with organisms of the previous one. However, for bigger nets, such as those used to collect micronektonic organism, it is difficult to use multiple net units, and multiple cod ends are preferred because of their simplicity, but with the problem of sample contamination from having a common net passageway. We present here a cod-end Multisampler design, VERDA, that uses a carrousel-like system. Similar to some sediment traps, the system works like a revolver with six or eight compartments whose turning mechanism is triggered when the net arrives to a programmed depth level. This prototype was built with inexpensive and recycled components and electronics similar to Arduino® and Teensy PCB to carry out electronic control. The net we used for testing the equipment was a mid-size midwater trawl of ca. 30 m^2 and total length of 58 m that works with a single towing cable and no doors. The overall system is useful for all type of ships, due to the relatively easy deployment operations and because the Multisampler does not need electrical cable or acoustics. In our case, we used a Marport® (Reykjavik Iceland) and Scanmar (Åsgårdstrand, Norway) sensors for real-time depth monitoring and opening distance.

Keywords: stratified sampling; micronekton; midwater trawls; Mesopelagos

Citation: Castellón, A.; Olivar, M.P. VERDA: A Multisampler Tool for Mesopelagic Nets. *J. Mar. Sci. Eng.* **2023**, *11*, 72. https://doi.org/10.3390/jmse11010072

Academic Editor: Gualtiero Basilone

Received: 17 November 2022
Revised: 30 November 2022
Accepted: 24 December 2022
Published: 2 January 2023

1. Introduction

The micronekton community is composed of organisms between 2 and 20 cm that swim freely and, unlike plankton, can avoid being carried away by ocean currents. These types of organisms are relatively good swimmers, and are able to avoid plankton nets. Conversely, they are too small for conventional commercial pelagic nets, which results in their extrusion through the large mesh sizes. Another feature of these organisms is that they tend to aggregate in different layers of the water column, with important differences in their vertical location throughout the day cycle. Therefore, for scientific purposes, it is imperative to collect them by discriminating the different layers of the water column. Much effort has been made in this regard for meso- and macrozooplankton [1], and similar prototypes are widely used by the scientific community. There is far less equipment development and a lack of consensus to sample micronekton.

Multinet gears have a rigid frame where various nets are encased, and by means of an electronic or mechanical device, are sequentially dropped (by time or by pressure), trapping the organisms in different collectors, allowing for sampling stratified by depth. Among these types of plankton nets, the most commonly used are MOCNESS [2], BIONESS [3], RMT [4], Tucker [5] or HYDROBIOS [6] multinets. Some of the gears to collect larger micronekton organisms are similar in structure to commercial pelagic gears, with a structure that consists of ropes and twines that hold a mesh net. These nets use doors or hydrofoils, kites, depressors and ballast and floats, which, by offering resistance to water, open the mouth of the gear. Among the latter, the IKMT (Isaacs Kit Midwater Trawl) [7], Cobb Midwater trawl [8], Mesopelagos trawl [9] and others generally do not use multisampling systems, and when they do, it usually consists of a cod-end Multisampler placed at the end of the nets. Sample contamination due to the organisms that may be entangled inside the meshes and eventually released to the cod ends is the main challenge for these type of systems that have a single net passageway.

The objective of our work was to develop a system to sample micronekton organisms at discrete strata and that could be used with midwater trawls of small to medium size and in a variety of vessels. A cod-end Multisampler was designed and tested in combination with an already developed midwater trawl of medium size (Mesopelagos). Here, we present the Multisampler design and discuss some of the results obtained in several surveys.

2. Materials and Methods

2.1. Midwater Trawl

This investigation is framed within several projects focused on the study of mesopelagic communities (Table 1). It was necessary to determine the appropriate net to collect the target organisms, but also take into account the characteristics of the research vessels used in the different surveys, which do not always permit the use of trawl doors. Commercial fishing trawls use doors for opening the mouth of the net. The operation, maneuvering for launch and recovery of doors need gears on deck and supports on the ship aft. The use of kites instead of doors makes the operations easier, with few gears involved, allowing its use in research ships of all sizes.

Table 1. Acronyms of projects and surveys (Survey). Research vessels (RV). Geographical limits of the survey (Region). Month and year of the survey (Period). Vertical sampling ranges (Depth). Type of acoustic equipment (Echosounder). Number of stations sampled with the system (N. Stn).

Survey	RV	Region	Period	Depth (m)	Echosounder	N. Stn
MAFIA	*Hesperides*	Tropical and Equatorial Atlantic	April 2015	0–800 m	EK60	14
Bathypelagic	*Sarmiento de Gamboa*	Northeastern Atlantic	May–June 2018	0–1800 m	EK60	12
CONECTA	*García del Cid*	Western Mediterranean	July 2016		-	4
WINFISH	*García del Cid*	Northwestern Mediterranean	March 2017	0–200	-	10
SUMMER	*Sarmiento de Gamboa*	Off Iberian Peninsula	October 2020	0–700 m	EK80	19

The Mesopelagos net designed by Ifremer (Fisheries Biology and Technology Laboratory, LTBH, Lorient, France) [9] and manufactured by Le Drezen (https://www.ledrezen.com/, accessed on 1 March 2022) was used in our surveys (Table 1). This gear works with a single traction cable, and uses kites on its wings as well as floats on the headline and a 120 kg ballast on the footrope to open the mouth (Figure 1). The gear has a total length of 58 m, and consists of graded-mesh netting starting with 30 mm and ending with 4 mm. The net sensors Marport® Trawl Explorer (Reykjavik, Iceland) and Scanmar opening distance sensor (Åsgårdstrand, Norway) were used in our surveys to know, in real time, the depth of the equipment and the horizontal and vertical mouth aperture.

(**a**) Surface view (**b**) Underwater view

Figure 1. Surface (**a**) and underwater (**b**) images (Félix Descalzo) of the mouth of Mesopelagos net. Observe the kites in the wings and floats in the headrope.

According to the manufacturer, the nominal opening of the rig is 7 m vertical between headrope and footrope, and 8 to 9 m horizontal between kites [9]. The geometry of the mouth varied depending on the ship and winch speeds, and may be also modified by adding floats, ballast and kites. Other authors give mouth areas of ca. 65 m^2 and vertical (v) and horizontal (h) openings between 5 and 6 m and 10 and 12 m, respectively [10,11]. During our hauls, almost all the records of opening measures were approximately 4 to 5 m horizontal (h) $\times$ 6 to 7 m vertical (v). From the image presented in Figure 1b, taken during a haul at 2 knots, and taking the floats and kites as references, the distance was 5 (h) $\times$ 9 (v) meters. This difference can be attributed to the position and posture of the height sensor that was located on the headrope. A tilt of this sensor produces lower readings as it receives the echo from the sides of the gear instead of from the footrope. Being a very light gear, the sensor location is important, and floats had to be added to compensate for sensor weight.

Trawling speed varied between 1.8 knots and 2.5 knots. As much as possible, both the ship's heading and speed were kept constant during the hauls. The ratio between the cable released and depth in the water column varied from a value of 3 for hauls down to 200 m to a value of 1.7 for hauls reaching 2000 m depth. Other authors [10] give a ratio between 2.3 (for 600 m) and 2.6 (for 540 m), at a speed of 2.5 knots. In our case, these ratios are somewhat lower: 2.1 (for 600 m) at a speed of 2 knots. The ratio between the length of the cable and the depth also depends on other factors such as the speed of the cable release, the total weight of the gear, mouth opening, sea state and currents, etc.

2.2. Sampling Protocol

To determine the diel vertical migration processes of micronekton species, it was necessary to sample the water column at different depths. Details of the sampling design are presented in Section 3. In order to sample mesopelagic communities in the water column, both day and night hauls were carried out, discriminating strata that ranged from 20 to 400 m width and from the surface to a maximum of 2000 m depth, depending on the survey's particular objectives.

2.3. Organisms Studied

We present here a brief overview of the main zoological groups collected during our last survey (SUMMER, October 2020), i.e., mesopelagic fishes, decapods, euphausiids and gelatinous organisms. Abundances are analyzed in terms of numerical abundance or biomasses, standardized according to the water filtered. The volume of water filtered was estimated considering an average mouth opening of 30 m^2 multiplied by the distance traveled during each haul.

3. Results

3.1. Design of the VERDA Multisampler

VERDA Vehicle

Our Multisampler consists of a vehicle transporting and encasing several electronic and mechanical devices. It is built in fiber glass and was designed with a funnel shape (Figures 2 and 3) starting with a cylindrical section (250 mm in diameter) that was tied to the end of the Mesopelagos net. From that point on, it widened until reaching a diameter adequate to support the toothed plate that held the collectors. A single 100 mm diameter opening in the stern of the vehicle matched the current collector. An internal net communicated the mouth with the opening of the collector. In this way, it was intended that the water that entered through the mouth could reduce it speed inside the vehicle so as to prevent organisms from being too damaged within the vehicle. Two openings were made in the sides to force the water flow-through. In the lower part of the vehicle, a casing with the motor and the electronics was placed. A pinion transmitted the rotation to the plate. The plate rotation moved the collecting bags so that before each haul, a collector was placed in front of the aft opening (Figure 2).

The buoyancy of the entire vehicle including the casing had to be neutral. During the first trials, the vehicle rotated and caused the net to close at the torsion point, preventing the samples from reaching the collectors. Two wings were installed to prevent this turning. The casing containing electronics, the depth sensor and the motor was placed in the lower part, as a ballast, and two floats were placed in the upper part to prevent the vehicle from turning (Figures 2 and 3). In addition, a sea (drift) anchor was placed to stretch the entire net. In this way, the previous problem was avoided.

An important point is that the geometry of the net cannot twist during the launching of the equipment. To ensure this, the upper part of the net, marked with a green line on the mesh, must coincide seamlessly with the upper part of the vehicle. Another problem that appeared was that the collecting bags, on occasion and due to the turbulence at the stern of the ship, became entangled between the pinion and the plate. To avoid this, all the ends of the collecting bags were tied together and these, in turn, were tied to the rope that held the sea anchor (Figure 2). With this configuration, the problem was solved.

Figure 2. View, in inverted position, of vehicle without the casing. Observe the way collector bags are attached to sea anchor rope.

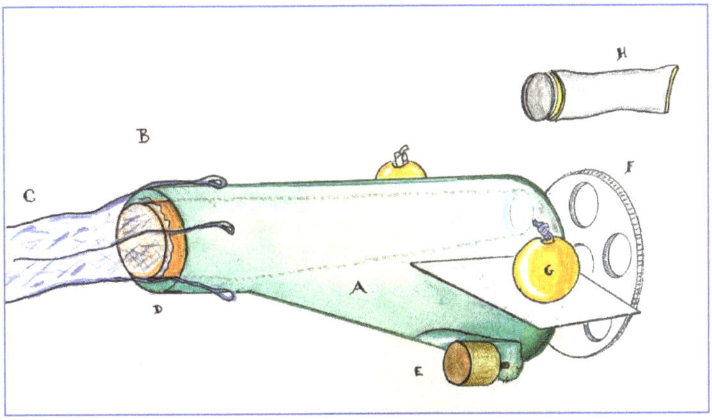

Figure 3. Scheme of the VERDA Multisampler. (A) Fiber glass body. (B) Mouth. (**C**) Mesopelagos net. (D) Attachment. (E) Motor. (F) Plate. (G) Floats. (H) Collector (net).

3.2. Plate and Pinion

Two plates with toothed gears were designed. The diameters of the plates were 400 mm and 398 mm. Both have 264 teeth. The first may contain six collectors (six holes for cylindrical bushings) (Figure 4a). The second was for eight collectors (Figure 4b). In the second case, the collector area was somewhat less than that of the first. In this way, the two plates could be used interchangeably on the same vehicle and the same pinion, avoiding major modifications. The only modification in this case was the number of turns the shaft had to make for positioning the next collector. The pinion has 11 teeth; then, for the first plate, three turns are needed (33 teeth) and for the second, four turns (44 teeth).

(**a**) Round bushings (**b**) Triangular bushing

Figure 4. Plate designs, with round (**a**) and triangular (**b**) bushing.

The non-cylindrical bushings (Figure 4b) of the second plate were made with a 3D printer using PLA filament (polylactic acid with a degradation time of two years) in solid printing. To check their reliability, they were lowered to a depth of 3700 m without any damage due to pressure. However, depending on the PLA filament used (different color), some suffered damage during some sampling, but the result was positive for most of the bushings until the end of the cruise.

3.3. Control Unit

The control unit consisted of a Technicap housing (La Turbie, France https://www.technicap.com/, accessed 1 March 2022) used for the sediment traps and built in aluminum anodized for 5000 m depth. In this case, only the casing and the motor were used. In addition, the carter (casing containing motor and electronics) was designed 10 cm longer. The original electronics were not used and one of our own, described here, was designed for the needs of our sampling. The depth sensor was a Keller Piezoresistive OEM Pressure Transmitter Type PA-20D/100 and 200 bar (Keller AG Winterthur, Switzerland (https://keller-druck.com/en/products/, accessed 1 March 2022)) This sensor has an I2C interface protocol (https://en.wikipedia.org/wiki/I%C2%B2C, accessed 1 March 2022). The casing was machined to house the sensor. The thread pitches of this sensor are the gas/hydraulic standard, so a pressure gauge was used to carry out the tests and calibrations in the laboratory without having to go out to sea or use hyperbaric chambers for haul simulations.

The microcontrollers were: Teensy 3.1/3.2 microcontroller development system (https://www.pjrc.com/teensy/, accessed 1 March 2022) and Arduino® Micro microcontroller (https://www.arduino.cc/en/Main/Products, accessed 1 March 2022). Arduino Micro had the software for controlling the whole system, and the Teensy microcontroller interface with the Keller pressure sensor through I2C protocol sent the data to the Micro using a serial interface. The Teensy I2C line is a 3.3-volt line. The final design was using only Teensy for the whole control.

The system controls the motor electronically with a MOSFET. Technicap's system uses a stem to count the turns of the shaft. In our case, three and four turns were necessary for the pinion to align with the collector openings. A CNY 70 Reflective Optical Sensor with Transistor Output was used to detect stem passages.

To record the data, a µSD card was used. When the device is deployed, the clock of the system is reset and the pinion is placed in its starting position. Time and Keller sensor data, temperature and pressure were recorded in the µSD card. The motor activation time, the number of each sampling collector and the time of its rotation, as well as the setup files are also stored in this card. The setup file consists of a maximum depth to reach, and the desired depths to be sampled (those where the collectors must turn). The procedure consists of lowering the system to exceed by ca. 20 m the maximum desired sampling depth (i.e., the depth of the first sample) and then starting the ascent. The system is triggered when sensors perceive that the net is rising and the maximum pre-programmed sampling depth is reached. From here on, the different samples are triggered as the depth configured in the setup file is reached.

The software loop works by continuously reading the Keller pressure sensor. A buffer stores previous depth data and calculates the average slope of vehicle depth to know if the equipment is descending or ascending. The system uses a system of flags (Booleans):

- In the water: start to submerge (5 m deep);
- Arrived at the maximum depth (in setup);
- Arrived at layer "*n*";
- Sinking or rising.

After several trials, it was decided that the best configuration for the VERDA Multi-sampler was to leave the first hole empty, without the cod end in the first bushing ("0"), so that the vehicle does not fish when sinking. The rest of the holes were used with the corresponding collecting bags. The end of the hauls was set when reaching 10 m on the ascent, and in this moment, the plate rotates to place it back in hole "0". In this way, it is avoided that, in the recovery of the device, with turbulence produced in the stern of the boat, the last sample can be emptied.

3.4. Collection of Organisms

The different surveys in which this system was used (Table 1) provided samples where many of the micronekton groups were represented, from small fishes and crustaceans of sizes ca. 2 to 20 cm, to gelatinous organisms (Figure 5). Gelatinous items were mainly represented by jelly fish, siphonofores and salps (Tunicata), and were the most abundant taxonomic group both in number and biomass in most of the samples. Within crustaceans, euphausiids dominate in terms of number of individuals, while decapods were dominant in terms of biomass. Fishes were represented by several mesopelagic families, among which Stomiiformes and Myctophiformes were the most abundant and frequently represented [12]. Nevertheless, size frequency distributions within species were skewed towards small size classes. On the other hand, fishes of the genus *Cyclothone*, although being abundant in our surveys, were underrepresented by these hauls when comparing to concurrent collection with finer meshes [13]. Finally, cephalopods were poorly represented in the samples, and most of their collections corresponded to paralarval stages.

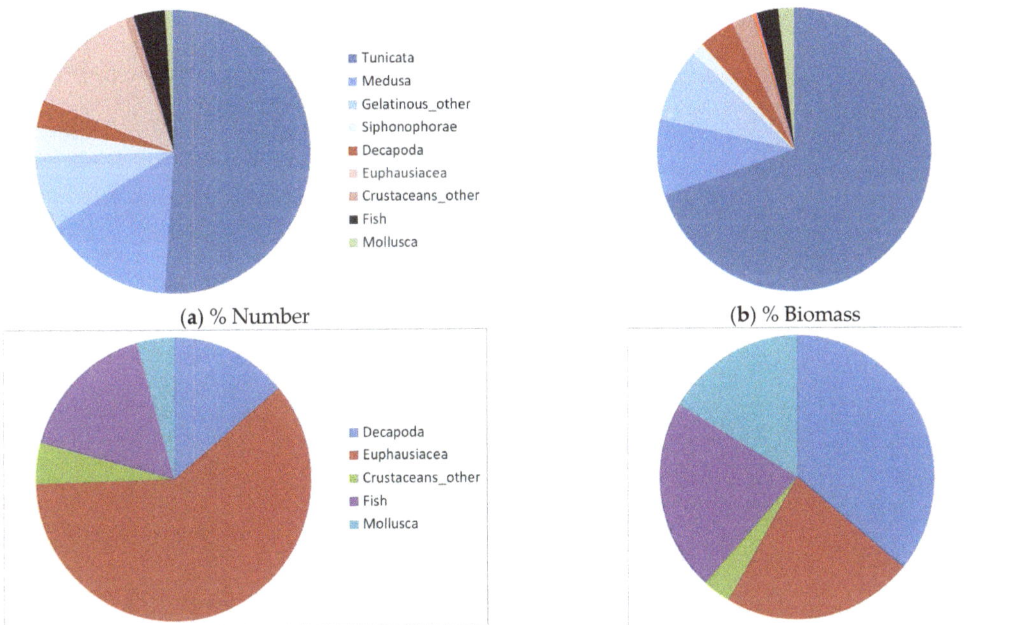

(c) % Number excluding gelatinous organisms (d) % Biomass excluding gelatinous organisms

Figure 5. Relative contribution, in number (**a**) and biomass (**b**), of the main micronekton and macrozooplankton groups collected during the SUMMER-2020 cruise around the Iberian Peninsula.

The vertical resolution of the VERDA system allowed for the discrimination of samples obtained from vertical strata (Figure 6), from several hundreds of meters to just 20 m. In our surveys, samples obtained on a day/night basis at different layers in the water column offered information on the night vertical displacements of fishes (almost all myctophids and some species of stomiiforms) from the mesopelagic layers (200–1000 m) where they are located during the day to the upper 100 m at night [12,13].

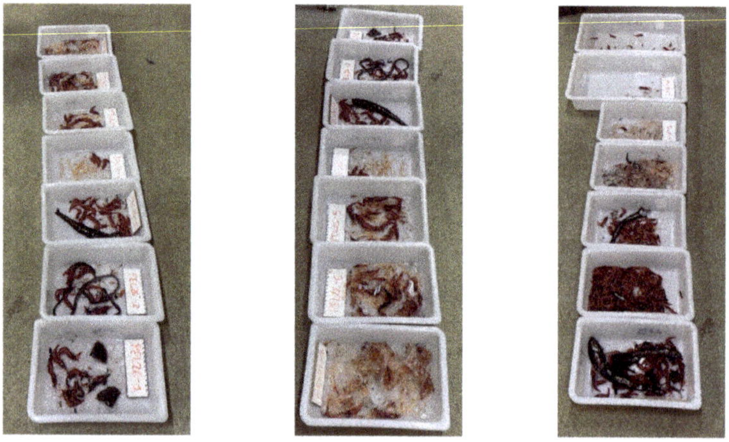

Figure 6. Photographs of the results of three hauls. Observe the difference in the catches for the different strata. **Left** and **right**: deeper strata in the front. **Middle**: deeper strata at the bottom.

4. Discussion

4.1. Investigations on Distribution Patterns

Investigations on mesopelagic organisms' distributions and abundances have a number of limitations and constraints related to both the type of organism and their behavior, and the difficulties in accurately sampling them from the different layers of the water column. The system we present here departs from the scientific midwater trawls previously designed [9] combined with the VERDA Multisampler we described in this paper. The system efficiently discriminates samples obtained at different layers of the water column (from several hundreds of m to 20 m), being particularly valuable to investigate daily vertical migratory displacements of mesopelagic organism, and has been the subject of a few earlier publications [12–15].

The system catches a variety of mesopelagic fishes, decapods, euphausiids and gelatinous organisms that were well represented in terms of number of species and relative abundances. For mesopelagic fishes, decapods and euphausiids, our assessment is that they are well represented in terms of diversity [12]. Abundance and biomass estimations are, however, important issues yet to be solved due to net escape and mesh extrusion. Net avoidance has been advised as an important problem in the assessment of biomass of mesopelagic fishes from net hauls [16]; however, there are no studies of the Mesopelagos net selectivity for these organisms, and therefore, our abundance estimations cannot be directly used as the actual abundance at sea. In general, in our hauls, juvenile fish were better represented than the older and larger adults, indicating higher net avoidance from these more advanced developmental stages with higher swimming and vision capability. These comments can be extended to the very low capture of cephalopods, mostly represented by the low swimming paralarval stages. One possibility to reduce net avoidance may be to increase haul speed, but for an efficient Mesopelagos net deployment, there is a speed constraint of 1.8 to 2.0 knots [9]. The use of larger nets will enable the possibility of hauling at higher speeds, which points to the use of a combination of these types of nets to obtain a wide spectrum of sizes [17].

A word of caution must also be given regarding the collection of small and thin fishes, such as those of the genus *Cyclothone* (which has one of the greatest abundances of individuals of any vertebrate genus [18]), as well as for the assessment of metamorphic stages of most lanternfish, which are thinner that the mesh size of the Mesopelagos net and are extruded through the meshes. In our surveys, we overcame this by using a combination of the Mesopelagos midwater trawl and a smaller plankton net, MOCNESS-1 m^2 (mesh size 0.2 mm) [12,19].

4.2. Future Research Directions

The use of this system, Mesopelagos + VERDA Multisampler, in combination with acoustic sampling using the on-board biological echo sounders (as in some of our surveys), may be used to decide, beforehand, the layers to be sampled, and eventually to correlate the acoustic information with the biological samples. The next step is to use an acoustic/USBL (Ultra Short Base Line) modem to be able to position the equipment on the echo sounder's echogram. Through the acoustic modem, it will be possible to "trigger" the VERDA Multisampler remotely to obtain samples on selected scattering layers observed on the live echogram. In this way, we trust that the acoustic information (Target strength, especially) can be correlated with the fished individuals. However, this has a difficulty since the fished individuals travel almost 60 m from the mouth of the gear until they reach the VERDA Multisampler, and this time/distance must be taken into account.

With the present system, the gear depth is currently obtained in real time using the Marport Trawl Explorer acoustic sensor. This sensor is located in the mouth of the gear. With the USBL, which will be located in the vehicle, the depth of the VERDA Multisampler (vehicle) will be obtained through the modem. With this information, it is possible to know what the geometry of the gear is and if the vehicle goes higher or lower than the mouth of the gear. Being able to have real-time information on the relative location of the vehicle and the mouth of the net could be useful to analyze the behavior of the entire system, and will allow placing the equipment more reliably in the echograms or even to sample, for example, near the seabed.

The next step in the control electronics will be to incorporate software to receive and send information through an acoustic modem. On the one hand, it will allow us to receive the "fire" order from the ship and also to send control data to the ship to monitor the operation of the system. A microcomputer (Raspberry Pi zero) and Python programming will be included to achieve a more user-friendly interface.

Finally, another improvement will consist of designing a larger plate and holes and cod ends of a larger size to be able to house the gelatinous organisms in particular.

5. Conclusions

This gear uses kites on its wings as well as floats on the headline and a 120 kg ballast on the footrope to open the mouth; therefore, it is an easy gear to deploy on any vessel that has a suitable winch and cable (Figure 1). It works with a single traction cable, so there is no need for heavy fishing doors to open the mouth.

Ideally, the vehicle should be at the same height as the mouth. The use of a sea (drift) anchor helps this purpose.

The final VERDA Multisampler design allows us to sample at seven different strata with a resolution from 10 to 20 m. The system works well both for horizontal and oblique hauls. Our recommendation, however, is to prevent long sampling times when gelatinous organisms are dominant, such as in the presence of jelly fish and salps. The accumulation of large quantities of these heavy organisms in the collectors compromises the plate rotation.

Author Contributions: A.C. was responsible for the design, development and validation of the VERDA Multisampler system and necessary software, as well as optimization of the entire sampling system. M.P.O. was responsible for the investigation of micronekton distribution patterns, decision of sampling schemes and data curation and analyses. Both authors contributed to the draft preparation. All authors have read and agreed to the published version of the manuscript.

Funding: This research was funded by the projects "Migrants and Active Flux in the Atlantic Ocean" (MAFIA, CTM2012-39587-C04) and "Biomass and Active Flux in the Bathypelagic Zone" (Bathypelagic, CTM2016-78853-R) from the Spanish Ministry of Economy and Competitiveness, and SUMMER (grant agreement number 817806) from the European Union (Horizon 2020 Research and Innovation Programme), and the support through the "Severo Ochoa Centre of Excellence" accreditation (CEX2019-000928-S).

Institutional Review Board Statement: The study was approved by the Ethics Committee of the Spanish National Research Council (CSIC) on the bases that this investigation did not include any procedure with animal experimentation and in accordance with the provisions set forth in Royal Decree 53/2013 (1 February 2013).

Data Availability Statement: No new data were created or analyzed in this study. Data sharing is not applicable to this article.

Acknowledgments: We would like to thank to the crew of R/V Hespérides, R/V Garcia del Cid and R/V Sarmiento de Gamboa, scientists and technicians of the Marine Technology Unit (UTM.CSIC) for their support and help during the cruises.

Conflicts of Interest: The authors declare no conflict of interest.

References

1. Wiebe, P.H.; Benfield, M.C. From the Hensen net toward four-dimensional biological oceanography. *Prog. Oceanogr.* **2003**, *56*, 7–136. [CrossRef]
2. Wiebe, P.; Morton, A.W.; Bradley, A.M.; Backus, R.H.; Craddock, J.E.; Barber, V.; Cowles, T.J.; Flierl, G.R. New development in the MOCNESS, an apparatus for sampling zooplankton and micronekton. *Mar. Biol.* **1985**, *87*, 313–323. [CrossRef]
3. Sameoto, D.D.; Jaroszynski, L.O.; Fraser, W.B. BIONESS, a New Design in Multiple Net Zooplankton Samplers. *Can. J. Fish. Aquat. Sci.* **1980**, *37*, 722–724. [CrossRef]
4. Dimmler, W.; Klindt, H. A new electronic system for the rectangular midwater trawl (RMT). *Mar. Biol.* **1990**, *107*, 345–348. [CrossRef]
5. Davies, I.E.; Barham, E.G. The Tucker opening-closing micronekton net and its performance in a study of the deep scattering layer. *Mar. Biol.* **1969**, *2*, 127–131. [CrossRef]
6. Weikert, H. and John, H.C. Experiences with a modified Bé multiple opening-closing plankton net. *J. Plankton Res.* **1981**, *3*, 167–176. [CrossRef]
7. Besançon, H.C. The ISAACS-KIDD midwater trawl (IKMT): Experiences with a new way of rigging. In *ICES, Gear and Behavior Committee*; International Council for the Exploration of the Sea: Copenhagen, Denmark, 1973; C.M. 1973/B:5.
8. McNeely, R.L. Development of the John M. Cobb pelagic trawl–A progress report. *Commer. Fish. Rev.* **1963**, *25*, 17–27.
9. Meillat, M. Essais du chalut mésopélagos pour le programme MYCTO 3D—MAP de l'IRD, à bord du Marion Dufresne. *IFREMER Internal. Rep.* **2012**. R.INT.RBE/STH/LTH 2012-05.
10. Cotté, C.; Ariza, A.; Berne, A.; Habasque, J.; Lebourges-Dhaussy, A.; Roudaut, G.; Espinasse, B.; Hunt, B.P.V.; Pakhomov, E.A.; Henschke, N.; et al. Macrozooplankton and micronekton diversity and associated carbon vertical patterns and fluxes under distinct productive conditions around the Kerguelen Islands. *J. Mar. Syst.* **2022**, *226*, 103650. [CrossRef]
11. Béhagle, N.; Cotté, C.; Lebourges-Dhaussy, A.; Roudaut, G.; Duhamel, G.; Brehmer, P.; Josse, E.; Cherel, Y. Acoustic distribution of discriminated micronektonic organisms from a bi-frequency processing: The case study of eastern Kerguelen oceanic waters. *Prog. Oceanogr.* **2017**, *156*, 276–289. [CrossRef]
12. Olivar, M.P.; Castellón, A.; Sabatés, A.; Sarmiento-Lezcano, A.; Emelianov, M.; Bernal, A.; Yang, Y.; Proud, R.; Brierley, A.S. Variation in mesopelagic fish community composition and structure between Mediterranean and Atlantic waters around the Iberian Peninsula. *Front. Mar. Sci.* **2022**, *9*, 1028717. [CrossRef]
13. Olivar, M.P.; Hulley, P.A.; Castellón, A.; Emelianov, M.; López, C.; Tuset, V.M.; Contreras, T.; Molí, B. Mesopelagic fishes across the tropical and equatorial Atlantic: Biogeographical and vertical patterns. *Prog. Oceanogr.* **2017**, *151*, 116–137. [CrossRef]
14. Hernández-León, S.; Olivar, M.P.; de Puelles, M.L.F.; Bode, A.; Castellón, A.; López-Pérez, C.; Tuset, V.M.; González-Gordillo, J.I. Zooplankton and Micronekton Active Flux Across the Tropical and Subtropical Atlantic Ocean. *Front. Mar. Sci.* **2019**, *6*, 535. [CrossRef]
15. López-Pérez, C.; Olivar, M.P.; Hulley, P.A.; Tuset, V.M. Length–weight relationships of mesopelagic fishes from the equatorial and tropical Atlantic waters: Influence of environment and body shape. *J. Fish Biol.* **2020**, *96*, 1388–1398. [CrossRef] [PubMed]
16. Kaartvedt, S.; Staby, A.; Aksnes, D.L. Efficient trawl avoidance by mesopelagic fishes causes large underestimation of their biomass. *Mar. Ecol. Prog. Ser.* **2012**, *456*, 1–6. [CrossRef]
17. Pearcy, W.G. Quantitative Assessment of the Vertical Distributions of Micronektonic Fishes with Opening/Closing Midwater Trawls. *Biol. Oceanogr.* **1983**, *2*, 289–310. [CrossRef]
18. Nelson, J.S.; Grande, T.C.; Wilson, M.V.H. *Fishes of the World*, 5th ed.; John Wiley & Sons Inc: Hoboken, NJ, USA, 2016; p. 434.
19. Peña, M.; Olivar, M.P.; Balbín, R.; López-Jurado, J.L.; Iglesias, M.; Miquel, J. Acoustic detection of mesopelagic fishes in scattering layers of the Balearic Sea (western Mediterranean). *Can. J. Fish. Aquat. Sci.* **2014**, *71*, 1186–1197. [CrossRef]

Journal of *Marine Science and Engineering*

MDPI

Article

Barcelona Coastal Monitoring with the "Patí a Vela", a Traditional Sailboat Turned into an Oceanographic Platform

Inmaculada Ortigosa [1,*], Raul Bardaji [2], Albert Carbonell [3], Oriol Carrasco [4], Marcella Castells-Sanabra [1], Rafel Figuerola [3], Nina Hoareau [4], Jordi Mateu [1], Jaume Piera [4], Joan Puigdefabregas [5], Joaquín Salvador [4], Carine Simon [4], Ignasi Vallès-Casanova [4] and Josep L. Pelegrí [4,*]

1 Departament de Ciència i Enginyeria Nàutica, Universitat Politècnica de Catalunya, 08003 Barcelona, Spain; marcella.castells@upc.edu (M.C.-S.); jordi.mateu@upc.edu (J.M.)
2 Marine Technology Unit, Department of European Multidisciplinary Seafloor and Water Column Observatories, CSIC, 08003 Barcelona, Spain; bardaji@utm.csic.es
3 Club Patí de Vela de Barcelona, 08003 Barcelona, Spain; albert.ca89@gmail.com (A.C.); info@pativelabarcelona.com (R.F.)
4 Departament d'Oceanografía Física i Tecnològica, Institut de Cienciès del Mar, CSIC, 08003 Barcelona, Spain; carrasco@icm.csic.es (O.C.); nhoareau@icm.csic.es (N.H.); jpiera@icm.csic.es (J.P.); jsalvador@icm.csic.es (J.S.); csimon@icm.csic.es (C.S.); valles@icm.csic.es (I.V.-C.)
5 Departament de Física, Universitat de les Illes Balears, 07122 Palma de Mallorca, Spain; joan.puigdefabregas@uib.cat
* Correspondence: inmaculada.ortigosa@upc.edu (I.O.); pelegri@icm.csic.es (J.L.P.); Tel.: +34-934010873 (I.O.)

Citation: Ortigosa, I.; Bardaji, R.; Carbonell, A.; Carrasco, O.; Castells-Sanabra, M.; Figuerola, R.; Hoareau, N.; Mateu, J.; Piera, J.; Puigdefabregas, J.; et al. Barcelona Coastal Monitoring with the "Patí a Vela", a Traditional Sailboat Turned into an Oceanographic Platform. *J. Mar. Sci. Eng.* **2022**, *10*, 591. https://doi.org/10.3390/jmse10050591

Academic Editors: Georgy I. Shapiro and Mikhail Emelianov

Received: 31 January 2022
Accepted: 20 April 2022
Published: 27 April 2022

Abstract: Shelf waters near large cities, such as Barcelona, are affected not only by meteorological episodes but also by anthropogenic influence. Scientists usually use data from on-site coastal platforms to analyze and understand these complex water ecosystems because remote sensing satellites have low spatiotemporal resolution and do not provide reliable data so close to the coast. However, platforms with conventional oceanographic instrumentation are expensive to install and maintain. This study presents the scientific adaptation and initial measurements from a "patí a vela", which is a very popular unipersonal catamaran in Barcelona. This versatile sailing vessel has been adapted to contain several low-cost sensors and instruments to measure water properties. Here, we describe the setup of a multi-parameter prototype, and then focus on results obtained using a low-cost temperature profiler. First, the temperature data are compared and validated with another conventional oceanographic instrument used in monthly oceanographic cruises. Then, field measurements between July and November 2021 are used to explore the relationship between air and water temperature in the Barcelona coastal area, showing the seasonal evolution of the temperature profile. We conclude that citizen sampling from fully sustainable sailing boats may turn into an effective strategy to monitor the urban coastal waters.

Keywords: scientific sailing boat; low-cost sensors; temperature-depth profiler; citizen science

1. Introduction

The United Nations adopted, in 2015, the 2030 Agenda for Sustainable Development, a blueprint towards global welfare for current and future generations. Within this agenda, sustainable growth in the marine and maritime sectors requires the development of effective tools, in terms of decreased uncertainty and low cost, for seawater quality monitoring.

Coastal marine environments are particularly vulnerable to both climate change and global change. The coastline is indeed susceptible to major impacts associated with anthropogenic climate change, particularly the rising sea level and the more frequent extreme meteorological events [1]. It is also affected by increased contaminants and degraded benthic habitats, with far-reaching implications for marine ecosystems and human health and welfare [2]. However, coastal waters are under sampled both in time and space [3] due to the difficulty of setting oceanographic platforms that can carry out continuous sampling away from the coastline and the limited accuracy and spatiotemporal resolution of satellites in the shelf waters. In contrast, in recent years, there have appeared different innovative ways of sampling coastal waters that take advantage of the number of participants engaged in nautical recreation sports [4–7].

Barcelona, with a population of over 3.3 million people in its metropolitan area, has a high impact on its coastal marine system. This, in turn, causes Barcelona citizens to become negatively affected by unhealthy marine ecosystems, impacting other economic areas such as local fisheries and tourism. With heavy rain and high seas, stormy episodes are remarkable examples of this two-way interaction. They lead to high-polluted coastal waters (because of water runoff from the city) and beach-width reduction, which affect the coastal environment and shoreline at different time scales.

Given its social and ecological relevance, it is of fundamental importance to understand and preserve the marine environment of Barcelona. The first condition to understand this environment is to monitor its spatiotemporal variations. Despite the apparently easy access to the marine environment in Barcelona, coastal monitoring is limited by the high cost and limited flexibility of classical oceanography. Apart from the water quality control carried out in summer (bathing water status) and a daily examination of the surface temperature carried out by the Catalan Water Agency [8,9], the only systematic year-long offshore sampling has been along a line normal to the Somorrostro Beach (PUDEM line), carried out since 2002 [10,11]. These data are available from the Institut de Ciències del Mar (ICM) [12] and have led to several reports and scientific publications [13–15].

In order to improve the sampling frequency and spatial coverage, we propose a local strategy for monitoring the Barcelona inner-shelf waters with the participation of sailors that use the patí a vela (PV), a recreational sailing boat widely used in Catalonia. Nowadays, the PV fleet is also found far beyond Catalonia, with sailing boats in Valencia and Andalusia, and also reaching the French, Belgian and Dutch coasts.

According to the International Patin Sailing Association (ADIPAV) data, more than 200 PV sailing boats could be directly involved with citizen sampling near Barcelona. If part of these recreational boats were equipped with an oceanographic low-cost-sensors platform, the sampling capacity of environmental indicators in Barcelona and nearby coastal waters would be significantly enhanced. In this article, we first explain how the PV is adapted to become an oceanographic sampling platform and then illustrate its potential for monitoring the coastal waters by presenting temperature data gathered in the summer and fall of 2021.

2. Patí Científic Project

The project "Development of a citizen monitoring program for the Barcelona coastal waters: the Scientific Patí a Vela (PATI CIENTIFIC)" is a collaborative initiative of the ICM, the Club Patí de Vela de Barcelona (CPVB), and the Facultat de Nàutica de Barcelona, with funding from the Barcelona City Council. The main objective of PATI CIENTIFIC is to develop an inclusive and transformative low-cost resource that can effectively help monitor the coastal waters. This is undertaken by designing, constructing and testing two main sampling devices: a multi-parametric box adapted on-board the PV, which becomes a multi-platform instrument to sample the surface waters while the PV sails, and a temperature-depth profiler, which can be used in specific sites to observe the thermal structure of the water column. In this way, the PV is easily converted into a scientific PV (SPV); a sailboat ready to provide valuable environmental information about the nearby coastal ocean.

All data collected with the SPV are freely available thanks to the development of a Web portal (https://paticientific.org/ (accessed on 13 April 2021) where data are incorporated and retrieved in the context of the philosophy of Open Science, hence promoting citizen engagement in scientific data collection and turning research into a process of social participation.

3. Materials and Equipment

3.1. Scientific Patí a Vela Description

A PV is a lightweight sailing catamaran, usually wooden and with a crew of only one person. It originated in Barcelona and nearby Badalona during the middle of the 20th century, as citizens wanted to bath in clean waters, away from the near-beach pollution that was largely caused by factories along the coastline. The standard design currently in use follows the original one designed by the Mongé brothers in 1943 [16,17]. The PVs' main peculiarities are that it does not have a rudder, keel or boom, so it goes into the water directly from the beach (Table 1).

Table 1. Main characteristics of the Scientific Patí de Vela.

LOA	Beam	Displacement (Hull)	Sail Area	Hull Type
5.60 m	1.60 m	89 kg	12.60 m^2	Catamaran

The PV is maneuvered by modifying the relative distance between the hull and sail pressure centers by simply displacing the crew along the deck (Figure 1). When the crew moves, so does the boat's gravity center, and the PV trim is consequently modified. The sail pressure center can be controlled by adjusting the tension in the sail.

Figure 1. Picture of an SPV while departing from the CPVB premises, near downtown Barcelona, with the ICM facilities at the back. The insets illustrate the steering control of the PV, reprinted/adapted with permission of Ref. [17]. 1974, Editorial Hispano Europea.

The PV turns into the SPV when adapted and equipped with an on-board multiparametric box, which contains a water pump, battery, electronics and a sensor box (Figure 2). The sensor box is modular, so that sensors can be easily incorporated into or removed from the box. We have tested temperature (PT1000), conductivity (pro K10), pH (LabGrade) and dissolved oxygen (ENV-40-DOC) sensors provided by Atlas Scientific (Long Island City, NY, 11101); although, only the temperature and conductivity sensors are used routinely for monitoring. The sensors are connected to a Smart Citizen kit, which contains a central data logger, developed by FabLab Barcelona (https://fablabbcn.org/ (accessed on 13 April 2022)). Two flexible tubes reach in and out from the sensor box, pumping surface water through the several sensors located inside the box and releasing it back to the sea. In this way, we can continuously sample the surface oceanographic variables, which can either be stored internally in the data logger or transmitted with a smartphone. The overall cost of the multiparametric box with temperature and conductivity sensors is about EUR 550.

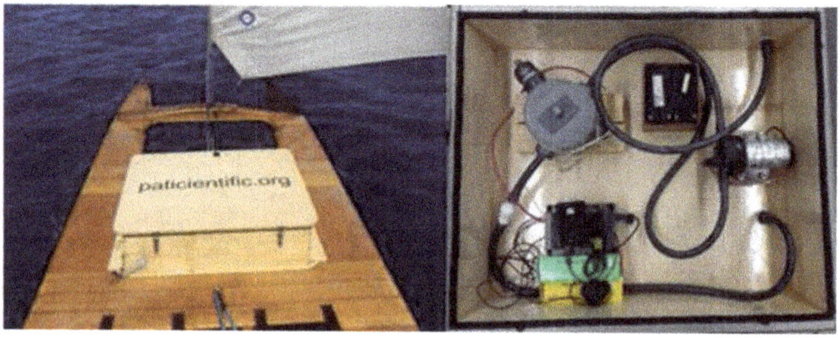

Figure 2. (**Left**) Removable on-board multiparametric box installed in a PV. (**Right**) Detail of the interior configuration of the removable on-board multiparametric box.

The on-board multiparametric box, which is placed between the second and third bench of the catamaran (Figure 2), modifies the displacement, gravity center and trim of the boat. If the skipper wants to maintain the direction of the boat, he/she has to sit closer to the stern than without the instrumented box installed onboard. However, these changes have little effect on the boat's stability, resistance and maneuverability. Ortigosa et al. [18] analyzed the effects of this prototype box on the hydrodynamics of the sailing boat, both numerically and at sea, and determined the best configuration that preserves the essential stability characteristics so that it does not interfere with the seakeeping performance of this recreational boat in the water.

Figure 3 illustrates the changes in sea-surface temperature (SST) and sea-surface salinity off downtown Barcelona as sampled with the multiparametric box on the SPV on 22 July 2021. The data, gathered in less than two hours, illustrate substantial temperature and salinity changes in relatively small distances, emphasizing the potential of the SPV to quickly sample the short-scale spatial variability.

3.2. SPV TD Instrument and Sensors

The SPV is also equipped with a hand-held open-source temperature-depth (TD) profiler that can be deployed down to depths of about 100 m and stores data on a uSD memory card. The low-cost sensors reduce the price of the entire setup by a factor of ten and open the possibility that several SPVs simultaneously sample the coastal waters, significantly improving the spatial and temporal monitoring. Further discussions in this article focus on the profiles obtained by this SPV TD profiler.

The TD profiler needs to be simple enough for use by sailors; additionally, it has to be small so that it does not interfere substantially with the center of gravity and maneuverability of the boat (Figure 4). The TD profiler consists of temperature and pressure sensors connected to an Arduino Pro Mini, with all electronics and sensors assembled in a waterproof housing made of commercial PVC components [19].

The sensors used are from Blue Robotics: the Celsius Fast Response temperature sensor with 0.1 °C accuracy that samples between −5 and 50 °C and has an operating depth range of 0–975 m; and the Bar30 pressure sensor with an accuracy of 200 mbar and operating range of 0–300 m. Connections are made with a custom-designed PCB, and activation is performed with an external magnetic switch. When the profiler is switched on, it starts recording data continuously at 1 Hz on a uSD memory card. The instrument is lowered by hand and has been tested down to 100 m depth. The overall cost for the components of this SPV TD is less than EUR 300.

Figure 3. (**Top**) Sea-surface temperature and (**bottom**) sea-surface salinity on 22 July 2021, along the navigation track of the SPV.

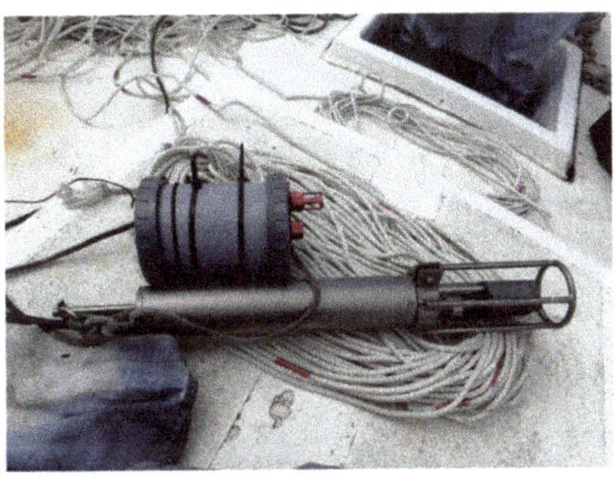

Figure 4. SPV TD aside the Sea & Sun CTD 48Mc probe; the respective lengths are approximately 20 cm and 45 cm long.

During the field campaign, the SPC sails to every study point, where the temperature vs. depth profile is recorded. As the PV has no electronics or navigation systems, the positioning is set using a smartphone. The procedure of recording the temperature profile at every point consists of the following steps: the boat is stopped in the desired location; the low-cost TD profiler is switched on and kept on the surface for 60 s to acclimatize the temperature sensor; the profiler is released until reaching the desired depth and then it is recovered and switched off.

3.3. SPV TD Sensors Validation

The open-source TD profiler was validated by comparison to data from a Sea & Sun 48Mc CTD, gathered down to 100 m in a station located in Estartit, along the Catalan coast and north of Barcelona. For the comparison, both datasets were first collocated in time, then interpolated at 1 Hz, and finally the temperature and pressure data were fitted with a linear regression. Figure 5 compares the temperature profiles from both instruments, and

Figure 6 presents the dispersion graphs for both sensors, with regression coefficients very close to one.

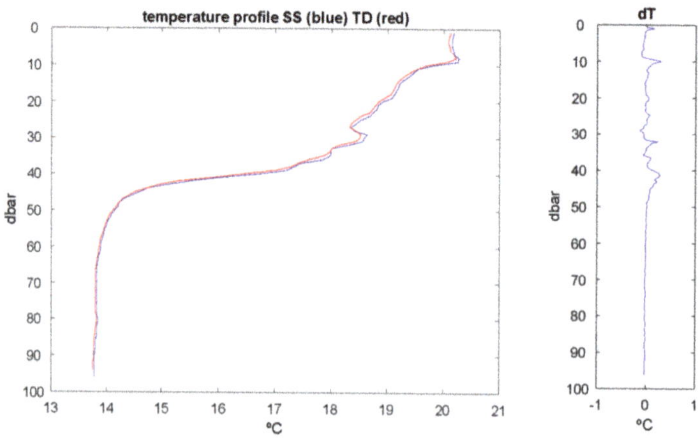

Figure 5. (**Left**) Temperature profiles obtained with the SPV TD profiler (red curve) and with the Sea & Sun CTD 48Mc (blue) and (**right**) their difference (CTD minus TD values).

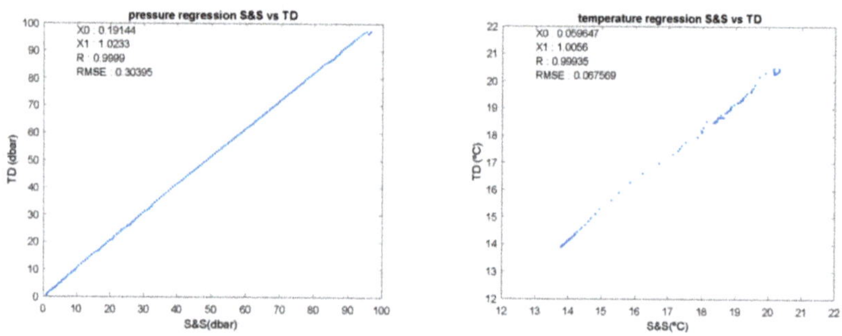

Figure 6. Regression plots for (**left**) pressure and (**right**) temperature between the PV TD and Sea & Sun CTD 48Mc probes.

The SPV TD produces a slightly smoother profile than the Sea & Sun CTD, which is consistent with a slightly slower response from the Celsius Fast Response temperature sensor. Nevertheless, the downcast temperature profiles match very well, with a global difference of -0.054 ± 0.055 °C, below the nominal resolution of both sensors. Note that in the deeper and more stable part of the profile (50 to 100 m) the difference is smaller (-0.024 ± 0.005 °C) while differences are larger in the upper 50 m of the profile (-0.080 ± 0.066 °C), possibly due to a slower response of the SPV TD or partly to turbulence generated by the Sea & Sun profiler, which was placed 1 m below the SPV TD.

3.4. Data Acquisition and Quality Control in the SPV Surveys

The data acquisition is quite simple and can be carried out by non-specialized sailors. Once the SPV arrives at the study point, the TD profiler is switched on, held for one minute at the sea surface for sensor stabilization, and lowered with a rope at a rate of about 0.5 m s^{-1} down to the maximum desired depth. Once the profiler is recovered onboard, it is switched off and the SPV sails to the next sampling point. As the ship has no electronic navigation, positioning is recorded on a mobile phone and stored in an independent dataset to be merged with oceanographic data in the post-processing steps.

A specific quality control is applied to the raw TD measurements, adapted from common oceanographic CTD data processing [20,21] and from specific observations during the SPV cruise. During the data processing, a quality flag value is assigned at each measurement (Table 2).

Table 2. Quality control flag values applied during the data processing.

Flag Value	Definition Term	Abbreviated Term
0	No quality control procedures have been applied to the data value. This is the initial status for all data values entering the working archive.	No quality control
1	Good quality data value that is not part of any identified malfunction of the instrument and has been verified as consistent with real phenomena during the quality control process.	Good
2	This value has been derived by interpolation from other values in the data object.	Interpolated
3	Data value is recognized as unusual during quality control, possibly inconsistent with real phenomena.	Suspect
4	An obviously erroneous data value	Bad
9	The data value is missing.	Missing

Data are considered "bad" (flag = 4) if one of the following occurs: (i) data correspond to the first 10 s after switching on, as the temperature sensor needs times to stabilize; (ii) data are located between the sea surface and 1 m depth; (iii) a temperature value is too far from the mean nearby values, and the criteria chosen being that the difference in data with the mean value is greater than four times the standard deviation. Data are flagged as "suspect" (flag = 3) if one of the following happens: (i) the downcast velocity is less than 0.2 m/s, potentially leading to multiple values at one single depth; (ii) we observe a loop in the profile due to the movement of the boat. Finally, the good data (flag = 1) are averaged every 0.2 m in order to create a temperature profile with regular sampling. When data are missing, an interpolation is applied and the data are flagged (flag = 2). Figure 7 shows a sample dataset in a station after applying the quality control procedure.

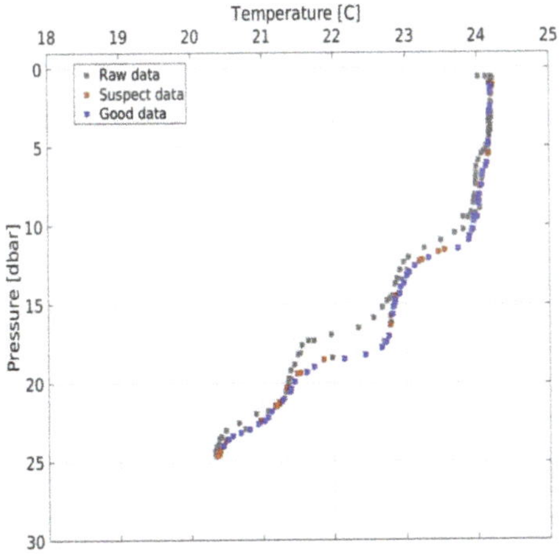

Figure 7. TD profile after applying the data quality procedure (station TD5, 1 October 2021). Both the downcast (colored points) and upcast (grey points) profiles are shown. Blue dots represent the good data (flag = 1) while orange dots correspond to data flagged as suspect data (flag = 3).

4. Methods

4.1. Study Area

The coastal zone studied is in front of the Somorrostro beach, near downtown Barcelona. This area is also close to the Port Olympic, a Marina with remarkable yachting activity during the summer season (Figure 8). Most of the Barcelona coastline becomes a fairly crowded beach during summer, and the study area is also affected by runoff water from the city collected by two of the city outfalls. In the case of high rainfall, the city outfalls may discharge untreated street water into the sea [15]. Therefore, a strong anthropogenic pressure affects the natural conditions of the study area.

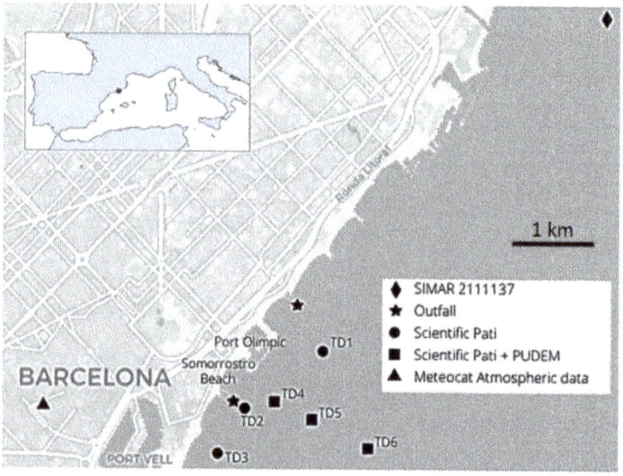

Figure 8. Study area off Barcelona with the location of the TD, wave and meteorological stations. Stations TD4, TD5 and TD6 coincide with three of the stations of the monthly PUDEM monitoring [15].

The area of study is the inner-shelf in front of the city of Barcelona, in the NW Mediterranean Sea (Figure 8). This is an area where the atmosphere–ocean interaction modifies the water masses throughout the year, with a mean monthly range of 14 °C. From spring to autumn, a seasonal thermocline is formed; in winter this thermocline vanishes and the water column turns homogeneous [15]. The zone is characterized by a predominant current from the northeast that drives the relatively low-density shelf water [22].

Six measurement stations (TD1 to TD6) were used to assess the complexity of the area, which goes from 41.3749° N to 41.3856° N and from 2.1951° E to 2.2162° E (Table 3). Station TD1 allows the study of the area located east of Port Olympic. This point receives the direct flow of water, which enters the study zone from the northeast. It may also influence one of the outfall points that can be seen on the map. TD2 is located in front of one of the water discharge points of the city and in front of the Port Olympic mouth. TD2 and TD3 are the points closer to the beach.

TD4, TD5 and TD6 are stations along a line normal to shore, which permit studying the evolution of the vertical temperature structure with increasing depth and distance to shore. These three points coincide with the study points during the oceanographic PUDEM cruises. In Figure 8, we also indicate the location of a SIMAR model point, where wave period and height are available from Puertos del Estado [23], and a nearby city meteorological station, where wind velocity, air temperature and solar radiation are available from the Catalan Meteorological Service [24].

Table 3. Position of the six measurement stations, showing the maximum depth (Mdepth) reached by the TD profiler. The bottom depth in stations TD5 and TD6 is about 30 and 40 m, respectively.

	Latitude (N)	Longitude (E)	Mdepth (m)
TD1	41.3856°	2.2099°	23
TD2	41.3798°	2.1986°	12
TD3	41.3749°	2.1951°	11
TD4	41.3806°	2.2029°	20
TD5	41.3785°	2.2084°	28
TD6	41.3757°	2.2162°	28

4.2. Period of Study

For real-time coastal ocean monitoring, field campaigns with the SPV were carried out in the study area in July, August, September, October and November 2021, for a total of 15 days (Table 4). This represents nearly one weekly cruise, designed to record the temperature profile in depth using the low-cost temperature profiler synchronized with meteorological data. Data are missing for some weeks, as the SPV requires some minimum wind conditions for sailing.

Table 4. Dates between July and November 2021 when TD data were acquired with the SPV.

Date	TD1	TD2	TD3	TD4	TD5	TD6
28 July	X	X	X			
3 August	X	X	X			
17 August	X	X	X			
31 August				X	X	X
6 September				X	X	X
17 September	X	X	X	X	X	X
21 September	X	X	X	X	X	X
28 September	X	X	X	X	X	X
1 October	X	X	X	X	X	X
5 October	X	X	X			
7 October	X	X	X	X	X	X
18 October	X	X	X			
28 October	X	X	X	X	X	X
8 November	X	X	X	X	X	X
18 November				X	X	X

Not every field campaign obtained data at all stations. If the meteorological conditions changed during the field campaign, caused by an extreme rise or fall in the wind, it may not have been possible to complete the sampling in all stations.

4.3. Other Datasets

We use SST data from the WST-non time-critical Sentinel-3 satellite products [25], available from the Puertos del Estado web platform (www.puertos.es (accessed on 20 November 2021)). This satellite delivers daily data at a spatial resolution of 1 km (0.3 °C precision when cloudiness is below 20%). A single SST value is derived along the swath of the satellite, using the best-performing sensor algorithms and supporting data field. Only data flagged as being of "excellent quality" are used. This causes that only 25 satellite SST values are available for our 4-month study period.

Hourly wave period and height data come from the coastal model SIMAR [23], developed by Puertos del Estado (www.puertos.es (accessed on 20 November 2021)). The closest point to the sampling area used in this study is SIMAR 2111137 (2.25°E, 41.42°N). Hourly observations of wind velocity, air temperature and solar radiation come from the Catalan Meteorological Service [24], concretely from the station located at 41.38° N 2.17° E, near downtown Barcelona. Daily averages of each variable are calculated, except for the maximum radiation and accumulated daily rainfall.

Meteorological and wave data during the period of our study are shown in Figure 9. Figure 9b shows the squared wind speed, as an indicator of the wind energy available to mix the water column, either in the form of waves or as turbulence that can increase the depth of the surface mixed layer. This figure reveals that wind remains relatively low during the summer months (July, August and early September) and that there are four events where the values exceed 40 $m^2\ s^{-2}$: 22–23 September, 4 October, 20 October and particularly 10–11 November, when values exceeded 120 $m^2\ s^{-2}$. Figure 9d presents the wave energy per unit length (proportional to the squared significant wave height) for this same period. This figure suggests a good correspondence between intensified winds and increased waves, indicating that most of the wave episodes are locally generated.

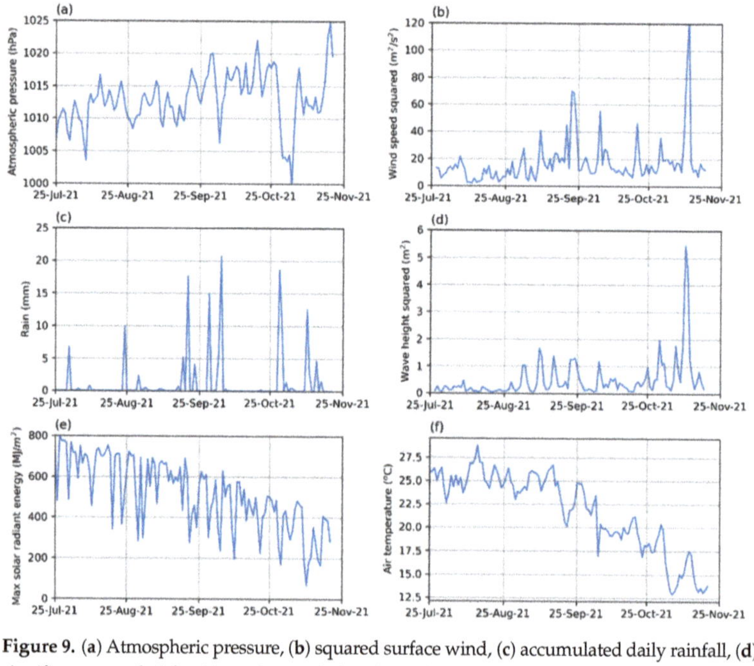

Figure 9. (**a**) Atmospheric pressure, (**b**) squared surface wind, (**c**) accumulated daily rainfall, (**d**) squared significant-wave height, (**e**) maximum daily solar radiation, and (**f**) surface air temperature.

Figure 9 also shows several variables that illustrate the meteorological forcing over the coastal ocean. Figure 9a presents the atmospheric pressure, with fairly stable values between 1007 and 1020 hPa, except for a minimum during the first days of November, which reached 1000 hPa. Figure 9e presents the maximum daily radiation, with a decreasing trend after the peak values in late July and intermittent minima that last one or two days and represent cloudy conditions. Figure 9c presents daily rainfall, with little rain during July and August and increasing in fall. High rainfall events correspond to periods of low radiation and high winds, except for the early-November period of low pressure that does not show wind intensification. Finally, Figure 9f shows the characteristics of the surface air temperature, with maximum values until mid-September followed by a fast-decreasing trend, with some intermittency that is tallied to atmospheric pressure and solar radiation.

5. Results and Discussion

In order to illustrate the potential of the SPV for monitoring the inner shelf, in Figure 10 we present the evolution of the temperature distribution in the water column as obtained with the TD profiler during the field campaigns, carried out roughly about once per week between 28 July and 18 November (Table 4). In these profiles, we have removed the upper few meters of the water column such as to remove the temperature variations associated with the diurnal cycle.

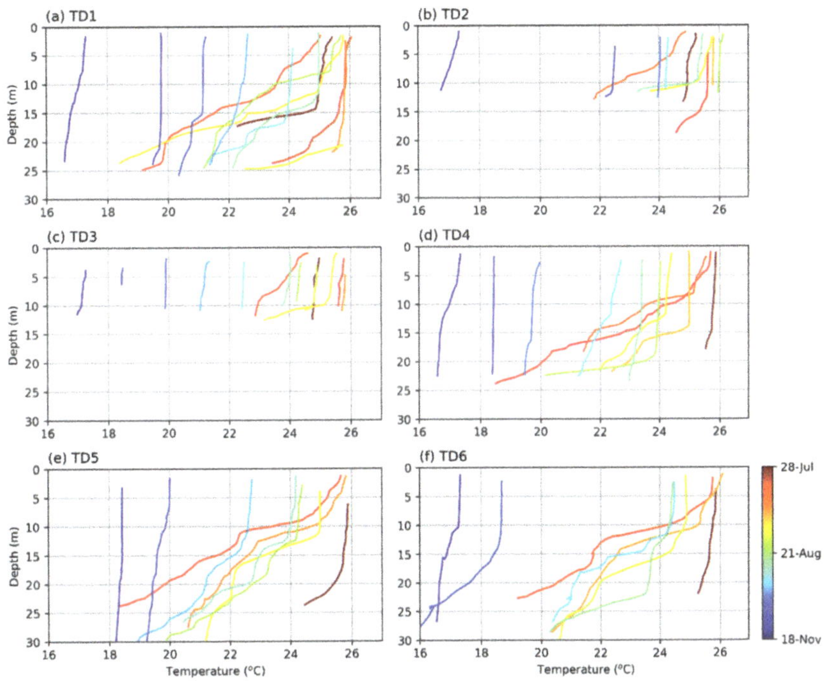

Figure 10. Temperature profiles (**a–f**) are colored as a function of time. Note that the start and end dates for stations (**a–c**) (top color bar) are not the same as for stations (**d–f**) (bottom color bar). The reader is referred to Table 4 for the exact sampling dates.

Stations TD2 and TD3 correspond to nearshore and relatively shallow waters (depths of 12 m and 11 m, respectively), which are fully affected by the influence of the surface gravity waves. For the short waves that characterize the nearshore platform off Barcelona, with typical periods of 4 s, waves feel the seafloor at water depths less than about 25 m. It is remarkable that some samples of these points exceed the depths of 12 m and 11 m. The sea floor in this area is irregular and the SPV is a floating device, so that it is difficult to sample the same exact location each time. The temperature profiles in TD2 and TD3 show that, during August and September, the upper 10 m of the water column remained most of the time over 24 °C and nearly homogeneous. The single exception is the August 17 profile (light red), which does show a clear thermocline, reflecting that sampling took place at a time of very low winds and waves, high solar radiation and maximum air temperature (Figure 9b,d,e,f). One relevant issue is that there are substantial differences between the temperature in these two sites, despite them being separated by only about 500 m, perhaps related to the location of TD2 just in front of a city runoff-water outlet. In particular, TD3 illustrates a gradual cooling of the water column throughout October and November as a result of the continuous decrease in solar radiation (Figure 9e).

At the deeper stations (TD1, TD4 and particularly TD5 and TD6) there is a substantial vertical structure except during late October and November. As at the shallower stations, the peak sea-surface temperatures take place in July and August and the temperatures progressively decrease during October and November. Station TD5 illustrates that the bottom waters (about 30 m) remain between about 18 and 21 °C during the entire period.

Rain events in excess of 10 mm per day occurred on 24 August, 20 and 28 September, 2 and 29 October, and 10 November. These rain events brought some runoff water from the city outlets but their effect was very limited to the surface of most waters. Finally, solar radiation was intermittent, caused by some cloudiness and rain, but until the end of September it remained nearly always above 400 Watts m^{-2}. Hence, we may infer that the observed decrease in sea-surface temperature was due to the negative trend in radiation and the air temperature drop rather than to the intermittent cloudiness.

One main mechanism for air–sea heat exchange is the flux of latent heat, which is proportional to water–air temperature difference near the sea surface. To analyze the water–air temperature difference during the period under study, we use daily-mean air temperatures as calculated from the hourly values provided by the Catalan Meteorological Service [24], and the SST values as derived from a combination of the TD near-surface data and the Sentinel-3 data flagged as excellent quality. With this strategy, we have produced a fairly continuous time series (values every few days) for daily temperature differences at the water–air interface (Figure 11). We are aware that the combination of the TD near-surface data and the satellite can introduce some unresolved daily variability, as the Sentinel data likely does not properly resolve the diurnal cycle. Nevertheless, Figure 11 shows that this strategy allows for resolving the autumn trend of increasing temperature differences.

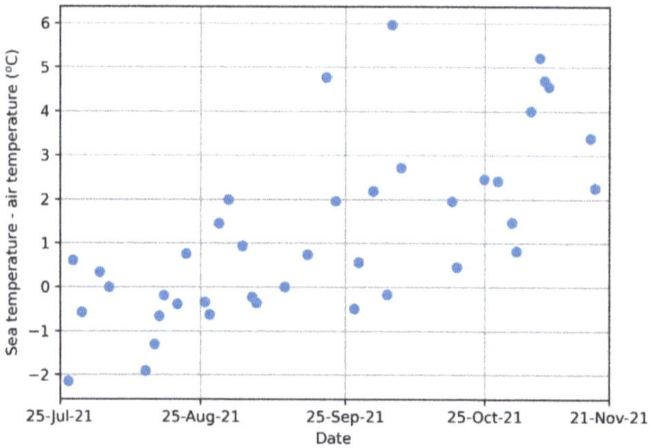

Figure 11. Time series of sea-surface water–air temperature differences in the study area.

The water–air temperature-difference time series also shows that there were three events of maximum differences, on 22 September, 5 October and 10 November (Figure 11), which correspond to windy and cloudy conditions that brought a substantial decrease in the surface air temperature while the surface waters remained much less affected (Figure 9b,e,f).

We close this article with a simple analysis of the TD data that shows the evolution of the water column and its relation with meteorological forcing, hence illustrating the SPV possibilities for monitoring the coastal ocean. We restrict our analysis to those stations with bottom depths of 20 m or greater, i.e., stations TD1 and TD4 close to shore and stations TD5 and TD6 further offshore (Table 3 and Figure 8).

For our purposes, we define two additional variables: the overall temperature gradient $\Delta T/\Delta z$, simply calculated as the temperature difference between the surface and bottom waters divided by the water depth, and the (surface) mixed layer depth d, defined with a simple criterion that sets the base of the mixed layer where the temperature gradient exceeds 0.15 °C m^{-1} (which considers the accuracy of our temperature sensor, 0.1 °C) These two variables are useful to discuss some of the key mechanical and thermodynamic aspects in the evolution of the mixed layer depth.

The overall temperature gradient and the mixed-layer depth are inversely related. The surface mixed layer deepens through the entrainment of subsurface cold waters, which leads to a decrease in the overall temperature gradient. Both variables are also subject to changes in air–sea heat exchange, directly through changes in solar radiation or indirectly through an increase in the water–air temperature difference that enhances the latent and sensible heat fluxes.

The incorporation of subsurface waters into the surface mixed layer causes an increase in the potential energy of the water column that requires a source of kinetic energy, which has to come from wind- and wave-induced turbulence. Hence, we expect that there will be an inverse relation between mixed-layer depth and overall temperature gradient, and that both $\Delta T/\Delta z$ and d will respond to changes in the mechanical stirring (by waves, winds or both) but conditioned (possibly to a lesser degree) by variations in the ocean–atmosphere heat exchange.

In order to assess these relations, we present scattered plots of these two oceanographic variables ($\Delta T/\Delta z$ and d) as a function of three variables that represent the different types of daily-mean meteorological forcing: water–air temperature difference ΔT, squared significant wave-height ${h_s}^2$ and squared wind-speed W^2 (Figure 12).

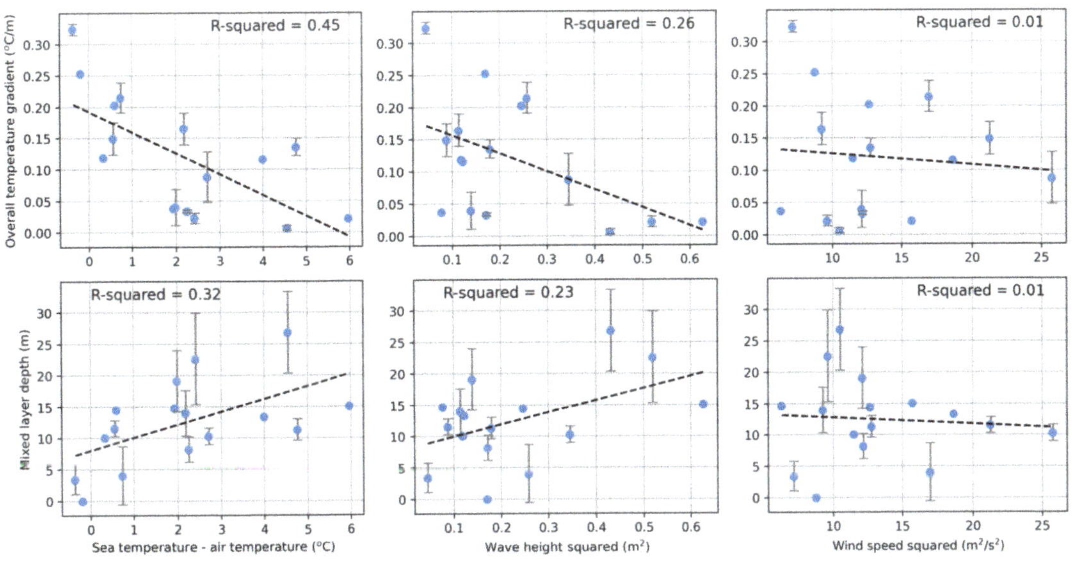

Figure 12. Scattered plots of (top panels) overall temperature gradient and (bottom panels) mixed-layer depth as a function of (from left to right) sea–air temperature difference, squared significant wave-height and squared wind-speed. The correlation coefficients are shown.

We observe that $\Delta T/\Delta z$ is inversely correlated to ΔT and d is directly correlated to ΔT (Figure 12, left panels). This is consistent with d increasing through the entrainment of subsurface waters into the mixed layer, hence causing a decrease in $\Delta T/\Delta z$.

Finally, we may explore if the energy required for vertical mixing comes from wave- or wind-induced turbulence, or both; notice that unstable vertical convection does not take place during summer-fall because the water column is stable as a result of the large increase in temperature towards the sea surface (Figure 10) [15]. We perform this analysis with scattered plots of $\Delta T/\Delta z$ and d as a function of $h_s{}^2$ and W^2 (Figure 12, center and right panels), and find that there are significant linear correlations for the wave energy while the relation is not significant for the wind energy. Hence, we are left with the conclusion that the seasonal deepening of the surface mixed layer, and consequent reduction in the overall temperature gradient, is caused by the progressive increase in the wave field (Figure 9d). This result is consistent with the homogeneity of the water column in shallow locations, where the bottom depth is less than the vertical influence of the wave regime (except for summer periods when there was little wave activity). Further, the lack of correlation of $\Delta T/\Delta z$ and d with W^2 tells us that those waves driving mixing are largely non-locally generated. These results, which are consistent with observations in the inner shelf of other shelf regions [26], illustrate the good potential of data gathered with the SPV.

6. Conclusions

We have presented the development of a low-cost sampling strategy for coastal waters with the patí a vela (PV), a sustainable historical Catalan sailboat. The strategy is based on the incorporation of both a multiparametric box for sampling surface waters and a profiler for measuring the structure of the water column, in both cases with low-cost sensors and open-source data logging, turning the PV into a scientific PV (SPV).

We have enumerated the main components of the multiparametric box (temperature and salinity sensors in its first phase) and have thoroughly described the depth profiler (temperature in this initial stage). Further, we have also shown that the proposed sampling setup is fully compatible with the standard sailing of this sailboat, and the good performance of the TD low-cost sensors was confirmed through a favorable comparison with a commercial instrument.

The two different types of low-cost open-source instruments, the multiparametric box and the depth profiler, complement each other finely, in a sense replicating the standard sampling strategy of research vessels. In the multiparametric box, surface water is continuously pumped through the sensors allowing for a swift sampling of the sea-surface conditions along the path of the sailboat. With one or a few SPV, the surface waters of a fairly large area (10–50 km^2) can be sampled in relatively short times (just a few hours); this is very appropriate for monitoring the water quality of surface waters, particularly after periods of intense rain and water runoff from the city. The depth profiler is the necessary counterpart for sampling the structure of the water column, particularly useful for assessing temporal changes associated with both natural and anthropogenic forcing.

As an example of the potential of the SPV, we have presented and analyzed the results obtained through 15 one-day campaigns that used a temperature-depth (TD) profiler, carried out between July and November 2021 in as many as six stations within an area of about 3 km $\times$ 3 km, in water depths from the coastline to about 40 m. The temperature-depth profiles show that the mixed-layer depth increases and the overall temperature depth-gradient decreases from summer into fall as the height of non-local waves progressively increases and the air temperature decreases.

The scientific patí a vela (SPV) turns out to be an inexpensive and maneuverable oceanographic platform to sample the coastal waters, used either individually for regular monitoring or with several units installed in a small fleet aimed at investigating the spatiotemporal ocean variability. Because of its wide utilization, as a very versatile and inexpensive sailboat, the SPV can grow as an inclusive and collaborative citizen science platform that can effectively contribute towards an improved monitoring and understanding of the inner continental shelf, particularly near densely populated areas.

Author Contributions: Conceptualization, J.L.P. and I.O.; data collection, mostly O.C.; data analyses, I.O. and I.V.-C.; software, R.B., I.V.-C. and N.H.; methodology, validation, formal analysis, investigation, writing, M.C.-S., J.M., R.B., A.C., R.F., O.C., N.H., J.P. (Jaume Piera), J.S., C.S., I.V.-C., J.P. (Joan Puigdefabregas), I.O. and J.L.P.; funding acquisition, J.L.P. All authors have read and agreed to the published version of the manuscript.

Funding: This work has been carried out within the framework of the 2019 Barcelona Pla de Ciència, with funding from the Barcelona City Council through the project "Development of a citizen monitoring program for the Barcelona coastal waters: the Scientific Patí Vela" (PATI CIENTIFIC; references 19SO1645-006, 19SO1649-006, 19SO1651-006).

Institutional Review Board Statement: Not applicable.

Informed Consent Statement: Not applicable.

Data Availability Statement: Data supporting reported results can be found at https://paticientific.org/index.php/dades/ (accessed on 13 April 2022) and through https://zenodo.org/search (accessed on 13 April 2022) by searching "pati cientific".

Acknowledgments: We are thankful to all the staff and members of the CPVB for their continuous support in the SPV operations, and to Josep Pascual for his key help in the sensor validation exercise. We are very grateful to the Barcelona City Council for their support through projects PATÍ CIENTIFIC (Pla Barcelona Ciència 2019, ref. 19SO1645-006, 19SO1649-006, 19SO1651-006) and AULAMAR (Fundació Bit Habitat, ref. ID253). This article is a publication of the Unidad Océano y Clima of the Universidad de Las Palmas de Gran Canaria, a R+D+I CSIC-associate unit. R.B. is supported by the project EMSO—Laboratorios Submarinos Profundos. The ICM authors also recognize the institutional support of the Spanish Government through the Severo Ochoa Center of Excellence accreditation (CEX2019-000928-S).

Conflicts of Interest: The authors declare no conflict of interest.

References

1. Harley, C.D.G.; Hughes, A.R.; Hultgreen, K.M.; Miner, B.G.; Sorte, C.J.B.; Thornber, C.S.; Rodriguez, L.F.; Tomanek, L.; Williams, S.L. The impacts of climate change in coastal marine systems. *Ecol. Lett.* **2006**, *9*, 228–241. [CrossRef] [PubMed]
2. Amores, A.; Marcos, M.; Carrió, D.S.; Gómez-Pujol, L.L. Coastal Impacts of Storm Gloria (January 2020) over the Northwestern Mediteranean. *Nat. Hazards Earth Syst. Sci.* **2020**, *20*, 1955–1968. [CrossRef]
3. Alverson, K.; Baker, D.J. Taking the pulse of the oceans. *Science* **2006**, *314*, 1657. [CrossRef] [PubMed]
4. Brewin, R.J.W.; Hyder, K.; Andersson, A.J.; Billson, O.; Bresnahan, P.J.; Brewin, T.G.; Cyronak, T.; Dall'Olmo, G.; de Mora, L.; Graham, G.; et al. Expanding aquatic observations through recreation. *Front. Mar. Sci.* **2017**, *4*, 351. [CrossRef]
5. Brewin, R.J.W.; de Mora, L.; Jackson, T.; Brewin, T.G.; Shutler, J. On the Potential of Surfers to Monitor Environmental Indicadtors in The Coastal Zone. *PLoS ONE* **2015**, *10*, e0127706. [CrossRef] [PubMed]
6. Wright, S.; Hull, T.; Sivyer, D.B.; Pearce, D.; Pinnegar, J.K.; Sayer, M.D.J.; Mogg, A.O.M.; Azzopardi, E.; Gontarek, S.; Hyder, K. SCUBA divers as oceanographic samplers: The potential of dive computers to augment aquatic temperature monitoring. *Sci. Rep.* **2016**, *6*, 30164. [CrossRef] [PubMed]
7. Bresnahan, P.J.; Wirth, T.; Martz, T.R.; Andersson, A.J.; Cyronak, T.; D'Angelo, S.; Pennise, J.; Melville, W.K.; Lenain, L.; Statom, N. A sensor package for mapping pH and oxygen from mobile platforms. *Methods Oceanogr.* **2016**, *17*, 1–13. [CrossRef]
8. ACA. Quality of the Beaches and Swimming Areas of Catalonia. Available online: https://aplicacions.aca.gencat.cat/platgescat2/app.php/en/playa/157# (accessed on 21 October 2021).
9. BCN, Bany i Platges, Ajuntament de Barcelona. Available online: https://www.barcelona.cat/ca/que-pots-fer-a-bcn/banys-i-platges (accessed on 1 August 2019).
10. ICM, Estació Litoral de Barcelona, Institut de Ciències del Mar. Available online: http://coo.icm.csic.es/site-page/stations#Barcelona-ELB (accessed on 1 August 2019).
11. ICM, PUDEM SOS Application, Institut de Ciències del Mar. Available online: http://ide.cmima.csic.es/sos-js/web/examples/sos-apptest (accessed on 1 August 2019).
12. ICM. Oceanographic Data - Coastal Ocean Observatory. Available online: http://coo.icm.csic.es/site-page/oceanographic-data (accessed on 21 October 2021).
13. Arin, L.; Guillén, J.; Segura-Noguera, M.; Estrada, M. Open sea hydrographic forcing of nutrient and phytoplankton dynamics in a Med. coastal ecosystem. *Est. Coastal Shelf Sci.* **2013**, *133*, 116–128. [CrossRef]
14. Romero, E.; Peters, F.; Arin, L.; Guillén, J. Decreased seasonality and high variability of coastal plankton dynamics in an urban location of the NW Mediterranean. *J. Sea Res.* **2014**, *88*, 130–143. [CrossRef]
15. Guillén, J.; Arin, L.; Salat, J.; Puig, P.; Estrada, M.; Palanques, A.; Simarro, G.; Pascual, J. Coastal oceanographic signatures of heat waves and extreme events of dense water formation during the period 2002–2012 (Barcelona, NW Mediterranean). *Scientia Marina* **2018**, *82*, 189–206. [CrossRef]
16. ADIPAV. Reglamento de la Disciplina Patín a Vela. *Clase: Patín a Vela*. Available online: http://adipav.org/node/1503 (accessed on 31 January 2022).
17. Depoorter, G. *El Patín a Vela*; Editorial Hispano Europea: Sitges, Spain, 1974; p. 328.

18. Ortigosa, I.; Castells-Sanabra, M.; Llevadot, J.M.; Bardaji, R.; Hoareau, N.; Simon, C.; Pelegrí, J.; Casanova, I.V. Adapting the existing coastal Patí a vela fleet for scientific purposes. *J. Maritime Res.* **2020**, *17*, 88–94.
19. Beddows, P.A.; Mallon, E.K. Cave Pearl Data Logger: A Flexible Arduino-Based Logging Platform for Long-Term Monitoring in Harsh Environments. *Sensors* **2018**, *18*, 530. [CrossRef] [PubMed]
20. SeaDataNet. Quality Quality Control Procedures. &th Framework of Ec Dg Research, V2.0, May 2010. Available online: https://www.seadatanet.org/content/download/596/3118/file/SeaDataNet_QC_procedures_V2_(May_2010).pdf (accessed on 20 November 2021).
21. Software Manual Seasoft V2: SBE Data Processing. CTD Data Processing & Plotting Software for Windows. Sea-Bird Scientific. Available online: https://www.seabird.com/asset-get.download.jsa?code=251446 (accessed on 9 December 2017).
22. Grifoll, M.; Aretxabaleta, A.; Espino, M.; Warner, J. Along-shelf current variability on the Catalan inner-self (NW Mediterranean). *J. Geophys. Res.* **2012**, *117*, c09027. [CrossRef]
23. Gómez, M.J.C. Carretero. Wave Forecasting at the Spanish Coasts. *J. Atm. Ocean Sci.* **2005**, *10*, 389–405. [CrossRef]
24. Measurements of the Meteorological Stations of the City of Barcelona, Open Data BCN, Servei Meteorològic de Catalunya. Available online: https://opendata-ajuntament.barcelona.cat/data/ca/dataset/mesures-estacions-meteorologiques (accessed on 20 November 2021).
25. Sentinel 3, EUMETSAT. Dedicated Copernicus Satellite Mission Delivering a Variety of High-Quality Ocean Measurements. Available online: https://www.eumetsat.int/sentinel-3 (accessed on 20 November 2021).
26. Babanin, A.V. *Breaking and Dissipation of Ocean Surface Waves*; Cambridge Univ. Press: Cambridge, UK, 2011.

Journal of
Marine Science and Engineering

MDPI

Article

Laboratory Study of Turbulent Mass Exchange in a Stratified Fluid

Andrey G. Zatsepin *, Valerii V. Gerasimov * and Alexander G. Ostrovskii

Shirshov Institute of Oceanology, Russian Academy of Sciences, Nakhimovskiy Prospect 36, 117997 Moscow, Russia; osasha@ocean.ru
* Correspondence: zatsepin@ocean.ru (A.G.Z.); gerasimov.vv@ocean.ru (V.V.G.)

Abstract: In this study, a laboratory experiment was conducted to investigate quantitatively turbulent exchange between two quasi-homogeneous layers of equal thickness and different density (salinity), as well as the fine structure of the density transition zone (interface) between the layers. The fluid was continuously stirred by a system of horizontally oscillating vertical rods, piercing through both layers and producing vertically homogeneous turbulent impact in a two-layered fluid. In every experimental run, the stirring process was carried out continuously from certain initial state up to the complete mixing of the layers. The buoyancy flux between the layers was estimated using the data on time changes of the salinity in both upper and lower layers. The fine structure of density interface was measured by vertically profiling conductivity microprobe. The results were presented in a dimensionless form and analyzed depending on two dimensionless parameters as follows: the Richardson number, Ri, and Reynolds number, Re. It was found that if $Ri > Ri^*(Re)$ where Ri^* is the critical Richardson number, the interface exists in "sharpening" mode and in "eroding" (diffusive) mode if $Ri < Ri^*(Re)$. The maximum mixing efficiency was achieved at critical Richardson number, when the density interface was in a transition state between the sharpening and diffusive modes.

Keywords: laboratory experiment; two-layered fluid; turbulent stirring; mixing efficiency; interface thickness

Citation: Zatsepin, A.G.; Gerasimov, V.V.; Ostrovskii, A.G. Laboratory Study of Turbulent Mass Exchange in a Stratified Fluid. *J. Mar. Sci. Eng.* **2022**, *10*, 756. https://doi.org/10.3390/jmse10060756

Academic Editors: Mikhail Emelianov, Georgy I. Shapiro and Anatoly Gusev

Received: 11 April 2022
Accepted: 23 May 2022
Published: 30 May 2022

1. Introduction

It is well known that the ocean stratification is almost never "smooth" and is characterized by the fine structure (FS), which is manifested as alternation of layers and interlayers with different values of vertical gradient of temperature, salinity, density, etc. According to Konstantin N. Fedorov, the well-known researcher of the FS in the ocean, the FS can be treated as "signature" of the mixing processes, both vertical and horizontal [1,2]. It should be noted that sometimes such signature is "calligraphic" when a regular FS is formed, consisting of a system of quasi-uniform layers separated by sharp density interfaces. Examples of such structures can be found in the above-mentioned books [1,2].

From general considerations, it follows that the type of mixing (convection, double-diffusive convection, velocity shear and shear-free turbulence, etc.) should not have a decisive effect on the final result—the transformation of an initially smooth vertical density gradient into a step-like structure profile. Rather the theoretical and laboratory modeling data show that vertically homogeneous turbulent mixing under certain conditions transforms a continuous density stratification into a step-like profile. The basic process of thin layering of the stratified fluid due to turbulent stirring was proposed by Phillips and Posmentier [3,4]. In a strongly stratified fluid, the process involves instability of vertical turbulent exchange. It was shown that at the Richardson number greater than some critical value, any local increase in the vertical density gradient tends to grow further on. This is due to the weakening of the vertical turbulent mass flux in the region of the increased gradient. Below and above this layer the gradient decreases while the mass flux increases. Over time, in the areas of turbulent mass flux decrease, sharp density interfaces are formed (similar to that how traffic jams occur in places of narrowing of the road). They suppress turbulence, and in the vicinity of them, in the areas of the turbulent mass flux increase, quasi-uniform layers are formed.

The possibility that the initially linearly density-stratified fluid exposed to vertically homogeneous stirring of undergoes the FS layering was demonstrated in laboratory experiments [5–8] (see Figure 1) as well as by means of mathematical and numerical modeling [9–11].

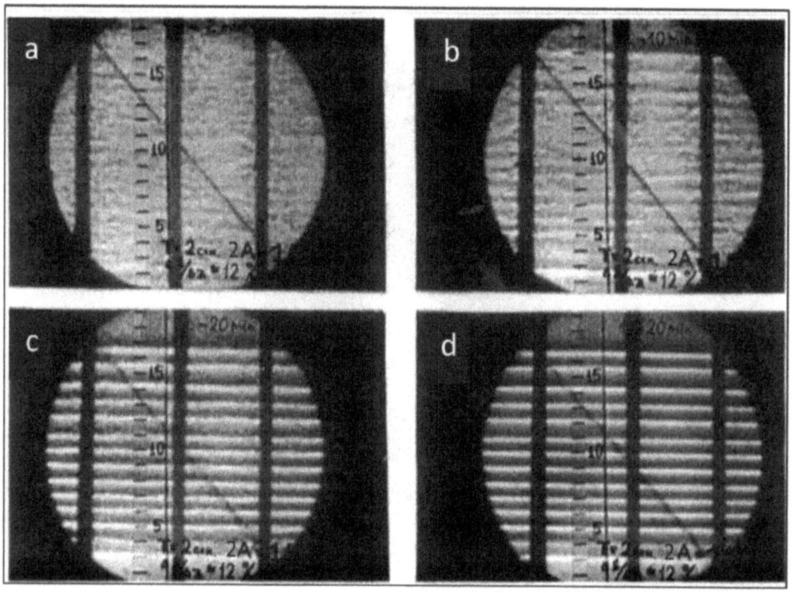

Figure 1. Consecutive shadowgraphs illustrate the formation of the step-like fine structure during long-term stirring of linearly stratified water column by horizontally oscillating vertical rods. The shadowgraphs were obtained during the experiment as follows: (**a**) top left at 10 min, (**b**) top right at 30 min, (**c**) bottom left at 240 min, (**d**) bottom right at 630 min. Experiment setup parameters were as follows: $A = 0.75$ cm, $d = 0.65$ cm, $T = 2$ s, $N^2 = (d\rho/dz)g/\rho = 8.4$ s^{-2}, where A is amplitude, d is diameter of the rod, T is period of rod oscillation, N is Brent-Väisälä frequency, $d\rho/dz$ is density gradient, ρ is water density, g is gravity acceleration).

In this study, which can be considered as a development of earlier work [8], the method of physical (laboratory) modeling is used to investigate quantitatively: (i) the turbulent exchange between two quasi-homogeneous layers of different salinity having equal thickness; and (ii) the structure of the density transition zone between the layers. The layers are continuously stirred by a fixed system of horizontally oscillating vertical rods piercing through both layers. The movement of the rods produces vertically homogeneous turbulent stirring of a two-layered fluid media. The experiment is different from the experiment of Linden [12] and Krylov and Zatsepin [13] which were focused on turbulent mass transfer through the density interface between the layers forced by the freely falling grid or vertically oscillating grids with square rods.

The objective for this work was to study the relationship between the structure of the water density boundary layer between two layers of different salinity and the vertical buoyancy flux induced by continuous and vertically uniform turbulent impact. The experiments were carried out in a wide range of the values of Richardson number in order to demonstrate abrupt transition from the mode of maintaining the density interface in a sharpened state to the mode of its erosion—diffusion expansion. The relevance of this transition to the Phillips-Posmentier model is discussed.

2. Materials and Methods

The experiments were carried out on a laboratory setup, the schematic diagram of which is shown in Figure 2. The laboratory tank made of organic glass was of an internal size of 36 × 15.5 × 25 cm^3. Horizontally oscillating metallic rod above the tank was oriented along the longer side of the tank. Six grids were fixed evenly at this rod with a $L = 5.5$ cm distance between the adjacent grids. Each grid consisted of six vertical glass rods with the diameter $d = 0.65$ cm with a distance $M = 2.8$ cm between each other (the mesh size). The glass rods were submerged in the water down to the 0.5 cm thick bottom layer of the tank. The metallic rod, on which the grids with the glass rods were fixed, was connected to a DC electric motor with an eccentric, which provided horizontal oscillation of the rod and the grids. By changing the supply voltage to the motor, the oscillation period T of the grids could be changed in the range from 1.3 to 6.7 s. By changing the attachment point of the metallic rod to the eccentric it was possible to vary the amplitude A of oscillations in the range from 0.5 to 1.8 cm. The measurement data were collected in a personal computer.

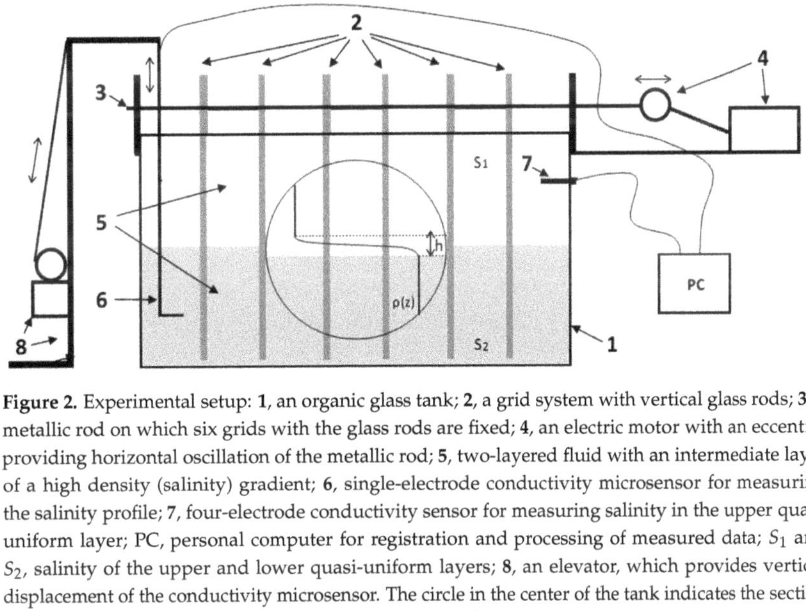

Figure 2. Experimental setup: **1**, an organic glass tank; **2**, a grid system with vertical glass rods; **3**, a metallic rod on which six grids with the glass rods are fixed; **4**, an electric motor with an eccentric providing horizontal oscillation of the metallic rod; **5**, two-layered fluid with an intermediate layer of a high density (salinity) gradient; **6**, single-electrode conductivity microsensor for measuring the salinity profile; **7**, four-electrode conductivity sensor for measuring salinity in the upper quasi-uniform layer; PC, personal computer for registration and processing of measured data; S_1 and S_2, salinity of the upper and lower quasi-uniform layers; **8**, an elevator, which provides vertical displacement of the conductivity microsensor. The circle in the center of the tank indicates the section of the plane-parallel beam of light by the shadowgraph device.

Before each experimental run, the tank was half filled with fresh distilled water, below which a water solution of NaCl of a given concentration was poured slowly to avoid mixing. As a result, a two-layered medium with layers of different salinity and equal thickness ($H_1 = H_2$, where H_1 is the thickness of the upper layer, and H_2 is the thickness of the lower one) was formed with a narrow density gradient zone $h \ll H_1, H_2$ between the layers. The temperature of the layers was the same with the accuracy of ±0.2 °C and close to room temperature.

After the tank was completely filled with two-layered water medium, the stirring with oscillating rods was switched on and this moment in time was taken as a start of an experimental run. An eccentric driven by the motor provided a sinusoidal oscillation of the metallic rod with the grids of glass rods for a given amplitude and period. During each experimental run, the current salinity S_1 in the upper layer was measured using a laboratory conductivity meter Expert 002. It is a four-electrode sensor with a diameter 0.5 cm, and with a length 3.0 cm. Using six subranges it provides the measurement of electrical conductivity in a range from 0.001 μS/cm up to 1000 mS/cm. The standard error of conductivity measurements does not exceed 2% (https://magazinlab.ru/konduktometr-jekspert-002.html, accessed on 25 May 2022).

A value of S_1 gradually increased with time due to the flux of salt from the lower layer across the density gradient layer. Throughout each experiment the thickness of the upper and lower layers remained practically unchanged. The four-electrode sensor of the conductivity meter Expert 002 was placed horizontally in the upper layer at a distance of 1.5 cm below the water surface in the tank. The measurements of the electrical conductivity with the four-electrode sensor were carried out with an interval of 1 s from the beginning to the end of the experimental run.

In all of experiments, along with the measurements of the salinity in the upper layer with the conductivity meter Expert 002, the vertical profiles of conductivity were measured using a single-electrode conductivity microsensor. It was attached to a mechanical yo-yo device providing vertical movement of the sensor at a speed of about 0.2 cm s^{-1}. Since the diameter of the electrode made from nichrome wire was only 0.01 cm, its vertical resolution was about of 0.1 cm. The signal registration frequency was 10 Hz. This made it possible to obtain almost non-smoothed vertical structure of the salinity distribution in the upper and lower layers and across the density gradient layer.

It should be noted that before each series of the experiments, the conductivity sensors were calibrated by immersing them together into plastic glasses filled with water of a given salinity. Initially, the salinity of the water in the glasses was set by the gravimetric method, and then it was determined with an accuracy of 0.01 ppm using the laboratory salinometer Autolab (Manly, Australia). The error of conductivity measurements using the conductivity meter Expert 002 did not exceed 2% of the measured value. The data of conductivity measurements by a single-electrode microsensor in the upper layer were referenced to the simultaneous readings of the conductivity meter Expert 002. The stability of salinity measurements with a single electrode microsensor was quite high; its drift did not exceed 0.1 ppm per day. In addition to the quantitative measurements described above, the Schlieren shadowgraph device was used to produce the picture of a density structure and fluctuations in the stratified fluid at the screen behind the tank (Figure 2). The shadowgraph images were photographed regularly during the experiments and compared with the salinity profiles measured at the same time when it was possible. Due to the use of a shadowgraph method, it was possible to monitor the density interface position even in the absence of measurements of the vertical salinity profiles by a single-electrode conductivity microsensor.

The laboratory study consisted of 39 experimental runs most of which were carried out at different values of the defining dimensional and dimensionless parameters (see Table 1). A full description of these options is provided in the next section of the paper. The experiment was carried out during the time period from June 2021 to February 2022.

Table 1. The values of the defining dimensional and dimensionless parameters of the experiment: $\Delta S^{initial}$, the initial value of the salinity difference between the layers; Re, the Reynolds number; $Ri^{initial}$, the initial value of the Richardson number; $2A$, double amplitude of the rods oscillation; T, the period of the oscillations of the rods.

Experiment	$\Delta S^{initial}$, ppm	Re	$Ri^{initial}$	$2A$, cm	T, s
1	10	135	84	2.6	4.0
2	10	132	89	3.3	5.2
3	10	134	86	3.3	5.1
4	12	147	85	3.3	4.6
5	14	159	84	3.1	4.0
6	14	160	84	3.3	4.3

Table 1. *Cont.*

Experiment	$\Delta S^{initial}$, ppm	Re	$Ri^{initial}$	$2A$, cm	T, s
7	14	159	86	3.3	4.3
8	14	158	86	3.3	4.3
9	16	170	85	3.3	4.0
10	18	180	86	3.3	3.8
11	19	168	102	3.3	4.1
12	20	190	84	3.7	4.0
13	20	194	81	3.3	3.5
14	24	207	86	3.3	3.3
15	24	211	83	3.3	3.2
16	24	207	86	3.3	3.3
17	24	204	89	3.3	3.3
18	24	204	89	3.3	3.3
19	28	223	86	3.3	3.1
20	28	223	87	3.3	3.1
21	30	214	100	3.3	3.2
22	30	214	100	3.3	3.2
23	30	214	100	3.3	3.2
24	32	238	87	3.3	2.9
25	36	254	86	3.3	2.7
26	38	238	103	3.3	2.9
27	42	274	86	3.3	2.5
28	46	287	86	3.3	2.4
29	46	288	85	3.3	2.4
30	48	267	104	3.3	2.6
31	56	293	96	3.3	2.3
32	56	317	86	3.3	2.2
33	64	341	84	3.3	2.0
34	75	364	87	3.3	1.9
35	76	370	85	3.3	1.8
36	88	365	102	3.3	1.9
37	90	402	86	3.3	1.7
38	108	441	85	3.3	1.5
39	139	498	86	3.3	1.4

3. Results

3.1. Dependence of Entrainment Velocity on Richardson Number

During every experimental run the stirring of the two-layer water medium using oscillating vertical rods was carried out from initial state until complete mixing of the layers (i.e., until achieving the homogeneity of the water column). The buoyancy flux F_b between the layers was calculated using the time series of the salinity values in the upper layer:

$$F_b = g\beta(dS_1/dt)H_1 \quad (1)$$

where S_1 is the instant salinity of the upper layer, t is the time, dS_1/dt is the rate of the salinity change in the upper layer, g is the acceleration of gravity, β is coefficient of salinity compression. Neglecting the thickness h of the density transition zone between the upper and lower layers, one can rewrite the buoyancy flux between the upper and lower quasi-homogeneous layers in the form:

$$F_b = g\beta U_e \Delta S \tag{2}$$

where U_e is the rate of turbulent entrainment of the water from the lower layer to the upper layer, and $\Delta S = S_2 - S_1$ is the instant salinity difference across the density interface between the layers, and $S_2 = S_2^{initial} - S_1$. Therefore, $\Delta S = S_2^{initial} - 2S_1$, where $S_2^{initial}$ is the initial salinity value of the lower layer. Equating the right-hand sides of (1) and (2), we obtain an expression for U_e:

$$U_e = (dS_1/dt)H_1/\Delta S = (dS_1/dt)H_1/\left(S_2^{initial} - 2S_1\right) \tag{3}$$

Following [6–8] the experimental data were presented in dimensionless form and analyzed based on the dimensionless Richardson and Reynolds numbers. The dimensionless form of U_e was obtained by dividing it on the velocity scale to some extent corresponding to the velocity scale of energy-carrying eddies in the turbulent layers.

In [7,8], this velocity scale was taken as an average speed of sinusoidal oscillation of the rods $U = 4A/T$, where A is the amplitude and T is the period of the rod oscillations. However, the zone of turbulence generation occupied by the oscillating rods engages a minor part of the horizontal section of the tank. Therefore, we came to conclusion that for the scale of the velocity of turbulent fluctuations caused by energy-carrying vortices, it is reasonable to choose the normalized velocity of the rod oscillations, which can be expressed as:

$$U = 4AT\cdot(\Sigma_r/\Sigma_0) \tag{4}$$

Here $\Sigma_r = NB2dA$ is the turbulence generation horizontal area, N is the number of vertical rods in the tank, Σ_0 is the area of the horizontal section of the tank, and B is an empirical coefficient that should be greater than one ($B > 1$). The basis for this assumption is the fact that the area Σ_r should be larger than the valence $2Ad$ due to presence of the viscous boundary layer around the rod and the mass of fluid dynamically attached to it. Taking into account that the efficiency of turbulent mixing, which is the fraction of the kinetic energy of turbulence spent on increasing the potential energy of a stratified fluid, reaches a maximum of about 20% [12,14], we obtained that the value of the coefficient B is approximately equal to 2.4, which seems to be rather realistic. The justification for such a parametrization of the turbulence velocity scale, as well as the length scale of energy-containing turbulent pulsations (eddy diameter) $l \approx M$, where M is the mesh size of the grid will be discussed in Section 3.5.

Based on the chosen scales of turbulent fluctuations in mixed layers, one can introduce two key dimensionless parameters of the experiment as follows: the Reynolds number, $Re = UM/\nu$, where ν is the kinematic viscosity of water, and the Richardson number, $Ri = g\beta\Delta SM/U^2$.

In the logarithmic coordinates, the dependence of U_e/U on the Ri number makes possible to verify the concept of the power-law dependence of the dimensionless entrainment rate on the Richardson number in a two-layered stratified aqueous medium [12]:

$$U_e/U = CRi^{-n} \tag{5}$$

Here n is an exponent to be determined as a result of the experiments. In all experimental runs, the behavior of the experimental points approximated by (5) was similar since there existed two pronounced quasi-linear parts in the plot (Figure 3).

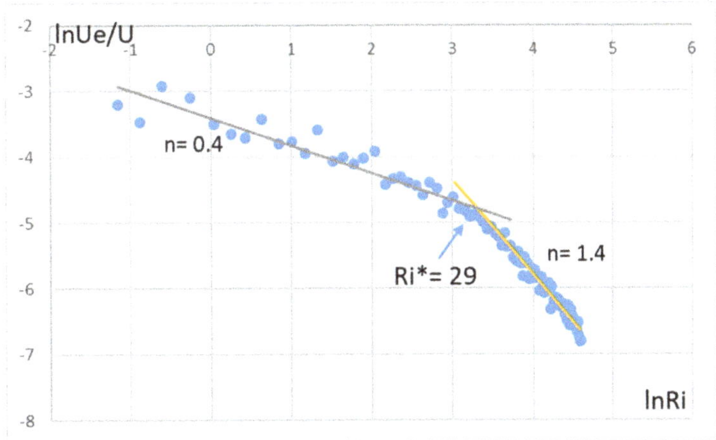

Figure 3. Dependence of $\ln u_e/u$ on $\ln Ri$ for the experimental run with the initial value of $Ri^{initial} = 100$ and $Re = 214$. The straight lines show two different asymptotic power dependences. Their intersection point at Ri^* indicates to the abrupt transition from one dependence to the other.

At vicinity of the point of intersection of straight lines at Figure 3 there is an abrupt transition from one turbulent mass transfer mode to the other. It should be noted that both values of the exponent n and the coefficient C in (5), as well as the value of Ri^*, are the functions of the Reynolds number Re. More detailed information on the dependences of C, n, and Ri^* on Re for $Ri > Ri^*$ will be given in Section 3.4.

3.2. Structure of the Density Transition Layer

A pattern of time evolution of the salinity profiles was obtained in the experimental runs. Since each experimental run began with a certain maximum value of Ri (with the greatest difference in salinity (density) between the layers), and ended when $Ri = 0$ (no density difference between the layers), it was possible to obtain the dependence of the density transition zone thickness and its structure on the Richardson number. In a wide range of its variation, the behavior of the density interface was found to be as follows. At the beginning, the thickness of the interface zone after a rather short stage of its adaptation to the turbulent stirring maintained in a sharpened state (Figure 4a–f) and then at $Ri \approx Ri^*$ it starts to expand (Figure 4g,h). These findings qualitatively agreed with the results of the experiment in which the density transition zone was formed under the impact of turbulence generated by vertically oscillating horizontal grids in the upper and lower water layers of different salinity and temperature [13].

Figure 5 shows the typical vertical salinity structure of the fluid column divided into three sublayers: two quasi-homogeneous sublayers (top and bottom) with thicknesses h_1 and h_3, correspondingly, and a middle sublayer with a maximum salinity gradient (density interface) with a thickness h_2. The thickness h_2 was determined as the vertical distance between the points of separation of the tangents from the real profiles.

Figure 6 shows an evolution of h_1 and h_2 with time in details, taking into account all of the salinity profiles obtained during the experimental run.

It follows from Figures 5 and 6 that h_2 characterize the thickness of the sublayer of the maximum salinity gradient (density interface) displaced between the upper and lower quasi-homogeneous sublayers. The density interface plays a key role in suppressing turbulence generated by the oscillating grids of vertical rods and preventing the penetration of eddies from one quasi-homogeneous sublayer to the other. Therefore, the characteristics of this sublayer most clearly reflect the change in the regime of mass transfer between the layers when the critical value of the Richardson number is reached.

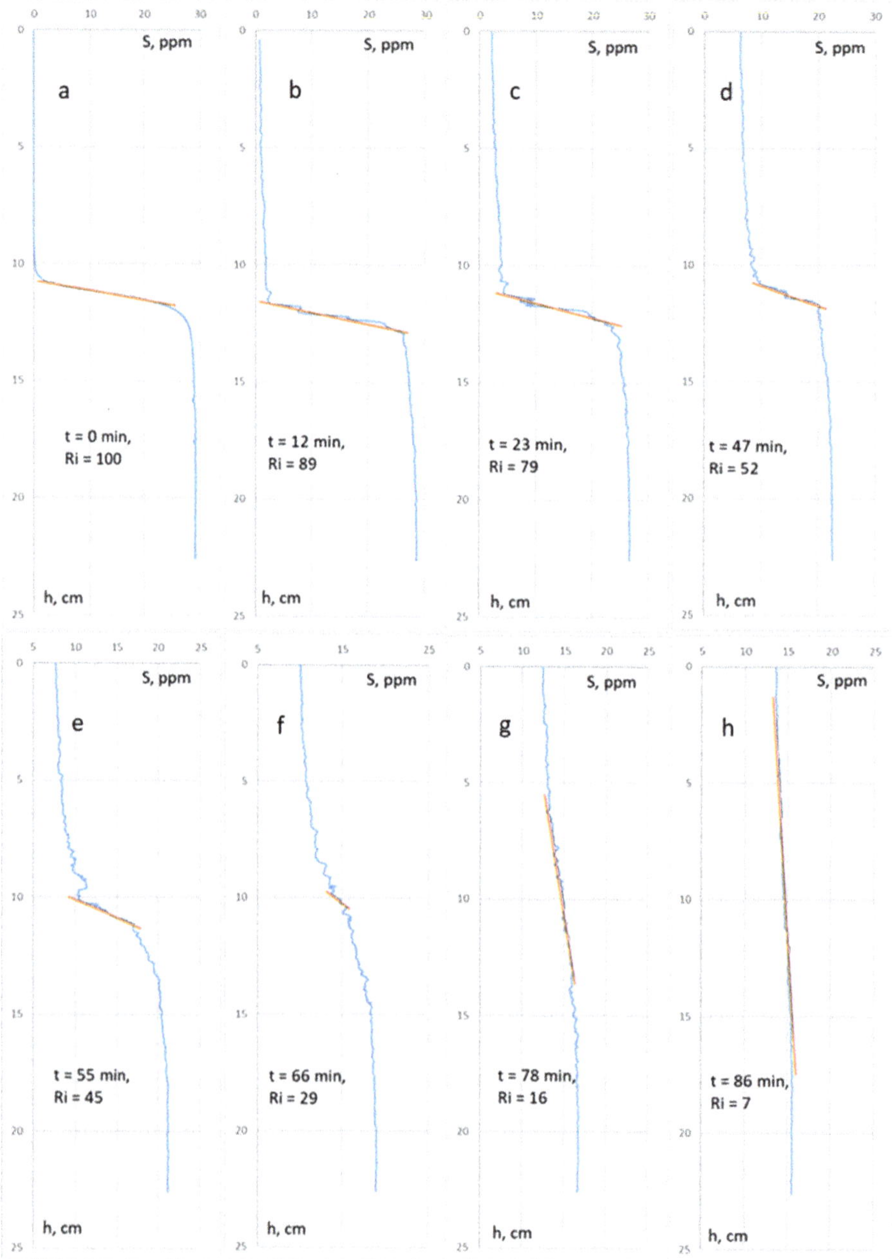

Figure 4. The sequence of salinity profiles measured during one of the experimental runs with initial conditions as follows: $\Delta S^{initial} = 30$ ppm, $Ri^{initial} = 100$, and $Re = 214$. Tangent lines (marked in orange) are added to the quasi-linear sections of the profiles with the maximum salinity gradient. The subfigures are shown during the experimental run at the time and the Richardson number as follows: (**a**): $t = 0$ min, $Ri = 100$; (**b**) $t = 12$ min, $Ri = 89$; (**c**) $t = 23$ min, $Ri = 79$; (**d**) $t = 47$ min, $Ri = 52$; (**e**) $t = 55$ min, $Ri = 45$; (**f**) $t = 66$ min, $Ri = 29$; (**g**) $t = 78$ min, $Ri = 16$; (**h**) $t = 86$ min, $Ri = 7$.

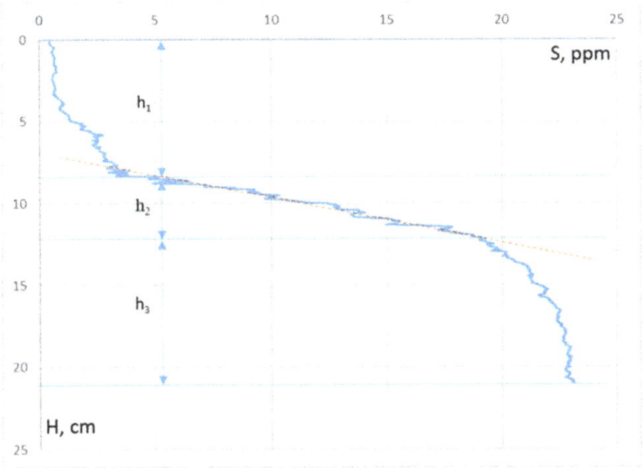

Figure 5. Vertical distribution of salinity in a two-layer fluid with an $\Delta S = 23$ ppm, $Ri = 86$ and $Re = 204$. Tangent line (marked in orange) is added to the quasi-linear sections of the profiles with the maximum salinity gradient.

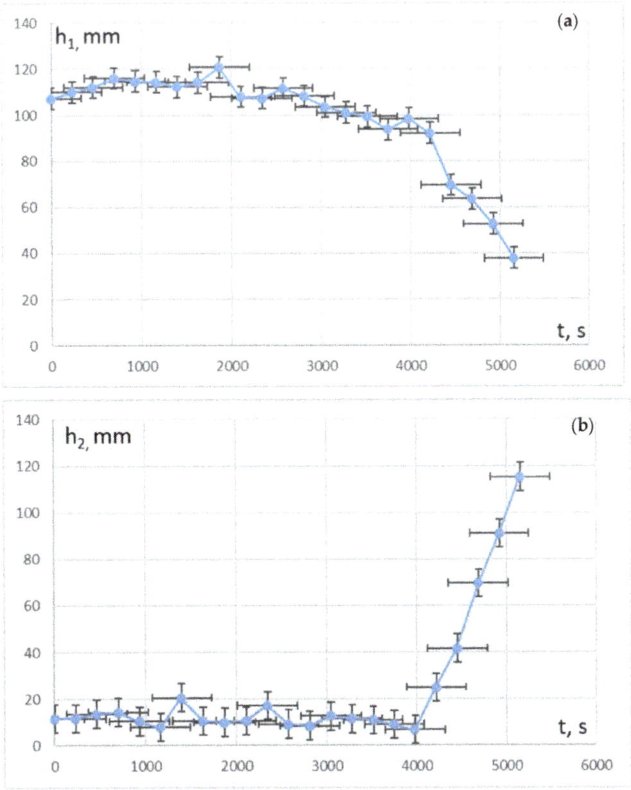

Figure 6. Variations of the sublayers thickness versus time: (**a**) the quasi-homogeneous sublayer thickness h_1; (**b**) the salinity interface sublayer thickness h_2, $Ri^{initial} = 100$; $Re = 214$. The bars display the standard deviation of the data.

The point at which the derivative of the $h_2/M(Ri)$ function change characterizes the transition from the mode of the density interface "sharpening" to the mode of its "erosion" (turbulent diffusion). In Figure 7 this transition corresponds to the number $Ri = Ri^* = 29 \pm 7$ at $Re = 214$. This result support a conclusion that during a vertically homogeneous turbulent stirring of the initially continuously stratified fluid, the fine-structure layering (step-like structure formation) is possible only at $Ri > Ri^*$.

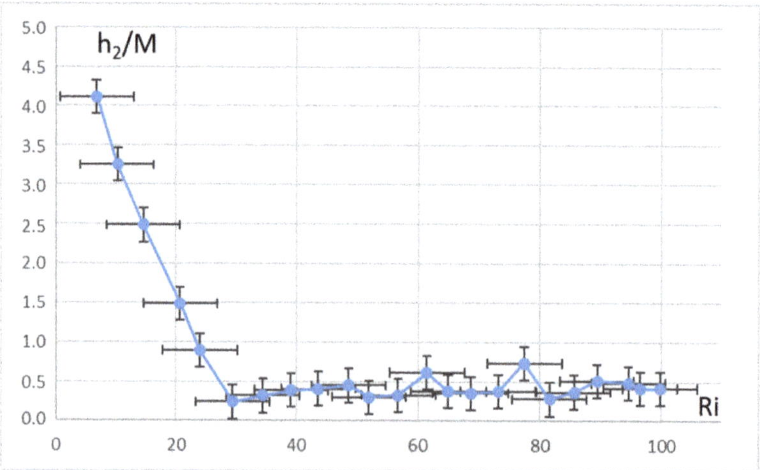

Figure 7. Dependence of h_2/M on Ri for the experimental run with initial salinity difference $\Delta S = 30$ ppm, $Ri^{initial} = 100$, and $Re = 214$. The bars are standard errors of measurements.

It should be noted that at $Ri < Ri^*$ the dimensionless thickness of the density interface is not only a function of Ri and Re, but also of time. Thus, it can be assumed that at $Ri < Ri^*$ the density boundary exists in a non-stationary regime and tends to expand over the entire thickness of the fluid layer in the tank with time. On the contrary, for $Ri > Ri^*$ the density interface exists in a quasi-stationary self-sustaining regime.

3.3. Mixing Efficience and It Dependence on Richardson Number

It is important to analyze the dependence of the Richardson flux number Rf as a function of Ri. By its physical meaning, Rf is the fraction of the kinetic energy of turbulence spent on mixing (i.e., on increasing the potential energy of a stratified fluid system) [12]. Here,

$$Rf = (U_e/U)Ri \tag{6}$$

The dependence of Rf on Ri (Figure 8), reaches a maximum value close to 0.2 at $Ri = Ri^* \approx 29$, when $Re = 214$. The maximum corresponds well to the point where the derivation of $h_2/H_1(Ri)$ change (Figure 7). Thus, it characterizes the transition from the sharpening interface mode to an eroding mode.

3.4. Dependence on the Reynolds Number

The experiments revealed a significant dependence of both C and n on Re at $Ri > Ri^*$ (see Formula (5)). These functions are shown in Figures 9 and 10, and by expressions (7) and (8), respectively.

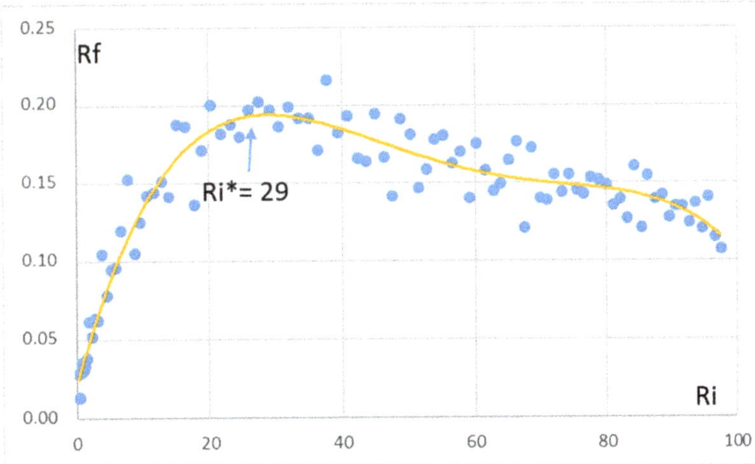

Figure 8. Dependence of Rf on Ri for the experimental run with $Ri^{initial} = 100$ and $Re = 214$.

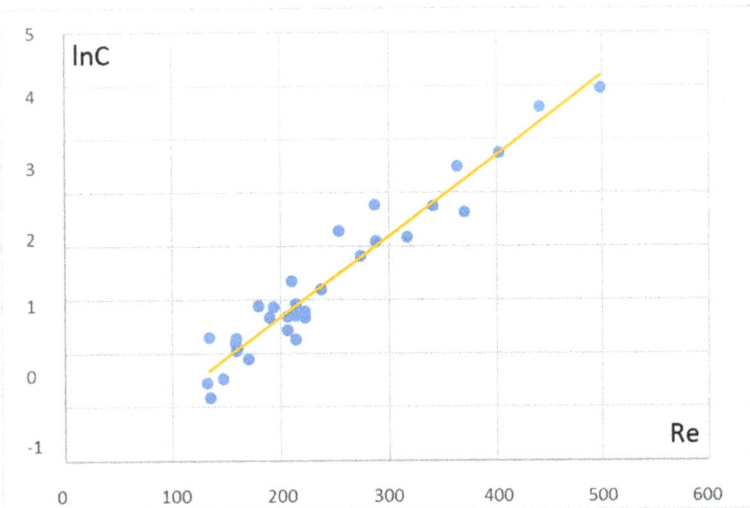

Figure 9. The dependence of $\ln C$ on Re for the case $Ri > Ri^{*}$.

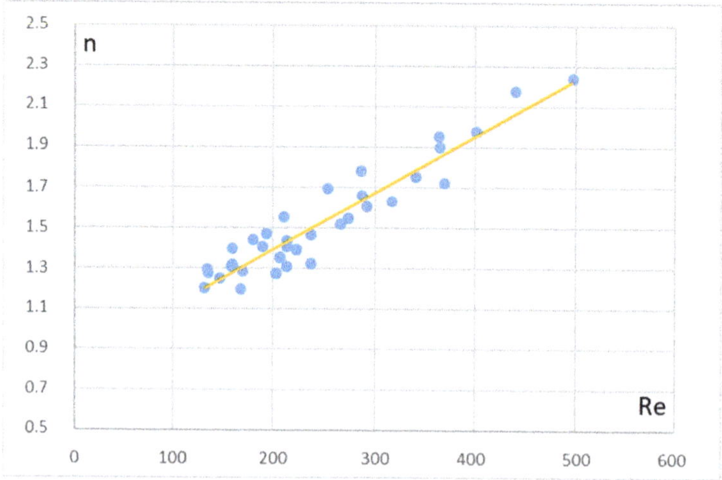

Figure 10. The function $n(Re)$ for the case $Ri > Ri^*$.

As follows from Figure 9, there was a quasi-linear dependence of $\ln C$ on Re given by

$$\ln C = 0.02Re - 3.33 \quad (7)$$

There was also a quasi-linear dependence of the exponent n on Re for the case $Ri > Ri^*$ as follows:

$$n(Re) = 0.003Re + 0.87 \quad (8)$$

As for the dependences of C and n on Re at $Ri < Ri^*$, they also exist, but their study is problematic, since in the mode of erosion of the density interface, the regularities of vertical turbulent mass transfer are also a function of time. Due to these circumstances, we did not consider it relevant to determine these dependencies.

At the critical Richardson number Ri^*, the regularities of the mass transfer between turbulent layers were changed. A dependence of Ri^* on the Reynolds number Re (Figure 11) was also observed.

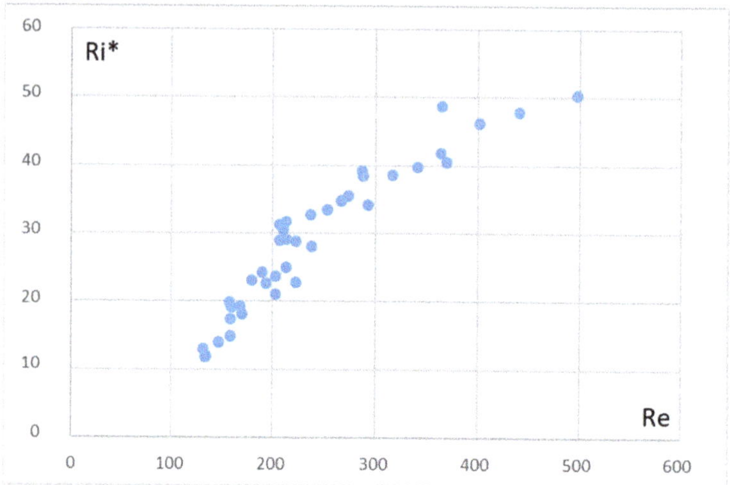

Figure 11. The critical Richardson number Ri^* as a function of the Reynolds number Re.

It follows from Figure 11 that Ri^* is a growing function of Re at least at low values of this parameter. An increase in the value of the critical Richardson number with the Reynolds number at which the mode of expansion of the density interface between layers is replaced by the mode of its sharpening was found previously in laboratory experiments on stirring of linearly stratified water medium [6,7]. However, there is a reason to expect that the increase of Ri^* slows down with increasing Re and it is possible that the function $Ri^*(Re)$ has a finite limit.

3.5. Analysis of Salinity Pulsations at the Density Interface and an Estimate of Energy-Containing Eddies Scale

Some useful information about the process of turbulent exchange between the layers of different density can be obtained from the analysis of salinity fluctuation records measured by electrical conductivity microsensors. For this purpose, two special experimental runs were carried out. In one of them, along with registration of shadowgraph images of the interface at different moments of time and measuring salinity profiles with an electrical conductivity microsensor, another conductivity microsensor was fixed inside the density interface to measure salinity fluctuations caused by turbulent mixing.

The salinity profile measured by the conductivity microsensor at one of the moments of time during the density interface sharpening mode is presented at Figure 12. The interface was very narrow: its thickness was about 1 cm and the value of the Brent-Väisälä frequency $N \approx 4\ s^{-1}$. This means that the period of free density interface oscillations $T \approx 1.6$ s was twice less than the period of vertical glass rods oscillation $T = 3.2$ s.

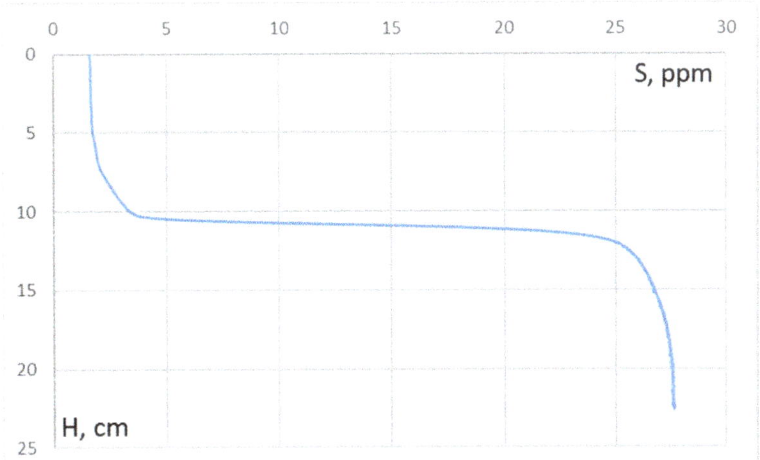

Figure 12. The snapshot of the salinity profile measured by the conductivity microsensor at a moment of time during the density interface sharpening ($Ri^{initial} = 100$; $Re = 214$).

The shadowgraph image of the density interface received nearly at the same time as the salinity profile at Figure 12 is presented at Figure 13. This image shows undulating curvature of the interface caused (possibly) by the impact on it of rather large (several centimeters in size) eddies that exist in the upper and lower layers together with the interfacial internal waves.

The readings of the electrical conductivity sensor located in the area of the density interface, is shown at Figure 14. Here, the red line is the smoothed data of salinity during the experimental run, and the blue line is the raw data indicating high-amplitude salinity fluctuations in a wide frequency range. An ensemble of the data shown on the graph by a double-sided arrow was taken for spectral analysis after removal of the linear trend.

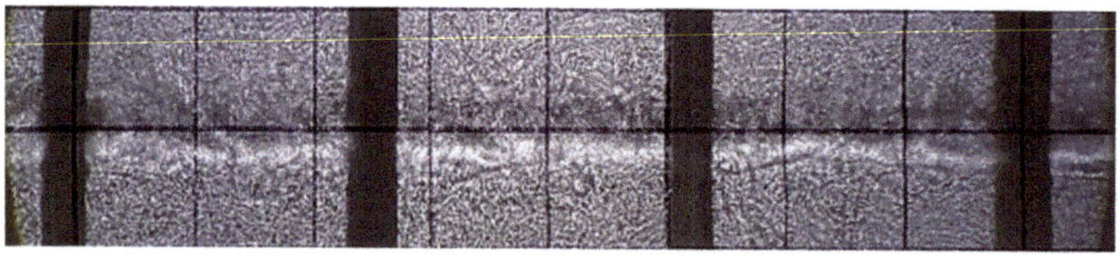

Figure 13. The shadowgraph image of the density interface received nearly at the same time as the salinity profile at Figure 12 ($Ri^{initial} = 100$; $Re = 214$).

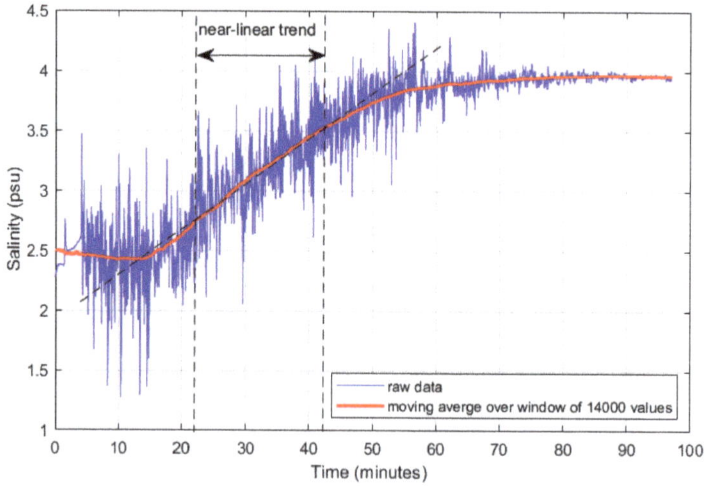

Figure 14. The salinity (blue line) as derived from the conductivity data registered by the microsensor of electrical conductivity with a frequency of 10 Hz in the density interface layer. Red line indicates moving average of the salinity. The vertical dashed bars indicate the data ensemble used for the spectral analysis.

The power spectrum of salinity fluctuations presented at Figure 15 was rather smooth without pronounced peaks. This was surprising since, in our opinion, the frequency of the rods oscillation, as well as its overtones, should have appeared. One of the possible reasons for the absence of these peaks, associated with the oscillation frequency of the rods, may be the mutual influence of neighboring grids of rods on turbulent fluctuations, causing a masking effect.

To test this hypothesis, an experimental run was carried out where only one grid with rods (instead of six grids) was oscillating in a tank while the other conditions were the same. As before, an electrical conductivity microsensor was located near the oscillating grid in the area of the density interface between the layers. The spectral energy of the measured salinity fluctuations is shown at Figure 16. The peaks of energy at $f = 3.3 \times 10^{-1}$ s^{-1} and 6.6×10^{-1} s^{-1} were clear-cut. These frequencies corresponded well to the grid oscillation period of 3.2 s and to its double frequency overtone. This confirmed the hypothesis that the absence of peaks associated with the oscillation frequency of the rods at Figure 14 was related to the mutual influence of neighboring grids of rods on the spectral composition of turbulent fluctuations. As for the threshold frequency at which the spectral energy reached its maximum values and a plateau (see Figure 15), it was about 3×10^{-2} s^{-1} which corresponded to the time scale $\tau \approx 30$ s. If we assume that this time scale characterized the overturn period of fluid particles in energy-carrying turbulent eddies, and the particle velocity U was determined by Formula (4), it was easy to obtain that $U\tau \approx \pi M$. Thus, the spectral analysis of salinity fluctuations indirectly confirmed the validity of the velocity scale U parametrization by Formula (4), as well as the fact that eddy diameter l was close to M i.e., to the mesh size of the grid. Still, it is necessary to refine these scales on the basis of direct measurements of turbulence parameters, which would be the task for the future studies.

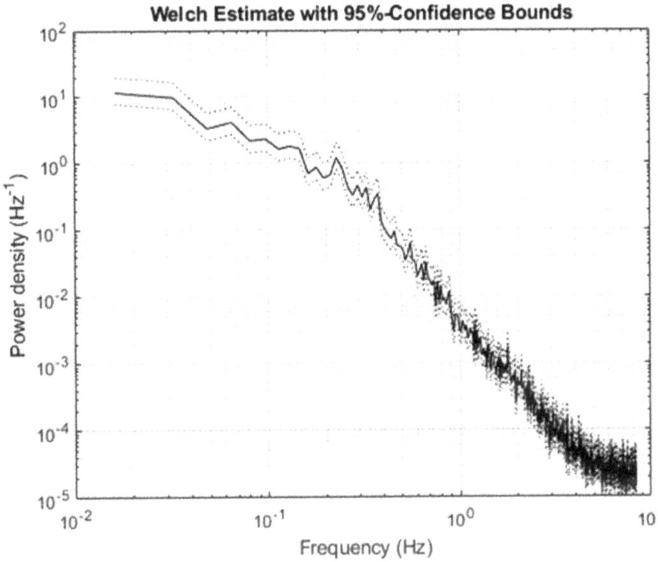

Figure 15. The variance-normalized power spectral density of salinity fluctuations shown in Figure 14 ($Ri^{initial}$ = 100, $Re = 214$).

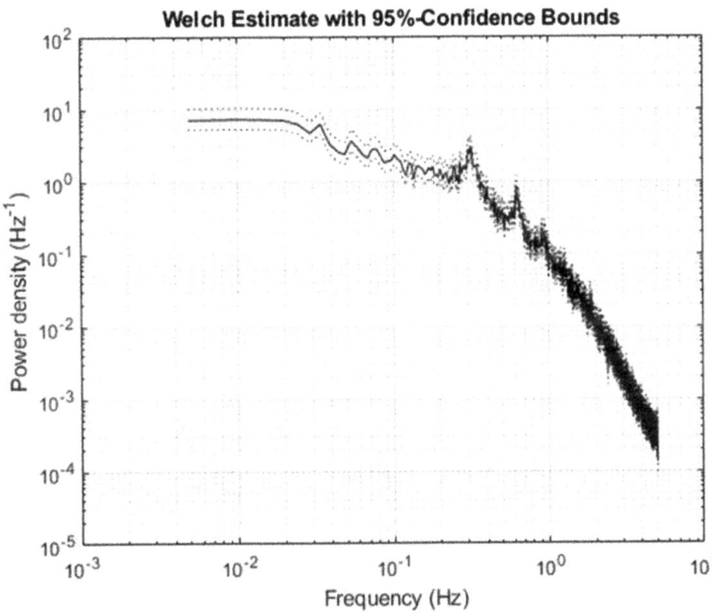

Figure 16. The variance-normalized power spectral density of the salinity fluctuations obtained in the experiment with the single oscillating grid.

3.6. Density Interface Structure, Mass Flux between the Turbulent Layers and Their Relation with Phillips-Posmentier Mechanism

The results of the experimental runs described above confirmed that the sharpening of the density interface between the turbulent layers of different salinity occurred in accordance with the Phillips-Posmentier model [3,4] as follows. In a turbulent stratified shear flow, the vertical mass flux can be represented as:

$$F = K(d\rho/dz), \tag{9}$$

where K is coefficient of turbulent exchange, $d\rho/dz$ is the vertical density gradient. The vertical exchange in a stratified fluid depends on the shear Richardson number $Ri_s = g(d\rho/dz)/\rho(dU/dz)^2$ [15], where dU/dz is the velocity shear. The coefficient K is represented by a decreasing power function of the Richardson number

$$K \sim C_1 {Ri_s}^{-n} \tag{10}$$

If $n > 1$, the mass flux F (9) is a decreasing function of the density gradient $d\rho/dz$. This means that if somewhere in the flow the density gradient increases locally then the mass flux through this region decreases. As a result, the gradient should increase further.

This situation is dynamically unstable, since a small deviation from equilibrium leads to a further increase in the deviation. Stratification tends to break up into turbulent layers of uniform density, separated by narrow interfaces with a large density gradient. For $n \leq 1$, such instability is absent, since the local inhomogeneities of the density gradient degenerate under the impact of the vertical mass flux.

Dealing with the obtained experimental data, we can move from describing turbulent mass transfer through the density interface via the entrainment velocity U_e to vertical turbulent exchange coefficient K. In our case, the integral estimate of turbulent exchange coefficient in the upper layer could be expressed as $K = U_e \cdot h_1$. Since both U_e and h_1 were measured in the experimental runs, it is possible to construct the dependence $K(Ri)$, or $Nu(Ri)$, where $Nu = K/k$ is the Nusselt number for salt and $k = 1.3 \times 10^{-5}$ cm^{2}s^{-1} is the salt molecular diffusion coefficient. The dependence $Nu(Ri)$ is expressed by:

$$Nu \sim C_2 Ri^{-n} \tag{11}$$

A plot of $\ln Nu$ versus $\ln Ri$ is shown in Figure 17. It follows from the figure that in the mode of the density interface sharpening ($Ri > Ri^*$), the value of the exponent in the Formula (11) $n = 1.4$ (>1) and at $Ri > Ri^*$, $n = 0.2$ (<1). It is obvious that these values of n agree well with the Phillips-Posmentier model. It should be mentioned that the molecular flux of salt in this experiment is much lower than the turbulent one even during the mode of interface sharpening ($Ri > Ri^*$).

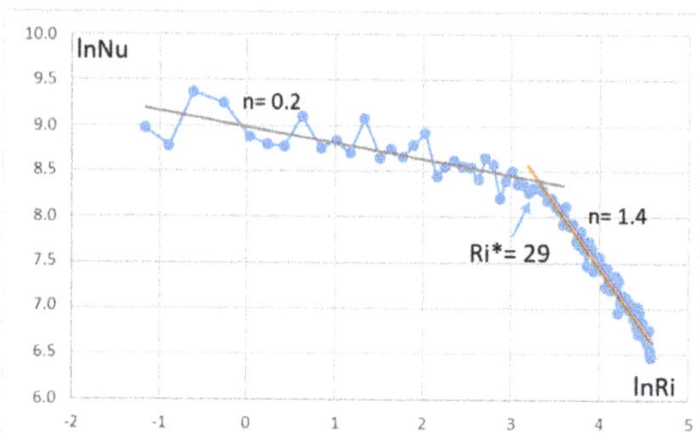

Figure 17. Dependence of $\ln Nu$ on $\ln Ri$ in the experimental run with $\Delta S^{initial} = 30$ ppm and $Re = 214$. $C_2 = 13$ for the asymptote at $Ri > Ri^*$.

4. Discussion

As it was shown in the Section 3.3, at $Ri \approx Ri^*$ the efficiency of the turbulent kinetic energy transition into the potential energy of stratification, reaches a maximum value, which is about 10–30%. Linden [12,14] analyzed the results of his experiments, as well as those carried out by other authors, and arrived at conclusions quite close to that. This means that we have chosen velocity and length scales of the energy-bearing turbulent eddies close to the real ones. At the same time, it is necessary to investigate how these parameters of turbulence are related to the parameters of vertical rods oscillation. Only after that will it be possible to apply findings of the laboratory experiment to natural conditions.

Of course, it can be argued that shear-free grid turbulence has nothing to do with natural conditions where turbulent mixing occurs in the regions of flow velocity shear, which is often observed at the density interfaces. In particular, Peligri and Sandra [15], proposed an alternative to Phillips-Posmentier model of mixing in stratified shear flow. According to [15], the maximum vertical mass flux occurs in the area of the maximum vertical density gradient, and was absent in the area of low-density gradient. As a result, a mixed layer is formed in the region of the initial density interface, whereas new density interfaces are formed above and below the mixed layer. For an experimental confirmation of this mechanism, Peligri and Sandra [15] referred to the experiment of Thorpe [16] conducted in an inclined sealed pipe filled with a two-layer stratified miscible fluid. As a result of the formation of a chain of turbulent Kelvin-Helmholtz billows along the length of the pipe and the mixing caused by them, the interface splits at two new density interfaces formed above and below the initial interface. Implementation of this process in natural conditions requires at least two conditions. Firstly, the region of development of the Kelvin-Helmholtz instability should have a large horizontal extension in order not to be collapsed by the stratification. Secondly, turbulence should persist for a long time to form a well-mixed region bounded above and below by sharp density interfaces. However, it is unclear so far how often such conditions can be realized in nature.

The close type of criticism on the realization of the Phillips-Posmentier mechanism was presented in [17]. The study concerned with the mixing in laboratory experiments in which an initially linearly stratified fluid was stirred with a rake of vertical bars. It was found that the flow evolution strongly depended on the Richardson number Ri. At low Ri, the buoyancy flux was a function of the local buoyancy gradient only, and could be modelled as gradient diffusion with a Ri-dependent eddy diffusivity. At high Ri, vertical vorticity shed in the wakes of the bars interacted with the stratification and produced well-mixed layers separated by interfaces. This process leads to layers with a thickness proportional to the ratio of grid velocity to buoyancy frequency for a wide range of Reynolds numbers Re and grid solidities. In this regime, the buoyancy flux was not a function of the local gradient alone, but also depended on the local structure of the buoyancy profile. It was proposed that, the layers were not formed by the Phillips-Posmentier model, but they resulted from vortical mixing previously thought to occur only at low Re.

The important factor of the impact of molecular effects on the turbulent diffusivity [18] and step-like structure formation in stratified shear flow was investigated in [11]. In particular, direct numerical simulation was used to investigate the process dependence on three non-dimensional parameters as follows the gradient Richardson number, the buoyancy Reynolds number and the Prandtl number. A simple expression for conditions favorable for the layer formation was obtained. It was shown that the layers were formed only in the case of rather large values of Prandtl number.

5. Conclusions

Overall, we would like to make two preliminary conclusions related to the regularities of the evolution of density interface between two water layers of different salinity and the buoyancy flux produced by continuous and vertically uniform turbulent impact at a wide range of Richardson numbers, rather low Reynolds numbers, and high Prandtl number (≈ 700).

1. The maintaining of a density interface between turbulent layers in a sharpening mode is possible only under the condition $Ri > Ri^*$. If $Ri < Ri^*$ the density interface exists in the eroding non-stationary mode. The transition between these two modes is abrupt and the physical explanation of such a kind of transition from one mode to the other seems to be consistent with Phillips-Posmienter mechanism.
2. The maximum mixing efficiency of a stratified fluid, is achieved at $Ri \approx Ri^*$, when the thickness of the density interfaces between the quasi-homogeneous layers is in a transition between the sharpening and eroding modes. Thus, the transformation of a continuously density-stratified aqueous medium into a step-like density structure should increase the efficiency of vertical turbulent exchange.

Additional research (laboratory, numerical, field) is needed to study further this complicated problem and its relation with the step-like FS formation in the turbulent stratified fluid—oceanic pycnocline.

Finally, a few words should be said about the applications of the results of this study. The possible practical use of the results is the prediction of the evolution of the oceanic fine structure (FS), formed by the action of various physical mechanisms. One of these mechanisms is related with cross-frontal intrusions: the water layers penetrating from one side of the oceanic front to the other along the isopycnal surfaces [1,2]. If the rate of turbulence in the stratified waters surrounding the intrusion is low (the Richardson number is higher than the critical one), the intrusion can exist for a long time and spread to a large distance from the front, maintaining its boundaries in a sharpened state. In the opposite case (the Richardson number is below the critical one), the intrusion will quickly disappear (erode) due to vertical turbulent diffusion. Evidently, the cross-frontal heat and mass transfer, the regularities of which are poorly understood, should depend on the lifetime of the intrusion. The list of such examples could be continued.

Author Contributions: Conceptualization, A.G.Z.; methodology, A.G.Z. and V.V.G.; laboratory setup, V.V.G.; formal analysis, V.V.G. and A.G.O.; resources, A.G.Z.; data curation, V.V.G.; writing—original draft preparation, A.G.Z.; writing—editing, A.G.O.; project administration, A.G.Z.; and funding acquisition, A.G.Z. All authors have read and agreed to the published version of the manuscript.

Funding: The work was carried out within the framework of the state assignment No. FMWE-2021-0002 and with the support of the Russian Foundation for Basic Research grant No. 20-05-000496.

Conflicts of Interest: The authors declare no conflict of interest.

References

1. Fedorov, K.N. *The Thermohaline Fine Structure of the Ocean*; Pergamon Press: Oxford, UK, 1978.
2. Fedorov, K.N. *The Physical Nature and Structure of Oceanic Fronts*; Springer: New York, NY, USA, 1986; 333p.
3. Phillips, O.M. Turbulence in a strongly stratified fluid: Is it unstable? In *Deep Sea Research and Oceanographic Abstracts*; Elsevier: Amsterdam, The Netherlands, 1972; Volume 19, pp. 7–81.
4. Posmentier, E.S. The generation of salinity fine structure by vertical diffusion. *J. Phys. Oceanogr.* **1977**, *7*, 298–300. [CrossRef]
5. Ruddick, B.R.; McDougall, T.J.; Turner, J.S. The formation of layers in a uniformly stirred density gradient. *Deep-Sea Res.* **1989**, *36*, 597–609. [CrossRef]
6. Park, Y.-G.; Whitehead, J.A.; Gnanadesikan, A. Turbulent mixing in stratified fluids: Layer formation and energetics. *J. Fluid Mech.* **1994**, *279*, 279–311. [CrossRef]
7. Dmitrenko, I.; Golovin, P.; Dehn, J.; Kassens, H.; Zatsepin, A. Influence of sea ice on under-ice mixing under stratified conditions: Potential impacts on particle distribution. *Estuar. Coast. Shelf Sci.* **1998**, *46*, 523–529. [CrossRef]
8. Zatsepin, A.G. On peculiarities and similarities of the coherent structure formation in stratified and rotating fluid. In *Turbulent Mixing in Geophysical Flows*; Linden, P.F., Redondo, J.M., Eds.; CIMNE: Barcelona, Spain, 2001; pp. 211–299.
9. Barenblatt, G.I.; Bertsch, M.; Dal Passo, R.; Prostokishin, V.M.; Ughi, K. A mathematical model of turbulent heat and mass transfer in stably stratified shear flow. *J. Fluid Mech.* **1993**, *253*, 341–358. [CrossRef]
10. Balmforth, N.J.; Smith, S.G.L.; Young, W.R. Dynamics of interfaces and layers in a stratified turbulent fluid. *J. Fluid Mech.* **1998**, *355*, 329–358. [CrossRef]
11. Taylor, J.M.; Zhou, Q. A multi-parameter criterion for layer formation in a stratified shear flow using sorted buoyancy coordinates. *J. Fluid Mech.* **2017**, *823*, R5. [CrossRef]
12. Linden, P.F. Mixing across a density interface produced by grid turbulence. *J. Fluid Mech.* **1980**, *100*, 691–703. [CrossRef]

13. Krylov, A.D.; Zatsepin, A.G. Frazil ice formation due to difference in heat and salt exchange across a density interface. *J. Mar. Syst.* **1992**, *3*, 497–506. [CrossRef]
14. Linden, P.F. Mixing in stratifled fluids. *Geophys. Astrophys. Fluid Dyn.* **1979**, *13*, 3–23. [CrossRef]
15. Pelegri, J.L.; Sangra, P. A mechanism for layer formation in stratified geophysical flows. *J. Geophys. Res.* **1998**, *103*, 30679–30693. [CrossRef]
16. Thorpe, S.A. Experiments on the instability of stratified shear flow: Miscible fluids. *J. Fluid Mech.* **1971**, *46*, 299–319. [CrossRef]
17. Holford, J.M.; Linden, P.F. Turbulent mixing in a stratified fluid. *Dyn. Atmos. Ocean.* **1999**, *30*, 173–198. [CrossRef]
18. Turner, J.S. The influence of molecular diffusivity on turbulent entrainment across a density interface. *J. Fluid Mech.* **1968**, *33*, 639–656. [CrossRef]

Journal of
Marine Science and Engineering

MDPI

Article

Fine Structure of Vertical Density Distribution in the Black Sea and Its Relationship with Vertical Turbulent Exchange

Oleg I. Podymov *, Andrei G. Zatsepin * and Alexander G. Ostrovskii

Shirshov Institute of Oceanology, Russian Academy of Sciences, 119991 Moscow, Russia
* Correspondence: huravela@yahoo.com (O.I.P.); zatsepin@ocean.ru (A.G.Z.)

Abstract: This paper is concerned with the analysis of the long-term regular time series of current velocity and conductivity, temperature, and depth (CTD) profiles, measured with the moored autonomous profiler Aqualog over the upper part of the continental slope at a fixed geographical location in the Northeastern Black Sea. This study focuses on the fine structure of the density profiles to show that the fine-structure Cox number (C) is a power function of the Richardson number (Ri). A similar inverse power relationship with the same exponent was found earlier for the coefficient of vertical turbulent mass exchange (K_ρ) and Ri. Based on those results, the analysis indicated a statistically significant correlation between C and K_ρ, which suggests that the estimations of K_ρ could be conducted from the CTD data only.

Keywords: Black Sea; vertical turbulent mass exchange; fine structure; Cox number; Richardson number

Citation: Podymov, O.I.; Zatsepin, A.G.; Ostrovskii, A.G. Fine Structure of Vertical Density Distribution in the Black Sea and Its Relationship with Vertical Turbulent Exchange. *J. Mar. Sci. Eng.* **2023**, *11*, 170. https://doi.org/10.3390/jmse11010170

Academic Editors: Mikhail Emelianov and Georgy I. Shapiro

Received: 23 November 2022
Revised: 26 December 2022
Accepted: 7 January 2023
Published: 10 January 2023

1. Introduction

One of the important tasks of physical oceanography is an assessment of vertical turbulent exchange in the stratified waters of seas and oceans. In particular, the vertical flux of nutrients from the deep waters into the photic layer, where most biological production takes place, depends on the intensity of the turbulent exchange [1,2]. Knowledge of the vertical turbulent mass diffusion coefficient and the vertical gradients of nutrients (first of all, nitrates) would make it possible to quantify this flux in the sea and assess its significance for bioproductivity and the carbon cycle [3].

The most reliable methods for estimating the coefficient of vertical turbulent mass diffusion are based on direct measurements of the turbulent energy dissipation rates and other parameters of hydrophysical microstructures on scales close to the Kolmogorov scale [4]. Such studies involved special equipment, including free-falling or towing streamline apparatuses with microscale hydrophysical probes [5,6], and were carried out all over the world ocean [7], including the Black Sea [8–10]. As a result of these experimental studies, the profiles of turbulent energy dissipation rate $\varepsilon(z)$, where z is depth, were determined and the coefficient of vertical turbulent diffusion $K_\rho(z)$ was calculated using several methods, including the semi-empirical formula proposed by Osborn [11]: $K_\rho \leq 0.2\ \varepsilon/N^2$, where N is the buoyancy frequency. However, direct measurements of turbulent exchange are rather difficult, and the vertical turbulent diffusion coefficient is often estimated on the basis of semi-empirical parameterizations and hydrophysical data with a much coarser resolution.

One such parameterization was proposed by Munk and Anderson [12]. It was based on the dependence of the coefficient of vertical turbulent diffusion K_ρ on the Richardson number Ri in a stratified shear flow:

$$K_\rho \sim Ri^{-n} \tag{1}$$

The exponent of the power function is often defined as n = 1.5, $Ri = N^2/Sh^2$, where $N^2 = -g/\rho_0 \cdot \partial\rho/\partial z$ is the squared buoyancy frequency, $Sh^2 = (\partial u/\partial z)^2 + (\partial v/\partial z)^2$ is the squared vertical shear of the current velocity, and u and v are the eastern and northern velocity components, respectively. Following this pioneering paper, many authors [13–17] applied modified formulas parameterizing the dependence of K_ρ on Ri. Basically, these formulas were defined as follows:

$$K_\rho = K_{0s}\,(1 + \alpha_s \cdot Ri)^{-ns} + K_{bs}, \quad (2)$$

where K_{0s}, K_{bs}, α_s, and n_s are constants that depend on the vertical averaging scales of Ri. Here, K_{0s} represents mixing in the unstratified ocean, and K_{bs} is the background mixing by sources other than shear-induced instability.

Forryan et al. [17] analyzed the parameterization (2) using the data of direct observations of vertical turbulent diffusivity from three separate ocean regions in the North Atlantic and Southern Ocean. It was found that equation (2) is robust for Ri > 1 at depths below the ocean near-surface boundary layer. In particular, when equation (2) was fitted to all the data for turbulent diffusivity from all three data sets simultaneously, then the best fit was for a smoothing window of 56 m with the following parameters: $\alpha_s = 1$, $n_s = 1.49$, $K_{os} = 3.62 \cdot 10^{-4}$ m^2/s, and $K_{bs} = 8.14 \cdot 10^{-6}$ m^2/s.

For the Black Sea in particular, Stanev [18] successfully used the parameterization proposed by Pacanowski and Philander [13] for the numerical general circulation model of the Black Sea.

Zaron and Moum [19] indicated that there are not enough dimensional groups in S^2, N^2, and K_ρ to provide a relation between K_ρ and Ri that is invariant to the dimensional units. In order to overcome this difficulty, they proposed a new parameterization, which is based not only on Ri but also included the shear length scale and the kinetic energy of the background flow. This parametrization appears to be very promisingand should be tried for the Black Sea.

Podymov et al. [20,21] estimated the values of K_ρ in the upper layer of the Black Sea (30–220 m) using the modified formula of Munk and Anderson and the Aqualog data of synchronously measured density and current velocity profiles from 2013 to 2016. It was shown that K_ρ varied from 10^{-6} to $8.5 \cdot 10^{-3}$ m^2/s, and its average value was $8.8 \cdot 10^{-5}$ m^2/s, in agreement with the previous estimates of K_ρ for the same depth layer in the Black Sea [22,23] based on ship measurements using the LADCP current velocity profiles and data of the CTD casts. In the latter case, the dissipation rate of turbulent energy was calculated using the values of velocity shear and buoyancy frequency, smoothed with a 10 m vertical window. These values were close to the estimates of K_ρ obtained for the upper 20–80 m stratified layer in the Black Sea, based on direct measurements of the turbulent energy dissipation rate [9].

Despite progress in the estimation methods for the coefficient of vertical turbulent diffusion in a shear flow of stratified fluids by using the parameterizations based on the Richardson number, there are at least two obstacles. First, the constant n in (1) can be less than 1.5 for small values of the Richardson number (Ri < 1) [17]. Laboratory experiments with a mixing of stably stratified fluid resulted in a similar conclusion [24]. Therefore, small values of the Richardson number could lead to uncertainty in the estimate. It is currently unclear which formula better suits such conditions. The second issue is that the Richardson number is a function of the vertical velocity shear and the buoyancy frequency, both of which depend on the averaging procedure, so the coefficient K_ρ should also depend on the size of the averaging window.

A possible approach to determine the vertical turbulent diffusion coefficient is based on the analysis of the fine structure (FS) of stratified ocean waters [25]. Usually, an FS appears in vertical profiles of temperature, salinity, density, velocity, concentrations of passive and non-passive tracers, etc. It is characterized either by the presence of layers (whose thickness is up to several tens of meters) with low vertical gradients of these parameters, separated by high-gradient interlayers (step-like type), or by layers with temperature and salinity inversions compensated in density (intrusive type). For the latter, a change in the sign of the vertical gradient to the opposite, in comparison with the sign of the background gradient, is considered an inversion.

An FS can have a reversible or irreversible nature. The reversible FS is usually associated with internal waves and their kinematic effect, which leads to adiabatic deformations in temperature and salinity profiles. The irreversible FS occurs as a result of internal vertical mixing events or quasi-isopycnic intrusions, both causing non-adiabatic profile changes and producing vertical and lateral fluxes in mass and momentum. In contrast to fluctuations of an isotropic turbulent nature, ocean FS layers have horizontal extensions usually 10^2–10^4 times larger than their vertical thickness [25].

FSs in different parts of the world ocean were the subject of many detailed studies during the last 40–50 years. Much attention was paid to transfrontal thermohaline intrusions, whose dynamics largely depend on buoyancy fluxes caused by differential diffusion–convection. Among the publications on this subject, we would like to mention the pioneering work of Joyce [26] and a review by Ruddick [27], which focused on the theory of intrusions and their laboratory modeling. In the case of Black Sea regional oceanography, investigations were mostly concerned with the intrusive type of FS formed via the intrusions of modified waters of the lower Bosphorus current into the Black Sea pycno-halocline and deeper layers [28,29]. Reliable evidence for such intrusions was found recently, based on the analysis of the vertical profiles of temperature, salinity, and dissolved oxygen obtained with the Argo floats [30–33].

A new step in the study of the Black Sea intrusions was conducted by Stanev et al. [34]. It was shown that optical backscattering coefficient data from the bio-Argo floats with a vertical resolution of 1 m can be successfully used to indicate thermohaline intrusions in the Bosphorus region and to the east of it. In this area, this coefficient is more than five times larger than elsewhere and could be an indicator of bacterial abundance and biological involvement in the precipitation of manganese-containing particles. The observations clearly identify thermal and «particle» intrusions down to ~700–800 m in the Bosphorus inflow area.

More recently, isopycnic mesoscale (20–50 km) thermohaline anomalies were identified and analyzed in the upper 300 m layer of the Northeastern Black Sea based on the archive of the ship-born CTD-data along 100 nautical miles cross-sections between the coastal zone and the deep part of the sea [35]. It was shown that the FS in the temperature profiles was mostly of the intrusive type at the boundaries of the thermohaline anomalies where an absolute value of the product $\Delta T\ [(\varphi_{ro} - \varphi_T)^2]^{1/2}$ was sufficiently large. Here, ΔT is the value of the isopycnic temperature anomaly between two CTD-stations, and $\varphi_{ro} - \varphi_T$ is the angle between isotherms and isopycnals. The observed temperature inversions did not result in density inversions. Their formation was due to quasi-isopycnic intrusions at the thermohaline fronts at the boundaries of the mesoscale thermohaline anomalies. Yet the number of profiles with the FS temperature inversions at each cross-section was much less than the number of profiles with the step-like FS associated with multimode internal waves and vertical turbulent mixing processes. It should be noted that the conditions for double-diffusive convection are lacking in the upper 300 m of the Black Sea [36]. Therefore, it could not be attributed any role in the formation of the step-like FS.

Part of the step-like FS is formed due to vertical mixing processes caused by breaking internal waves and shear flow instability. The latter mechanism seems to prevail in the Black Sea Rim Current zone (see below), where significant vertical velocity shear is observed from time to time. In such cases, the Richardson number has low values and the intensity of vertical turbulent mixing increases [9,37]. As it was shown in [9], a decrease in the bulk Richardson number, estimated using geostrophic velocity shear and CTD measurements of density vertical distribution, was associated with the FS interleaving intensity (a magnitude of deviation in the smoothed temperature, salinity, and density profiles from the original measurements) in the seasonal thermocline and the permanent pycno-halocline. At the same time, the FS intensity increases along with the turbulent energy dissipation, as was revealed in experiments with the free-falling microstructure probe Baklan [5]. However, these preliminary results require further research based on more numerous observations.

Of course, there is difficulty in extracting a turbulent mixing signal from FS data, since non-breaking internal waves and intrusion processes cause deformation in the density profile [25]. Furthermore, a standard CTD sampling is usually performed using a probe lowered from a ship on a conducting wire cable. This results in data pulsations, caused by cable jerking and probe snaking because of ship pitching, thus producing an artificial fine-scale vertical distribution of the thermohaline parameters. A particularly strong noisiness in fine-structure data should be expected when using massive rosette (carousel) complexes with water sampling bottles. This kind of measurement is conventionally used in multidisciplinary ocean studies. To reduce the data noise as much as possible, a free-falling CTD probe with a streamlined shape should be used [38,39]. However, such measurement methods are far from being widespread.

This paper presents the results of the studies of the FS characteristics of density profiles in the Northeastern Black Sea and their dependence on the Richardson number, based on the large dataset composed of synchronized and accurate measurements of density and current velocity profiles using the moored profiler Aqualog [40]. The Aqualog was deployed at the upper part of the continental slope within the zone of the Rim Current, which is a basin-scale current that encircles the Black Sea in a counterclockwise direction [28]. The main reason for the formation of a cyclonic annular current in the Black Sea is the cyclonic vorticity of the wind stress on its surface [18,41,42]. As was shown with a drifter experiment, the typical speed of the Rim Current in the upper layer is about 0.15–0.2 m/s, and it can increase up to 0.8–0.9 m/s [43,44]. Due to the sharp pycno-halocline located at a depth of 100–200 m, the current velocity decreases rapidly with depth in the upper 250 m layer, resulting in substantial velocity shear [45].

The Rim Current core is usually observed further into the deep sea than the Aqualog location. However, due to meandering, from time to time, the current moves close to the shore, and the Aqualog registers a significant increase in the flow velocity directed to the northwest. This happens when the mooring station moves into a cyclonic meander of the Rim Current. When it moves into an anticyclonic meander (or a mesoscale anticyclonic eddy), a southeastern countercurrent is observed. A typical period of a change in the velocity direction in the study area is about 10 days [46].

The aim of this work was to find the interrelations between the Black Sea FS characteristics of density profiles and the vertical turbulent exchange based on the Aqualog measurements. The data and methods are described in Section 2. The evidence for the dependence of the FS characteristics of the density profiles on the Richardson number is demonstrated in Section 3.1 and their interrelation with vertical turbulent exchange is analyzed in Section 3.2, along with their application for estimating the coefficient of vertical turbulent mass diffusion. The discussion of the results is contained in Section 4 and the preliminary conclusions in Section 5.

2. Materials and Methods

The analyzed data were collected using the autonomous moored profiler Aqualog [40] located on the continental slope, five nautical miles from the Black Sea coast near the town of Gelendzhik (Figure 1). The mooring depth was about 240 m.

The Aqualog profiler is capable of moving up and down along the mooring line stretched between the bottom anchor and the top floatation. The latter was positioned about 30 m below the sea surface—deep enough to avoid any significant effect on it from the surface waves (Figure 2).

Because of the stable propulsive force and streamlined body of the device, its speed along the mooring cable was stable (about 20 cm/s on average, with an SD of about 0.7 cm/s), varying within a few cm/s (Figure 3). During the observational survey, the Aqualog was equipped with the CTD probe Idronaut Ocean Seven 316 to measure temperature and salinity with a vertical resolution of 8–10 cm and the Doppler acoustic current meter Nortek Aquadopp 3D to measure current velocity with a vertical resolution of about 1 meter. Hence, the Aqualog was a reliable tool for the accurate measuring of the hydrophysical parameters with a fine vertical resolution.

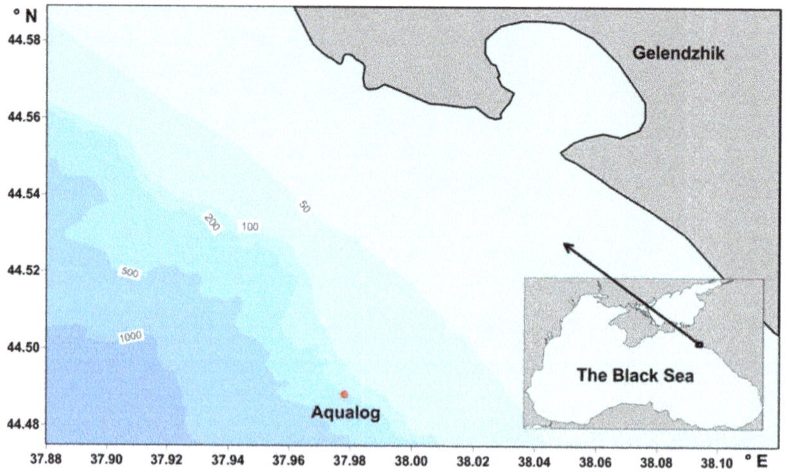

Figure 1. Location of the moored station with the Aqualog profiler on the continental slope of the Northeastern Black Sea. Shades of blue indicate bathymetry and numbers show the bottom depth in meters.

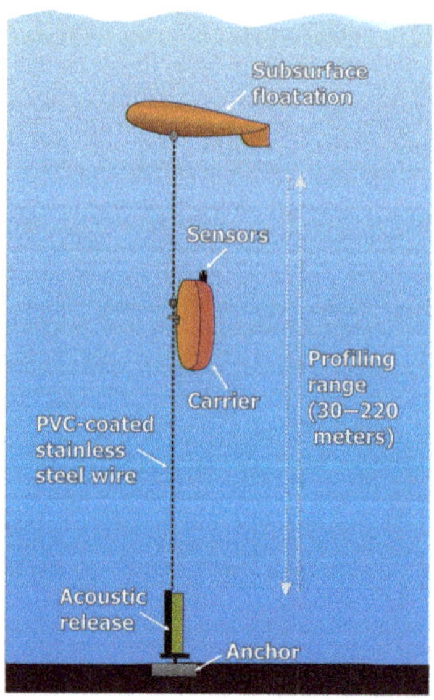

Figure 2. The Aqualog profiler mooring station.

Below, we analyze the data obtained with the Aqualog from 25 October 2015 to 1 March 2016. The profiler operated continuously for 129 days, measuring the vertical profiles of the hydrophysical parameters every 6 h from the subsurface floatation downward to a depth of about 220 m. All measured parameters, including the current velocity, were binned into 1 m vertical cells and considered as primary data.

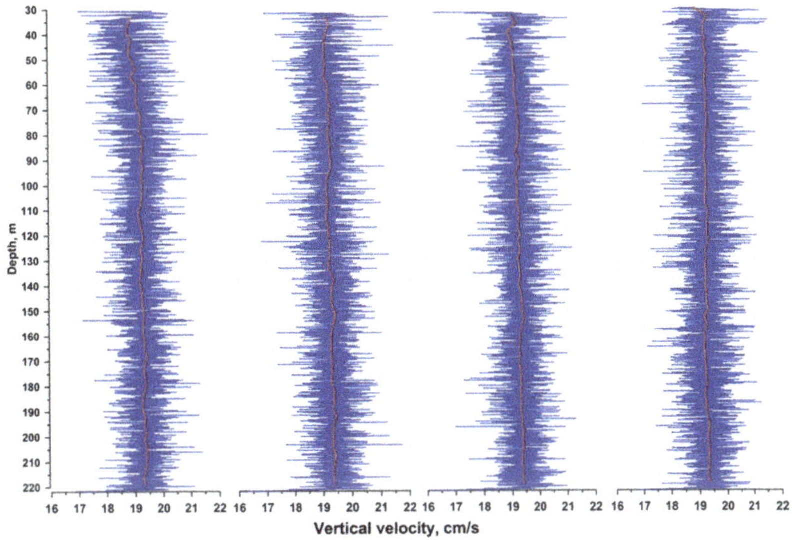

Figure 3. The Aqualog profiling speed measured with 2 Hz frequency. The figure shows first 4 profiles (1 day measurement). Orange lines are 3 m averaged values.

While the Doppler acoustic current meter Aquadopp provided us with high quality data, there was still a measurement error that could lead to extremely high values of Ri when velocity was rather low and density gradient was high, causing distortions in the Ri-based estimations. To remove these errors, the following procedure was used to calculate the vertically averaged Richardson number. The profiles of $N^2(z)$ and $Sh^2(z)$ were smoothed with a vertical moving average window of 25 m. The width of the window was chosen to overlay the FS layers, of which thickness usually did not exceed 10 m [9]. Henceforth, we shall use the angle brackets ($<...>$) to indicate that a certain parameter was smoothed in a 25 m averaging window, or its calculation was based on similarly smoothed components. The values of $<N^2>$ and $<Sh^2>$ were used to obtain the average Richardson number $<Ri> = <N^2>/<Sh^2>$ that characterized the stratified shear flow stability, the intensity of the vertical turbulent exchange, and the FS characteristics.

One of the most important methodological aspects of this work was the procedure of FS extraction from the primary data profiles. Let us note that the FS variability of the potential density profiles $\sigma(z)$ was determined with the variability of the density gradient $\Delta\sigma/\Delta z$, which characterized the density stratification of the aquatic environment. The density gradient was calculated from the primary vertical distribution of density as $\Delta\sigma/\Delta z$, where $\Delta\sigma$ is the density difference between the lower and upper points of the density profile with a distance of $\Delta z = 1$ m between them. The gradient magnitude may vary widely, and the gradient changes its sign in the layers in which the stable stratification is locally overturned. The FS variations appear around the averaged values of $\Delta\sigma/\Delta z$. To calculate the average gradient $<\Delta\sigma/\Delta z>$ we used the same moving average procedure with a 25 m window as that for the calculation of $<Ri>$. The fluctuating part (pulsation) of the density gradient was defined as the difference between the primary and the averaged gradient: $(\Delta\sigma/\Delta z)' = \Delta\sigma/\Delta z - <\Delta\sigma/\Delta z>$.

Typical profiles of the density gradient and its pulsations are presented in Figure 4. Let us consider an element of FS as a layer situated between consecutive odd crossings of the original profile with the smoothed profile (transition through zero for $(\Delta\sigma/\Delta z)'$). Such an element with the thickness h contains both positive and negative anomalies of $\Delta\sigma/\Delta z$ and could appear after an incomplete vertical mixing. The negative anomaly can be associated with a partly mixed portion of the profile and the positive anomaly, with the interface between the negative anomalies.

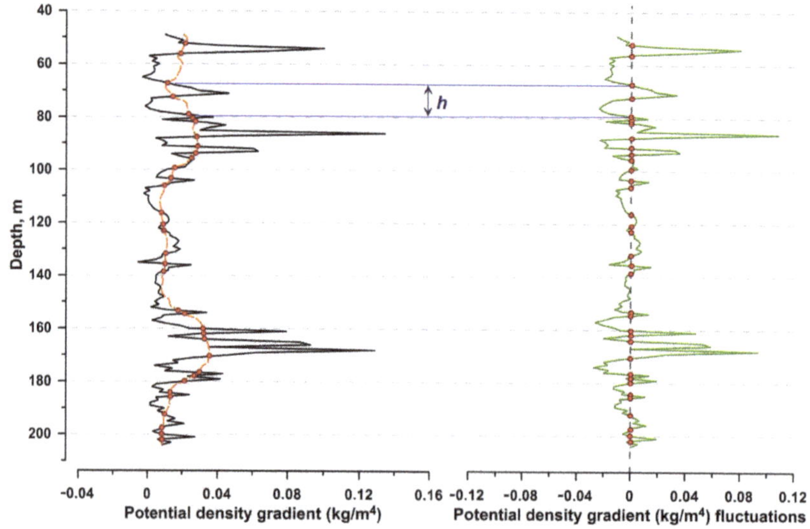

Figure 4. Left: a typical profile of the potential density vertical gradient $\Delta\sigma/\Delta z$. Orange dashed line shows the density gradient profile that was smoothed with the 25 m moving average. Red dots show the crossings of the original profile with the smoothed profile. **Right**: fluctuations in density gradient, i.e., the difference between the primary profile and the smoothed profile of the density gradient. Red dots show transitions through zero. The blue horizontal lines indicate a layer of a single element of FS, which includes one positive and one negative anomaly of the density gradient. Each pair of the consecutive anomalies constitutes a single FS element. The thickness of the FS element is depicted as h.

Along with the calculations of the thickness of all h-elements for all acquired profiles, we also calculated the thickness of the layers with positive and negative gradient anomalies (sublayers of h element, denoted as h^+ and h^-, respectively), as well as the thickness of the density gradient inversion layers, h_{inv}.

The following procedure was used to estimate the thickness of the density gradient inversion layers. Since these layers were relatively thin (their thickness in most cases was less than 1 m), the original CTD data were binned into 10 cm vertical cells and smoothed with a 1 m moving average. Since the density of water in the Black Sea generally increases with depth, the inversion layers were those where the density decreased with depth. All acquired profiles were scanned from top to bottom. The thickness h_{inv} of every inversion layer was calculated as an absolute depth difference between the bin, where density decreased with depth compared with the overlying (upper) bin, and the first underlying (lower) bin, where density increased with depth again. Therefore, the estimation error was ±10 cm. To cut off insignificant density fluctuations inside the step-like FS and minor probing errors, the density fluctuations smaller than 0.001 kg/m^3 were ignored.

Since the Richardson number was calculated with a 1 m vertical step, to find a connection between <Ri> and the h_{inv} intervals, the values of <Ri> were interpolated over a uniform 10 cm vertical grid, matching the 10 cm CTD data cells used for the h_{inv} calculation. After that, <Ri> was calculated for every h_{inv}, as a mean value of <Ri> over all the 10 cm cells that fell within a particulate h_{inv} layer.

To estimate the non-dimensional intensity of the FS density gradient pulsations we used a form of the Cox number [47]. Initially [48,49], this number was introduced as a dimensionless quadratic measure of microstructural fluctuations in temperature and salinity. In our case, the "fine scale" analog of this number has the following expression:

$$<C(z)> = <(\Delta\sigma/\Delta z)'^2>/<(\Delta\sigma/\Delta z)>^2 \quad (3)$$

To calculate $<C(z)>$, the profile of the squared values of density gradient pulsations $(\Delta\sigma/\Delta z)'^2$ was smoothed with the same 25 m window, as indicated by the angle brackets in (3). The values of $<C(z)>$ and their dependence on $<Ri>$ will be used to find the relationship between $<C(z)>$ and the turbulent mass diffusion coefficient K_ρ (see Section 3.2 below).

To calculate the turbulent mass diffusion coefficient K_ρ with Equation (2), we used the same "Aqualog" data, a 40 m smoothing window, and the previously obtained coefficients [20]: $\alpha_s = 0.2$, $n_s = 1.5$, $K_{0s} = 10^{-3}$, and $K_{bs} = 10^{-6}$.

3. Results

3.1. Density Gradient Fine-Structure Characteristics, and Their Dependence on the Richardson Number

The analysis of the 514 profiles of the sea temperature, salinity, and density collected using the Aqualog profiler showed that both the step-like FS and the intrusive FS were present in the study area (Figure 5).

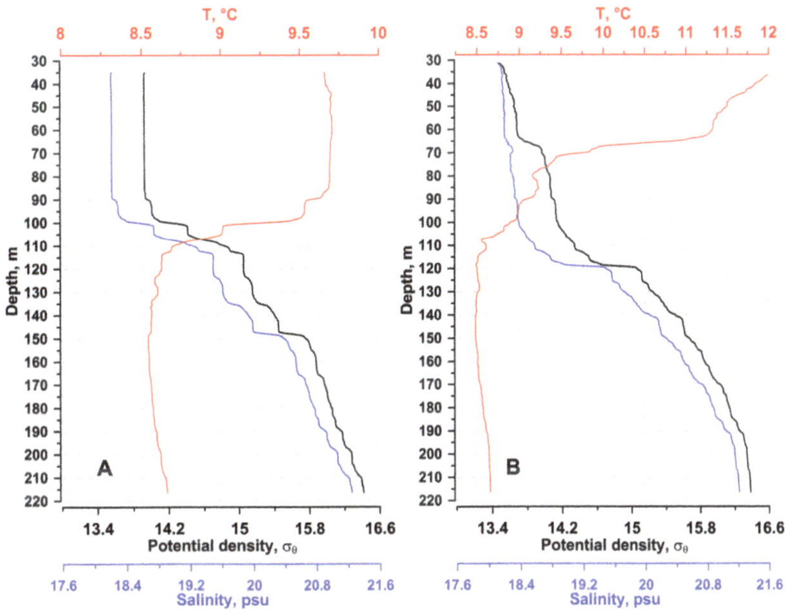

Figure 5. Profiles of the sea temperature, salinity, and potential density for the step-like (**A**) and intrusion (**B**) types of FS (the intrusion type is clearly observed in the temperature profiles only).

It should be noted that in the layer of 30–220 m (the Aqualog profiling range), potential density σ was mostly controlled by the salinity S, and the vertical density gradient was generally determined by the salinity gradient $\Delta S/\Delta z$. The temperature gradient below the thermocline was small and changed its sign at depths below 100–150 m. The temperature inversions caused by the intrusions at the thermohaline fronts [35] did not result in density inversions. These were always accompanied by salinity variations, which compensated for their contribution to the density.

The FS elements h were calculated (Table 1) on both a 10 cm vertical grid (the same grid that was used for the h_{inv} calculations) and the primary 1 m vertical grid. In both cases, the average density gradient was calculated using the same 25 m smoothing window.

When the data grid changed from 10 cm to 1 m, the average values of h, h+, and h^{-} almost doubled. While strong FS elements remained the same, with the same thickness, some smaller adjoined elements merged into one, and the vast background of tiny (less than 1 m) FS elements was substantially reduced. It appears that such an increase in calculating scale provides a better display of strongly pronounced FS elements and decreases the background level of weak h. For these reasons, we decided to use the results of the 1 m vertical grid calculations for further h-based analysis.

Table 1. Estimates of the mean and median thickness (in meters) of the layers with the density inversion (h_{inv}) and the FS element (h). h+ and h^{-} are the layers of the FS elements (their thickness), where density gradient is greater or smaller than its averaged value, respectively. SD denotes standard deviation.

Data Grid	h_{mean}	h_{median}	h^{-}_{mean}	h^{-}_{median}	h^{+}_{mean}	h^{+}_{median}	$h_{inv\ mean}$	$h_{inv\ median}$
10 cm	2.9	2.1	1.8	1	1.1	0.9	0.55	0.4
	SD = 2.4		SD = 2.1		SD = 0.93		SD = 0.4	
1 m	5.1	4	3.23	2	1.87	2	n/a	
	SD = 3.83		SD = 3.31		SD = 1.67			

As follows from Table 1 (the 1 m vertical grid calculations), the average thickness of the layers with the density gradient inversions was about 10 times smaller than the average thickness of the FS elements.

Figure 6 shows the distribution of the thickness h versus the Richardson number, <Ri>, averaged for every h-layer. The typical value of h was about 1–10 m and the maximum number of the FS elements was observed when <Ri> = 0.2−0.5.

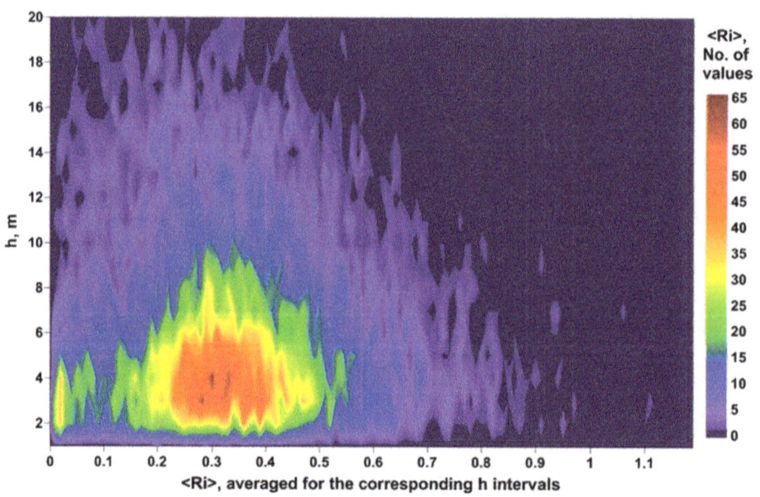

Figure 6. The thickness h of the FS elements versus the bulk Richardson number <Ri>.

The density inversions were observed mostly when <Ri> < 0.25 (Figure 7), in accordance with the criterion of Miles and Howard [50] for the instability of stratified flow with linear velocity shear.

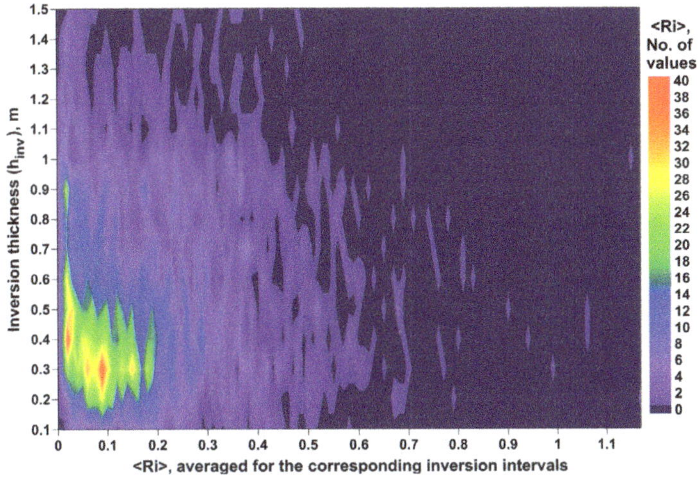

Figure 7. The distribution of the density inversion thickness h_{inv} versus the bulk Richardson number <Ri>.

There was evidence that the vertical velocity shear was the most important factor that determined the occurrence of the density inversions. Figure 8 shows the vertical profiles of density and velocity shear when the velocity shear was rather small, and Figure 9 shows the case when the velocity shear was sufficiently large.

In Figure 8, we can see the step-like FS with practically no inversions, while Figure 9 shows the density profile with many inversions.

A general picture of the time–depth distributions of <Ri>, <N^2>, and <S^2> is shown in Figure 10. As expected, there is a certain positive correlation between <Ri> and <N^2>, and a negative correlation between <Ri> and <S^2>.

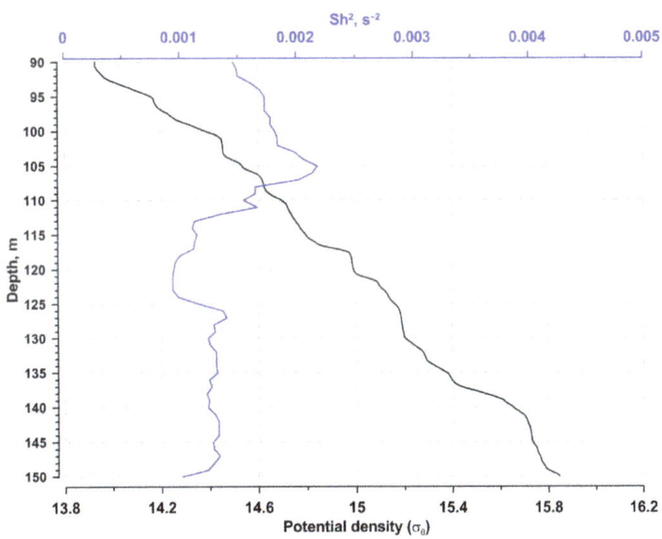

Figure 8. Example of the vertical profiles of the density and the squared velocity shear for the case when the velocity shear was small.

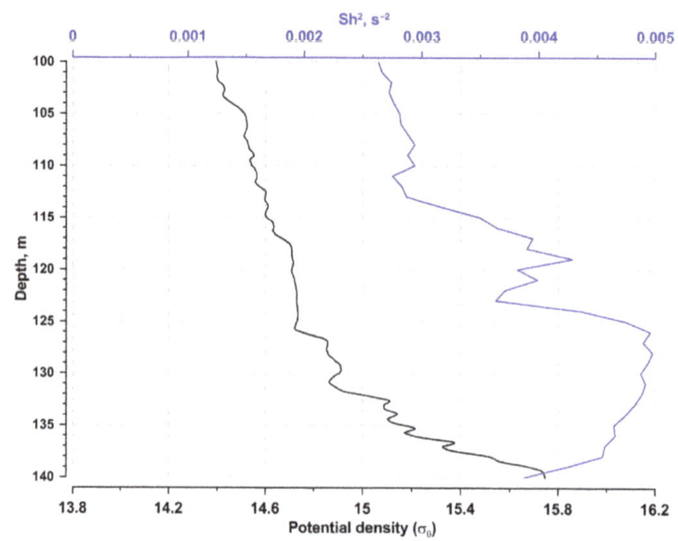

Figure 9. Same as Figure 8, but for the case when the velocity shear was large.

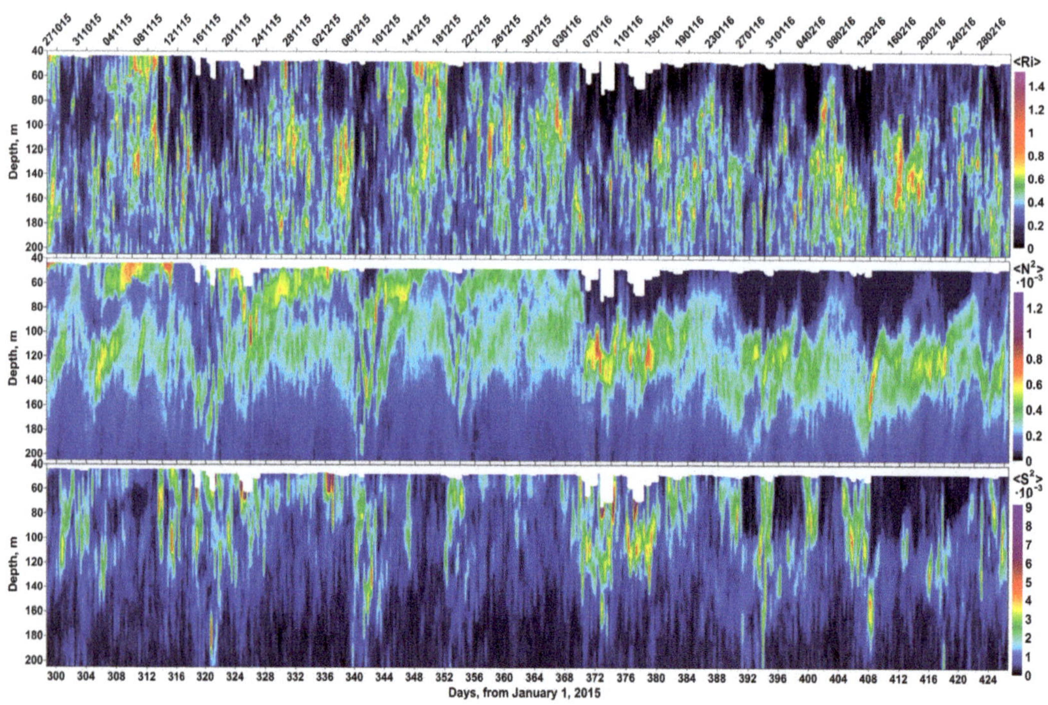

Figure 10. Time–depth diagrams of the bulk Richardson number, buoyancy frequency, and vertical velocity shear (N^2 and S^2 were smoothed with the 25 m moving average).

To estimate the non-dimensional intensity of density gradient pulsations, the FS Cox number <C> was calculated for all the data using (3). In addition to vertical smoothing, temporal averaging was applied to the data as well, using the isopycnal levels rather than depth coordinates. We used the isopycnals because the distribution of sea temperature, flow velocity, dissolved oxygen, etc., depends on $\sigma(z)$, and the gradient of these parameters over the density surface is very small in the Black Sea. Therefore, the temporal averaging of the FS characteristics should be more accurate when performed in the density domain [35]. For that purpose, the primary data, initially binned into the 1 m vertical cells, were re-binned into the isopycnal layers with a uniform vertical step of $\Delta\sigma = 0.01$ kg/m^3. An Akima spline [51] was used to interpolate the data over the thicker density grid. After this transformation, a 5-day moving average spanning 20 profiles was applied to each individual isopycnal within the potential density range of $\sigma = 14.1$–16.3 kg/m^3. The density limits were chosen to ensure that the selected density range was almost always within the depth limits of the studied layer. While the top and bottom depths of the Aqualog profiles remained mostly the same, the water density varied significantly, and a wider density interval would result in more data gaps near the endpoints of the profiles.

We shall use double angle brackets for further indication of both temporal and vertical smoothing (thus ending with <<C>> and <<Ri>>). The performed procedure provided us with the following diagram (Figure 11), indicating that there is a dependence of <<C>> on <<Ri>>, which is possible to represent as an inverse power function:

$$<<C>> = 0.2(<<Ri>>)^{-1.5} \quad (4)$$

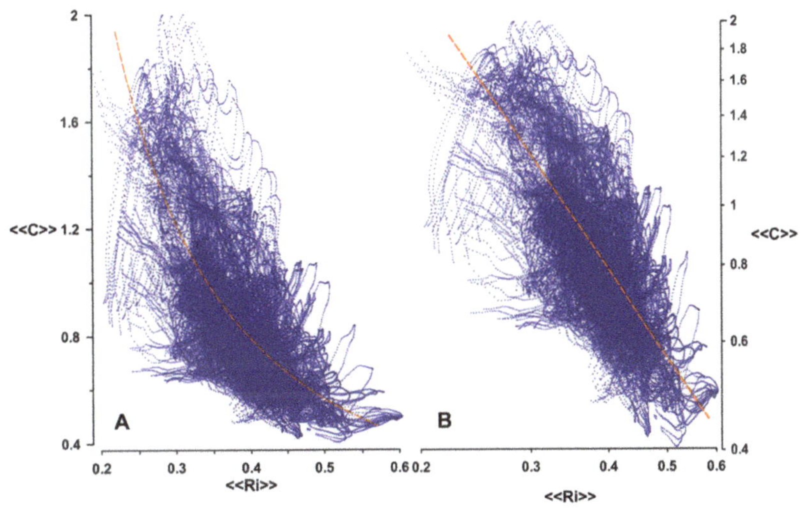

Figure 11. The values of <<C>> versus <<Ri>> for all measured data. (**A**) shows the data distribution of the linear coordinates and (**B**) shows the same data distribution of the decimal logarithmic coordinates. Dashed line represents the inverse power fit $<<C>> = 0.2(<<Ri>>)^{-1.5}$ or $lg<<C>> = -1.5\ lg<<Ri>> - 0.7$ for the decimal logarithmic scale.

An open-source software for non-linear curve-fitting, Fytik [52], was used to find the best approximation (4). The root-mean-square error (RMSE) for the approximation was 0.35, and the Pearson correlation coefficient was R = 0.68.

This result is very important since it indicates the existence of a statistically significant inverse power dependence of the non-dimensional FS pulsations on the Richardson number, which characterizes the stability of the stratified shear flow. A decrease in the Richardson number indicates the intensification of the process of vertical turbulent exchange, which contributes to the FS layering of the marine environment under strong stratification [53,54].

3.2. Interrelation between Intensity of the Fine-Structure Layering and Vertical Turbulent Mass Exchange

Since the established inverse power dependences of the FS Cox number <<C>> and the vertical turbulent mass diffusion coefficient K_ρ on the Richardson number are similar, it makes sense to look for a linear correlation between <<C>> and $<K_\rho>$.

Following Podymov et al. [20], we calculated $<K_\rho>$ using a 40 m averaging window by applying formula (2) with previously established values of the coefficients $\alpha_s = 0.2$, $K_{0s} = 10^{-3}$, $K_{bs} = 10^{-6}$, and $n_s = 1.5$. Then, the resulting values were smoothed with a 5-day moving average in the same manner as that for the calculations of <<C>>.

The time–depth diagram of $<<K_\rho>>$ and <<C>> (the calculations are performed on the time–density grid, but the diagram is drawn using the time–depth grid) indicates that there was indeed a certain positive correlation in the spatial-temporal dynamics of these parameters (Figure 12).

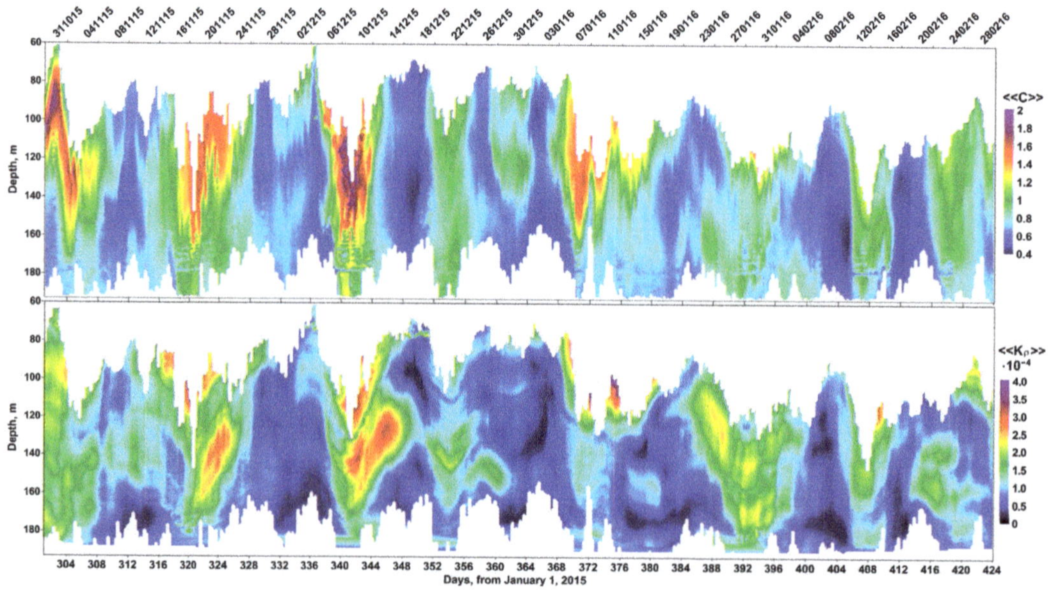

Figure 12. Time–depth diagram of <<C>> (**above**) and $<<K_\rho>>$ (**below**).

A scatter plot of $<<K_\rho>>$ against <<C>> was built to estimate the parameters for the linear correlation (Figure 13). The linear dependence of $<<K_\rho>>$ on <<C>> can be represented as follows: $<<K_\rho>> = (1.15 \cdot <<C>> + 0.18) \cdot 10^{-4}$ with Pearson correlation coefficient R = 0.54. Although the positive correlation between <<C>> and $<<K_\rho>>$ was not very high, it was still significant and indicated the existence of a quasi-linear dependence of the intensity of the FS layering on the vertical turbulent exchange. The root-mean-square error between the Ri-based $<<K_\rho>>$ values and the <<C>>-based approximation was $5 \cdot 10^{-5}$. This was about 1.5 times less than the median value of the turbulent vertical mass exchange coefficient itself.

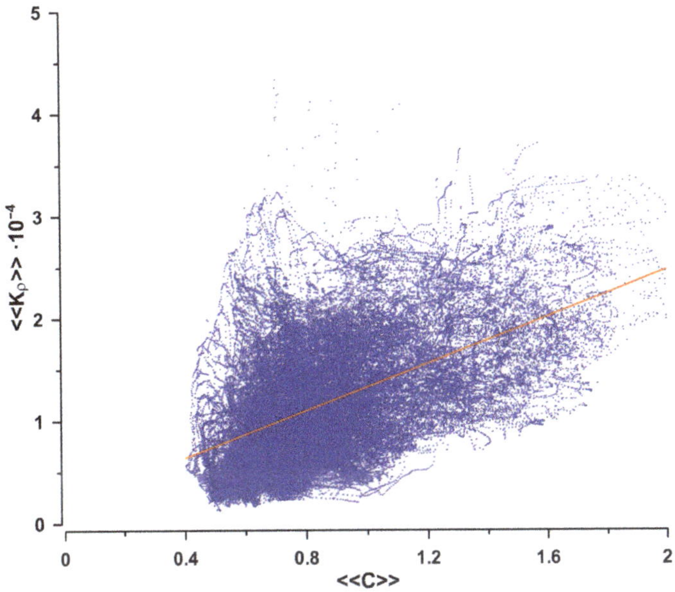

Figure 13. The values of <<K_ρ>> versus <<C>> and the linear regression fit.

4. Discussion

As was mentioned in the introduction, an estimation of the coefficient of vertical turbulent mass diffusion in a stratified water column is an important oceanographic problem. Ri-based parameterization of this coefficient is not always applicable since it involves both CTD data and data on the vertical distribution of the flow velocity or its vertical shear. Such data are present if the profiles of temperature, salinity, and current velocity are measured simultaneously with a CTD probe and either a ship-mounted ADCP or an LADCP in the profiling mode [22]. However, CTD data are usually not supplemented with the current velocity profiles. There is a lot of information on the vertical distributions of sea temperature, salinity, and density with the FS resolution missing the quasi-synchronous data on the current velocity shear. The question is whether the data of the CTD measurements alone could be successfully used to estimate the coefficient of vertical turbulent diffusion or not. We tried to address this challenge by analyzing the measurement data obtained using the moored profiler Aqualog in the Black Sea.

The estimates of the averaged Richardson number <Ri> based on the Ri values obtained for the high-resolution (one meter) data seem to be quite realistic. The smoothing of the N^2 and Sh^2 profiles with a 25 m moving average, before using them for the Ri calculation, made it possible to denoise the estimates and obtain a <Ri> that is a reliable characteristic of the stratified shear flow stability. It may serve as a criterion for the dynamical status of the flow. When the velocity shear is strong enough, the <Ri> values comply with the criterion of Miles–Howard for shear instability. Another piece of evidence in support of the reliability of this estimation is that almost all density inversions during the observed events of vertical mixing (overturning of stratified water layers) occur when <Ri> < 0.25 [50]. However, we should use the calculations of density inversions with caution, since they are usually performed at the lower limit of the sensors' vertical resolution and may be affected by measurement errors.

An important role in finding the correlation between the values of the FS Cox and Richardson numbers was the additional ensemble averaging of these parameters with a 5-day (20 profiles) moving average along the density surfaces. As it was shown earlier, a typical period of mesoscale variability is about 10 days in the zone of the Rim

Current [46,55]. Therefore, a 5-day average retains the mesoscale variability caused by meandering and eddy formation in the Rim Current area.

Another important factor is that the energy of internal waves in the Black Sea is relatively small compared with the ocean since the Black Sea belongs to non-tidal seas and is characterized by moderate wind forcing [22,23]. Thus, the processes of vertical mixing and FS formation depend mostly on the velocity shear produced by the highly baroclinic Rim Current and mesoscale eddies [9]. It should be noted that the strong dependence of the vertical mixing on the current velocity shear is characteristic not only of the Rim Current but also of other jet-like currents, for example, the western boundary currents [15].

The ensemble averaging of $<C>$ and $<Ri>$ provided us with an opportunity to look for a correlation between these parameters. It was found that $<<C>> \sim (<<Ri>>)^{-n}$ when n = 1.5 with a Pearson correlation coefficient R = 0.68 (RMSE = 0.35). Since the same power function with the same exponent n was successfully used by Podymov et al. [20,21] for the parameterization of the coefficient of vertical turbulent mass diffusion K_ρ, it was plausible to assume that a linear correlation exists between $<<C>>$ and $<<K_\rho>>$, where $<<K_\rho>>$ was calculated using the same Aqualog data and averaged in the same manner as $<<C>>$. We obtained a statistically significant linear correlation, although with a smaller correlation coefficient (R = 0.54).

The relationship found between the FS Cox number and the coefficient of vertical turbulent mass diffusion is only one step in a study of the FS characteristics and their dependence on the intensity of the vertical turbulent exchange in the Black Sea. In the future, it would be plausible to apply Thorpe's and Ellison's scales to the Aqualog data in order to estimate the turbulent energy dissipation velocity, as it was performed in, e.g., [56], and use it to calculate the turbulent mass diffusivity. The advanced methods for K_ρ calculation proposed by Zaron and Moum [19], or Gregg et al. [57] (as formalized by Kunze et al. [58]), also appear worth trying (e.g., [59]), using the high-resolution and long-term data obtained with the Aqualog profilograph.

5. Conclusions

We analyzed the data of continuous long-term measurements of the vertical distribution of density and current velocity in a depth range of 30–220 m sampled using the autonomous Aqualog profiler moored at the upper part of the Black Sea continental slope. The fine structure of the vertical density gradient and its relation to the averaged values of the Richardson number $<Ri>$ was estimated using high-resolution (one meter) data on the vertical distributions of water density and current velocity.

Two important fine-structure characteristics were calculated: the thickness of layers with increased and decreased density gradients and the thickness of layers with inversions of the density gradient. Based on the density gradient profiles binned into 1 m vertical cells, it was revealed that the thickness of a typical vertical fine-structure element (a layer of increased density gradient and an underlying layer of decreased density gradient) is about 10 times larger than the mean thickness of layers with density inversions calculated over a 10 cm vertical grid.

Most fine-structure elements corresponded to the values of $<Ri>$ from 0.2 to 0.5, while the density inversions were mostly observed when $<Ri> < 0.25$. Their amount was significantly less than the number of fine-structure elements. The existence of density inversions is fundamentally important because they indicate the ongoing process of vertical turbulent mixing, i.e., the overturning of stably stratified water layers, leading to the formation of irreversible elements in the fine structure. It also proves that the parameters of the fine structure are related to the processes of vertical turbulent mixing, which strongly depend on the Richardson number.

The Cox number C for the fine structure was calculated, and it was found that after vertical and temporal averaging, $<<C>>$ correlated with the Richardson number $<<Ri>>$ averaged in the same manner. Particularly, it was shown that $<<C>> \sim (<<Ri>>)^{-1.5}$ with Pearson correlation coefficient R = 0.68 and root-mean-square error RMSE = 0.35. The coefficient of vertical turbulent mass diffusion K_ρ had a similar power dependence on Ri, as was estimated earlier [20,21] for the Black Sea cold intermediate layer and pycnohalocline. Based on this result we found a linear correlation between <<C>> and <<K_ρ>>, with Pearson correlation coefficient R = 0.54 and RMSE = $5 \cdot 10^{-5}$ between Ri-based <<K_ρ>> values and a <<C>>-based approximation. This is about 1.5 times less than the median value of the turbulent vertical mass exchange coefficient itself.

The results of this work suggest that in the future, it might be possible to provide K_ρ estimations for the Rim Current zone of the Black Sea using the fine-structure characteristics of density profiles without any calculations of the Richardson number. It means that only the data of high-quality CTD measurements performed under certain hydrodynamic and meteorological conditions would be required for this procedure, and the estimation of K_ρ would be significantly facilitated. However, more work is required to refine the methods for a quality estimate of the vertical turbulent diffusion coefficient and to determine which of the fine-structure elements of density (temperature and salinity) profiles are best suited to reflect the processes of vertical mixing in a stratified fluid.

Author Contributions: Conceptualization, O.I.P. and A.G.Z.; data curation, O.I.P.; formal analysis, O.I.P. and A.G.Z.; funding acquisition, A.G.Z.; methodology, O.I.P. and A.G.Z.; project administration, A.G.Z.; resources, A.G.O.; writing—original draft, O.I.P. and A.G.Z.; writing—review and editing, O.I.P., A.G.Z. and A.G.O. All authors have read and agreed to the published version of the manuscript.

Funding: This work was carried out within the framework of the state assignment no. FMWE-2021-0002 and with the support of the Russian Foundation for Basic Research, grant no. 20-05-000496.

Institutional Review Board Statement: Not applicable.

Informed Consent Statement: Not applicable.

Data Availability Statement: The data presented in this study are available on request from the corresponding author. The data are not publicly available due to privacy.

Conflicts of Interest: The authors declare no conflict of interest.

References

1. Johnson, K.S.; Riser, S.C.; Karl, D.M. Nitrate supply from deep to near-surface waters of the North Pacific subtropical gyre. *Nature* **2010**, *465*, 1062–1065. [CrossRef] [PubMed]
2. Mikaelyan, A.S.; Mosharov, S.A.; Kubryakov, A.A.; Pautova, L.A.; Fedorov, A.; Chasovnikov, V.K. The impact of physical processes on taxonomic composition, distribution and growth of phytoplankton in the open Black Sea. *J. Mar. Sys.* **2020**, *208*, 103368. [CrossRef]
3. Ostrovskii, A.G.; Emelianov, M.V.; Kochetov, O.Y.; Kremenetskiy, V.V.; Shvoev, D.A.; Volkov, S.V.; Zatsepin, A.G.; Korovchinsky, N.M.; Olshanskiy, V.M.; Olchev, A.V. Automated Tethered Profiler for Hydrophysical and Bio-Optical Measurements in the Black Sea Carbon Observational Site. *J. Mar. Sci. Eng.* **2022**, *10*, 322. [CrossRef]
4. Monin, A.S.; Ozmidov, R.V. *Oceanic Turbulence;* Gidrometeoizdat: Leningrad, Russia, 1981; p. 320. (In Russian)
5. Paka, V.T.; Nabatov, V.N.; Lozovatsky, I.D.; Dillon, T.M. Oceanic microstructure measurements by Baklan and Grif. *J. Atm. Ocean. Technol.* **1999**, *16*, 1519–1532. [CrossRef]
6. Prandke, H.; Stips, A. Test measurements with an operational microstructure-turbulence profiler. Detection limit of dissipation rates. *Aquat. Sci.* **1998**, *60*, 191–209. [CrossRef]
7. Shroyer, E.L.; Nash, J.D.; Waterhouse, A.F.; Moum, J.N. Measuring Ocean Turbulence. In *Observing the Oceans in Real Time. Springer Oceanography;* Venkatesan, R., Tandon, A., D'Asaro, E., Atmanand, M., Eds.; Springer: Cham, Switzerland, 2018; pp. 99–122.
8. Gregg, M.C.; Yakushev, E.V. Surface ventilation of the Black Sea's cold intermediate layer in the middle of the western gyre. *Geophys. Res. Lett.* **2005**, *32*, L03604. [CrossRef]
9. Zatsepin, A.G.; Golenko, N.N.; Korzh, A.O.; Kremenetskii, V.V.; Paka, V.T.; Poyarkov, S.G.; Stunzhas, P.A. Influence of the dynamics of currents on the hydrophysical structure of the waters and the vertical exchange in the active layer of the Black Sea. *Oceanology* **2007**, *47*, 301–312. [CrossRef]
10. Samodurov, A.S.; Chukharev, A.M.; Kazakov, D.A. Basic Regularities of Vertical Turbulent Exchange in the Mixed and Stratified Layers of the Black Sea. *Phys. Oceanogr.* **2021**, *28*, 405–422. [CrossRef]
11. Osborn, T.R. Estimates of the local rate of vertical diffusion from dissipation measurements. *J. Phys. Oceanogr.* **1980**, *10*, 83–89. [CrossRef]
12. Munk, W.H.; Anderson, E.R. Notes on a theory of the thermocline. *J. Mar. Res.* **1948**, *7*, 276–295.

13. Pacanowski, R.C.; Philander, S.G.H. Parameterization of vertical mixing in numerical models of tropical oceans. *J. Phys. Oceanogr.* **1981**, *11*, 1443–1451. [CrossRef]
14. Peters, H.; Gregg, M.; Toole, J. On the parameterization of equatorial turbulence. *J. Geophys. Res. Oceans.* **1988**, *93*, 1199–1218. [CrossRef]
15. Pelegri, J.L.; Csanady, G.T. Diapycnal mixing in western boundary currents. *J. Geophys. Res.* **1994**, *99*, 18275–18304. [CrossRef]
16. Large, W.; McWilliams, J.; Doney, S. Oceanic vertical mixing—A review and a model with a nonlocal boundary-layer parameterization. *Rev. Geophys.* **1994**, *32*, 363–403. [CrossRef]
17. Forryan, A.; Martin, A.; Srokosz, M.; Popova, E.; Painter, C.; Renner, A. A new observationally motivated Richardson number based mixing parameterization for oceanic mesoscale flow. *J. Geophys. Res. Oceans.* **2013**, *118*, 3. [CrossRef]
18. Stanev, E.V. On the mechanisms of the Black Sea circulation. *Earth-Sci. Rev.* **1990**, *28*, 285–319. [CrossRef]
19. Zaron, E.D.; Moum, J.N. A new look at Richardson number mixing schemes for equatorial ocean modeling. *J. Phys. Oceanogr.* **2009**, *39*, 2652–2664. [CrossRef]
20. Podymov, O.I.; Zatsepin, A.G.; Ostrovsky, A.G. Vertical turbulent exchange in the Black Sea pycnocline and its relation to water dynamics. *Oceanology* **2017**, *57*, 492–504. [CrossRef]
21. Podymov, O.I.; Zatsepin, A.G.; Kubryakov, A.A.; Ostrovskii, A.G. Seasonal and interannual variability of vertical turbulent exchange coefficient in the Black Sea pycnocline in 2013–2016 and its relation to variability of mean kinetic energy of surface currents. *Ocean Dyn.* **2020**, *70*, 199–211. [CrossRef]
22. Morozov, A.N.; Lemeshko, E.M. The coefficient of vertical turbulent diffusion according to CTD/LADCP measurements in the northwestern part of the Black Sea in May 2004. *Morsk. Gidrofiz. Zh.* **2014**, *1*, 58–67.
23. Morozov, A.N.; Lemeshko, E.M.; Shutov, S.A.; Zima, V.V.; Deryushkin, D.V. Structure of the Black Sea currents based on the results of the LADCP observations in 2004–2014. *Morsk. Gidrofiz. Zh.* **2017**, *1*, 27–42.
24. Zatsepin, A.G.; Gerasimov, V.V.; Ostrovskii, A.G. Laboratory study of turbulent mass exchange in a stratified fluid. *J. Mar. Sci. Eng.* **2022**, *10*, 756. [CrossRef]
25. Fedorov, K.N. *The Thermohaline Fine Structure of the Ocean*; Pergamon Press: Oxford, UK, 1978.
26. Joyce, T.M. A note on the lateral mixing of water masses. *J. Phys. Oceanogr.* **1977**, *7*, 626–629. [CrossRef]
27. Ruddick, B.; Kerr, O. Oceanic thermohaline intrusions: Theory. *Prog. Oceanogr.* **2003**, *56*, 483–497. [CrossRef]
28. Ozsoy, E.; Unluata, U. Oceanography of the Black Sea: A review of some recent results. *Earth Sci. Rev.* **1997**, *42*, 231–272. [CrossRef]
29. Falina, A.; Sarafanov, A.; Volkov, I. Warm intrusions in the intermediate layer (150–500 m) of the Black Sea eastern gyre interior. *Geophys. Res. Lett.* **2007**, *34*, L22602. [CrossRef]
30. Falina, A.; Sarafanov, A.; Ozsoy, E.; Turuncoglu, U. Observed basin-wide propagation of Mediterranean water in the Black Sea. *J. Geophys. Res. Oceans* **2017**, *122*, 3141–3151. [CrossRef]
31. Stunzhas, P.A. About fine structure and the origin of the intrusions in the Black Sea. *J. Oceanol. Res.* **2019**, *47*, 152–163. (In Russian) [CrossRef] [PubMed]
32. Stanev, E.V.; Poulain, P.-M.; Grayek, S.; Johnson, K.S.; Claustre, H.; Murray, J.W. Understanding the dynamics of the oxic-anoxic interface in the Black Sea. *Geophys. Res. Lett.* **2018**, *45*, 864–871. [CrossRef]
33. Stanev, E.V.; Chtirkova, B.; Peneva, E. Geothermal convection and double diffusion based on profiling floats in the Black Sea. *Geophys. Res. Lett.* **2021**, *48*, e2020GL091788. [CrossRef]
34. Stanev, E.; Grayek, S.; Claustre, H.; Schmechtig, C.; Poteau, A. Water intrusions and particle signatures in the Black Sea: A Biogeochemical-Argo float investigation. *Ocean Dyn.* **2017**, *67*, 1119–1136. [CrossRef]
35. Zatsepin, A.G.; Podymov, O.I. Thermohaline anomalies and fronts in the Black Sea and their relationship with the vertical fine structure. *Oceanology* **2021**, *61*, 757–768. [CrossRef]
36. Mamayev, O.I.; Arkhipkin, V.S.; Tuzhilkin, V.S. TS-analysis of the Black Sea waters. *Oceanology* **1994**, *34*, 154–168.
37. Ostrovskii, A.G.; Zatsepin, A.G. Intense ventilation of the Black Sea pycnocline due to vertical turbulent exchange in the Rim Current area. *Deep Sea Res. Part I Oceanogr. Res. Pap.* **2016**, *116*, 1–13. [CrossRef]
38. Rudnick, D.L.; Klinke, J. The underway conductivity–temperature–depth instrument. *J. Atmos. Ocean. Technol.* **2007**, *24*, 1910–1923. [CrossRef]
39. Kuklev, S.B.; Zatsepin, A.G.; Paka, V.T.; Baranov, V.I.; Kukleva, O.N.; Podymov, O.I.; Podufalov, A.P.; Korg, A.O.; Kondrashov, A.A.; Soloviev, D.M. Experience of simultaneous measurements of parameters of currents and hydrological structure of water from a moving vessel. *Oceanology* **2021**, *61*, 132–138. [CrossRef]
40. Ostrovskii, A.G.; Zatsepin, A.G.; Soloviev, V.A.; Tsibulsky, A.L.; Shvoev, D.A. Autonomous system for vertical profiling of the marine environment at a moored station. *Oceanology* **2013**, *53*, 233–242. [CrossRef]
41. Korotaev, G.; Oguz, T.; Nikiforov, A.; Koblinsky, C. Seasonal, interannual, and mesoscale variability of the Black Sea upper layer circulation derived from altimeter data. *J. Geophys. Res.* **2003**, *108*, 19-1–19-15. [CrossRef]
42. Kubryakov, A.A.; Stanichny, S.V.; Zatsepin, A.G.; Krementetsky, V.V. Long-term variations of the Black Sea dynamics and their impact on the marine ecosystem. *J. Mar. Syst.* **2016**, *163*, 80–94. [CrossRef]
43. Zhurbas, V.M.; Zatsepin, A.G.; Grigor'eva, Y.V.; Eremeev, V.N.; Kremenetsky, V.V.; Motyzhev, S.V.; Poyarkov, S.G.; Poulain, P.M.; Stanichny, S.V.; Soloviev, D.M. Water Circulation and Characteristics of Currents of Different Scales in the Upper Layer of the Black Sea from Drifter Data. *Oceanology* **2004**, *44*, 30–43.

44. Poulain, P.-M.; Barbanti, R.; Motyzhev, S.; Zatsepin, A. Statistical description of the Black Sea near-surface circulation using drifters in 1993–2003. *Deep-Sea Res.* **2005**, *52*, 2250–2274. [CrossRef]
45. Morozov, A.N.; Zatsepin, A.G.; Ostrovskii, A.G.; Kuklev, S.B.; Fedorov, S.V. Vertical structure of currents in the upper part of the continental slope of the Black Sea in the region of Gelendzhik. *Izv. Atmos. Ocean. Phys.* **2017**, *53*, 632–640. [CrossRef]
46. Zatsepin, A.G.; Ostrovskii, A.G.; Kremenetskiy, V.V.; Piotukh, V.B.; Kuklev, S.B.; Moskalenko, L.V.; Podymov, O.I.; Baranov, V.I.; Korzh, A.O.; Stanichny, S.V. On the nature of short-period oscillations of the main Black Sea pycnocline, submesoscale eddies, and response of the marine environment to the catastrophic shower of 2012. *Izv. Atmos. Ocean. Phys.* **2013**, *49*, 659–673. [CrossRef]
47. Thorpe, S.A. *An Introduction to Ocean Turbulence*; Cambridge University Press: Oxford, UK, 2007; p. 240.
48. Gregg, M.C.; Cox, C.S. Measurements of the oceanic microstructure of temperature and electrical conductivity. *Deep-Sea Res.* **1971**, *18*, 925–934. [CrossRef]
49. Osborn, T.R.; Cox, C.S. Oceanic fine structure. *Geophys. Fluid Dyn.* **1972**, *3*, 321–345. [CrossRef]
50. Miles, J.W.; Howard, L.N. Note on a heterogeneous shear layer. *J. Fluid Mech.* **1964**, *20*, 331–336. [CrossRef]
51. Akima, H. Algorithm 526: Bivariate Interpolation and Smooth Surface Fitting for Irregularly Distributed Data Points. *ACM Trans. Math. Softw.* **1978**, *4*, 160–164. [CrossRef]
52. Vojdyr, M. Fytik, a Curve-Fitting and Data Analysis Program. Available online: http://fityk.nieto.pl (accessed on 31 October 2022).
53. Phillips, O.M. Turbulence in a strongly stratified fluid: Is it unstable? *Deep Sea Res. Oceanogr. Abstr.* **1972**, *19*, 7–81. [CrossRef]
54. Posmentier, E.S. The generation of salinity fine structure by vertical diffusion. *J. Phys. Oceanogr.* **1977**, *7*, 298–300. [CrossRef]
55. Ivanov, V.A.; Yankovskii, A.E. *Longwave Motions in the Black Sea*; Naukova Dumka: Kiev, Ukraine, 1992. (In Russian)
56. Cimatoribus, A.; van Haren, H.; Gostiaux, L. Comparison of Ellison and Thorpe scales from Eulerian ocean temperature observations. *J. Geophys. Res. Oceans* **2014**, *119*, 7047–7065. [CrossRef]
57. Gregg, M.C.; Sanford, T.B.; Winkel, D.P. Reduced mixing from the breaking of internal waves in equatorial ocean waters. *Nature* **2003**, *422*, 513–515. [CrossRef] [PubMed]
58. Kunze, E.; Firing, E.; Hummon, J.M.; Chereskin, T.K.; Thurnherr, A.M. Global abyssal Mixing inferred from lowered ADCP shear and CTD strain profiles. *J. Phys. Oceanogr.* **2006**, *36*, 1553–1576. [CrossRef]
59. Ostrovskii, A.; Stepanov, D.; Kaplunenko, D.; Park, J.-H.; Park, Y.-G.; Tishchenko, P. Turbulent mixing and its contribution to the oxygen flux in the northwestern boundary current region of the Japan/East Sea, April–October 2015. *J. Mar. Syst.* **2021**, *224*, 103619. [CrossRef]

Journal of
Marine Science and Engineering

Article

The Contribution of the Vendée Globe Race to Improved Ocean Surface Information: A Validation of the Remotely Sensed Salinity in the Sub-Antarctic Zone

Marta Umbert [1,2,*], Nina Hoareau [1,2], Jordi Salat [1], Joaquín Salvador [1], Sébastien Guimbard [3], Estrella Olmedo [1,2] and Carolina Gabarró [1,2]

[1] Department of Physical and Technological Oceanography, Institut de Ciències del Mar, CSIC, 08003 Barcelona, Spain
[2] Barcelona Expert Center on Remote Sensing, CSIC-UPC, 08003 Barcelona, Spain
[3] Ocean Scope, 29280 Brest, France
* Correspondence: mumbert@icm.csic.es

Abstract: The Vendée Globe is the world's most famous solo, non-stop, unassisted sailing race. The Institute of Marine Sciences and the Barcelona Ocean Sailing Foundation installed a MicroCAT on the One Ocean One Planet boat. The skipper, Dídac Costa, completed the round trip in 97 days, from 8 November 2020 to 13 February 2021, providing one measurement of temperature and conductivity every 30 s during navigation. More than half of the ship's route was in the sub-Antarctic zone, between the tropical and polar fronts, and it passed through areas of oceanographic interest such as Southern Patagonia (affected by glacier melting), the Brazil–Malvinas confluence, the Southern Pacific Ocean, and the entire Southern Indian Ocean. This sailing race gave a rare opportunity to measure in-situ sea surface salinity in a region where satellite salinity measurements are not reliable. Due to the decreased sensitivity of brightness temperature to salinity in cold seas, retrieving sea surface salinity at high latitudes remains a major challenge. This paper describes how the data are processed and uses the data to validate satellite salinity products in the sub-Antarctic zone. The sailing race measurements represent surface information (60 cm depth) not available from drifters or Argo floats. Acquiring measurements using round-the-world sailing races would allow us to analyse the evolution of ocean salinity and the impact of changes in the ice extent around Antarctica.

Keywords: sea surface temperature; sea surface salinity; ocean circumnavigation; ships of opportunity; SMOS validation; sub-Antarctic zone

Citation: Umbert, M.; Hoareau, N.; Salat, J.; Salvador, J.; Guimbard, S.; Olmedo, E.; Gabarró, C. The Contribution of the Vendée Globe Race to Improved Ocean Surface Information: A Validation of the Remotely Sensed Salinity in the Sub-Antarctic Zone. *J. Mar. Sci. Eng.* **2022**, *10*, 1078. https://doi.org/10.3390/jmse10081078

Academic Editor: Alfredo L. Aretxabaleta

Received: 23 June 2022
Accepted: 2 August 2022
Published: 6 August 2022

Publisher's Note: MDPI stays neutral with regard to jurisdictional claims in published maps and institutional affiliations.

1. Introduction

Sea Surface Salinity (SSS) is an important factor for understanding the ocean-ice-atmosphere interactions in the Southern Ocean, an area characterized by high winds, strong currents, the presence of fronts and under the influences of ice melting. For instance, the analysis of SSS allows us to monitor changes in sea ice distribution and concentration. The SSS anomalies are linked to sea ice production and water evaporation (positive SSS anomalies) and to melting and precipitation (negative SSS anomalies). Because of the distance and adverse weather conditions, acquiring in-situ salinity measurements in the sub-Antarctic zone is extremely difficult. As a result, satellite observations become a promising option for having continuous and synoptic monitoring of SSS in polar regions.

Since 2009, a new generation of satellites (SMOS from ESA and, later, Aquarius and SMAP from NASA) have provided radiometric measurements in L-band, allowing us to retrieve SSS. In Southern Ocean there are several aspects that hampers the SSS retrieval from L-band measurements. The reduction in sensitivity of brightness temperatures to SSS in cold seas degrades the accuracy of the SSS retrieval in polar regions. Moreover, L-band measurements are seriously affected by land–sea and ice–sea contamination effects.

Recently, these difficulties have been addressed over the Arctic Ocean, leading to satellite SSS measurements with enough quality to allow scientific studies [1–5]. However, in the Southern Ocean, SSS has lower variability range than in the Arctic Ocean, and roughness corrections mandatory to retrieve SSS are very challenging due to complex wind–current–wave interactions. Therefore having in-situ data as a ground truth is crucial for validating and enhancing the quality of satellite products in this region. Argo floats are usually used to validate SSS products, however the depth where Argo floats measure salinity is much deeper than the depth where the satellites observe salinity. The difference brings some biases, especially close to the coastal regions where much freshwater is supplied [6,7].

In 2010, the Barcelona World Race (BWR) organizers, scientists from the Institute of Marine Sciences (ICM-CSIC), and members of the Maritime Catalan Forum (FMC) agreed to equip one sailing boat participating in the BWR with a MicroCAT temperature and conductivity sensor, as well as an XCAT transmitter. The objective was to investigate if oceanic races could be used to measure global surface temperature and salinity in real time without interfering with navigation or penalizing the boats' race performance. Continuous surface monitoring was carried out for the first time during a round trip around the world's oceans in less than 4 months, contributing to a global study of ocean surface properties and were used to validate remotely sensed salinity products. This was of particular interest because the vessel's route included rarely sampled regions in the Southern Ocean [8]. After this successful first experience, the collaboration continued, allowing us to collect surface data of conductivity and temperature in the following competitions [9]. In this study we present a new data set, collected in the 2020–2021 Vendée Globe edition.

The Vendée Globe race follows a Clipper route circumnavigation, the fastest circumnavigation of the world. Taking advantage of the strong westerly winds of the Roaring Forties, the route runs from west to east through the Southern Ocean. The Vendée Globe is called after the Vendée département in France, where the event starts and finishes at Les Sables-d'Olonne. From Les Sables d'Olonne, participants onboard IMOCA 60 vessels (a monohull 18 m sailing yacht) navigate the Atlantic Ocean to the Cape of Good Hope; then clockwise around Antarctica, keeping Cape Leeuwin and Cape Horn to port; and finally returning to Les Sables d'Olonne crossing all Atlantic Ocean. The race usually lasts around 24,000 nautical miles (around 44,000 km) and takes place from November to February, aiming to sail through the Southern Ocean during the austral summer. The Vendée Globe is recognized as the greatest endeavour in ocean racing and an extreme challenge of individual endurance. The race presents significant difficulties, such as the strong wind and wave conditions in the Southern Ocean, the competition's lengthy unassisted duration, and the fact that the route takes racers far from any regular emergency service.

The 2020–2021 Vendée Globe was the ninth edition of the race. It started on 8 November 2020, with the first finishers arriving on 27 January 2021, and last boat arriving on 5 March 2021. There were a total of six female participants among the 33 starters and 25 finishers, which was a record high. The Spanish skipper Dídac Costa completed the race on board of the One Planet Ocean which sailed 24,500 nautical miles in 97 days, 6 h, 27 min and 3 s (Figure 1).

South of 40° S, the Vendée Globe route runs almost parallel to the Antarctic Circumpolar Current (ACC), the world's largest current, that flows eastward around the Antarctic continent, moving between 100 and 173 Sv [10,11]. The ACC is part of the Meridional Overturning Circulation, which ventilates the Pacific, Atlantic, and Indian Oceans' deep basins. Between the ACC's southern edge and the Subtropical Front to the north, three major fronts can be identified: the Southern ACC Front, the Polar Front, and the Sub-Antarctic Front in some places [12–14]. The Vendée Globe route crosses the Subtropical front, and the Sub-Antarctic Fronts, where few in-situ data are available.

The first objective of this study is to present a new set of in-situ data (later named as VG2020) obtained thanks to the collaboration between ICM, FNOB and with Vendée Globe organizers. On the other hand, we use this in-situ data to validate the current available remote sensed SSS products from SMOS and SMAP satellites in the sub-Antarctic zone, where few in-situ data are available.

Figure 1. Sailing racer Dídac Costa onboard the One Planet One Ocean, an Open 60 sailing boat arriving to Les Sables d'Olonne after a world tour during the ninth edition of the Vendée Globe race. Image from www.vendeeglobe.org, last accessed on 5 August 2022, with copyright of Jean-Louis Carli-Alea.

2. Data and Methods

2.1. Vendée Globe In-Situ Data

The Vendée Globe edition 2020–2021 (VG2020) in-situ data have been measured by using a Seabird SBE 37-SI MicroCAT installed in the sailing boat. The SBE 37-SI MicroCAT is a high-accuracy conductivity and temperature sensor equipped with a memory to save the data acquired. Externally powered, the MicroCAT is useful as a stand-alone monitoring device. The conductivity sensor has an accuracy of ±0.0003 S/m (0.003 mS/cm), a resolution of 0.00001 S/m (0.0001 mS/cm) and a stability of 0.0003 S/m (0.003 mS/cm) per month. The temperature sensor has an accuracy of ±0.002 °C (−5 to to 35 °C); ±0.01 °C (35 °C to 45 °C), a resolution of 0.0001 °C and an stability of 0.0002 °C per month. The sensor was calibrated in the factory just before the race started, in September 2020, and installed on the ship in October 2020. The instrument was connected to the ship's 24 volts batteries through a voltage regulator (7812 with TO220 capsule) so that the input voltage to the sensor remains always constant at 12 volts. This protects the sensor from power overloads. The sensor was installed and configured to autonomously record and store conductivity, temperature and timestamp with a frequency of one measurement every 30 s (two per minute). It has a temperature-compensated crystal oscillator (TCXO, 32,768 Hz) as a real-time-clock frequency source whose accuracy is ±1 min per year (0 °C to 40 °C). In this way timestamp of the data stored can be merged with the ship's GPS clock to retrieve the position and date of every sample. Once uninstalled after the race, the sensor has been checked in the laboratory to confirm the absence of possible drifts. Therefore no drift corrections were applied to the data.

The positions along the trajectory (Figure 2) where measured using a GPS installed in the boat. The sensor was installed inside the articulated keel which stabilized the boat (Figure 3). It was fixed in this space by means of two flanges anchored in the hull. In this way, the sensor remains nearly at 60 cm bellow the surface, stays in full contact with continuously renewed water. However, the sensor may be impacted by foam if the boat is sailing with the hull too close to the surface, during storm episodes, or when the boat sails too fast. In those cases, data can be affected by as the presence of bubbles that reduce the area of the conductivity sensor, underestimating the conductivity.

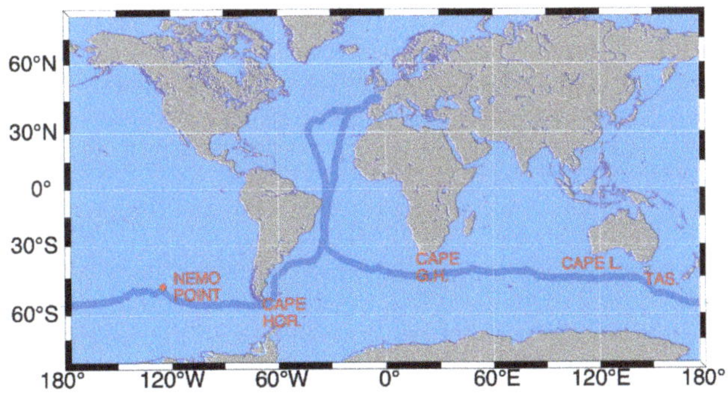

Figure 2. One Planet One Ocean trajectory during the 9th edition of Vendée Globe race. In red: TAS: Tasmania, CAPE G.H: Cape of Good Hope, CAPE HOR: Cape of Hornos, CAPE L.: Cape of Leeuwin, and the Nemo Point (124.39° W, 48.87° S) is indicated with a red cross.

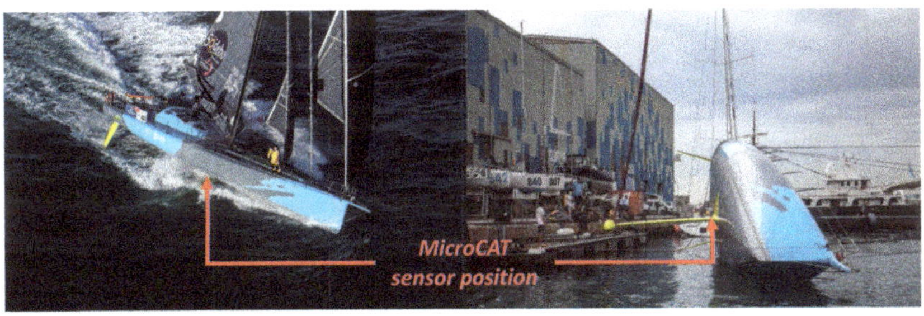

Figure 3. Position of conductivity-temperature sensor that was installed onboard the One Ocean One Planet boat.

2.2. Vendée Globe Data Filtering Process

The data obtained along the race course need specific assessment as they were measured with a non specialized oceanographic boat. As explained above, the sensor is located in the hull, therefore, the measurements are directly impacted by vessel speed conditions. Temperature (T) and conductivity (C), which are highly linked variables, are the sensor's primary data.

The filtering process applied to the data is the following:

1. Checking that temperature and salinity are within a geophysical values range. Temperature is filtered out when there are values outside the range of (0–35 °C), and conductivity when the resulting salinity lies outside the range of (30–40 psu). These filtered measurements include records obtained when sensor box was outside the water;
2. The temperature measurement is flagged when (a) there are short spikes denoting that the water might be stagnant, (b) when there are significant differences from the surrounding values and in the absence of strong temperature gradients.

Once the temperature record is considered acceptable, we then flag conductivity when (a) trends are significant opposite to those of temperature and (b) when producing salinity spikes, as detailed bellow. When only conductivity is flagged we considered that the temperature is good and salinity bad. As mentioned above, data are filtered when the local trends of T and C are opposite. In some cases, the water in the box is not renovated, causing an increase (or decrease) in the temperature and conductivity values that are not corresponding to the sea water crossed by the sailboat. If the water container gets stagnant, T rapidly rises (or decreases) and recovers once the water is replaced again. Moreover, under high-hull-speed conditions, the rise of the hull above the water causes air bubbles and foam getting the sensor box, and even empty the sensor box. Bubbles or foam only influence conductivity. These episodes of bubbles and foam entering the sensor box are typically related to the boat's heading in relation to the wind direction. A lot of foam may come under upwind sailing (i.e., wind blowing 45 to 90 degrees off the heading) especially under rough seas. A downwind sailing may rise the boat over the water, causing the sensor box to dry up. Under crosswind sailing, typically no foam enters the sensor box as boat is attached to the water, and is moving quickly.

A comparison of filtered and original measurements are shown in Figure 4. Abrupt peaks in temperature associated to situations of stagnation of the sensor are removed. Most of the salinity measurements affected by bubbles (abrupt decreases of salinity values) are filtered out. Note that, close to the equator, when the boat was coming back to France at the end of January 2021, the measurements exhibit a decrease in salinity that coincided with an increase in temperature that may reflect the effect of a heavy tropical rain.

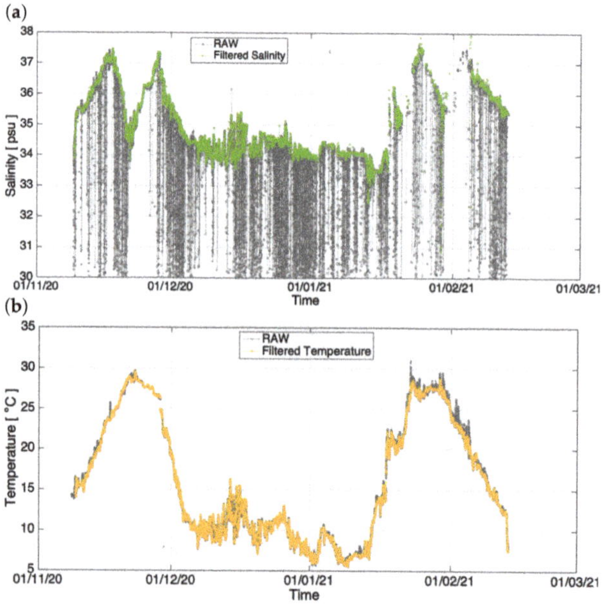

Figure 4. Temporal evolution of raw and filtered salinity data (**a**) and temperature (**b**) along the sailing race trajectory.

2.3. Remotely Sensed SSS Products

We use four different Level-3 SSS products. Two are derived from the Soil Moisture and Ocean Salinity (SMOS) ESA mission and two from the Soil Moisture Active Passive (SMAP) NASA mission. SMAP JPL v5 and SMAP RSS v4 correspond to 8-day running mean maps on 0.25 degree resolution grids provided by the Jet Propulsion Laboratory [15] and Remote Sensing Systems [16], respectively. SMOS BEC v2 and SMOS CATDS v5 are 9-day running mean product at 0.25 degree and 25 km resolutions respectively obtained from the Barcelona Expert Center on Remote Sensing of the ICM-CSIC [17] and the LOCEAN/IPSL (UMR CNRS/UPMC/IRD/MNHN) laboratory [18]. We have regridded all datasets to a common 0.25° regular grid. Each dataset was accessed on 20 January 2022 from the following links: https://doi.org/10.5067/SMP50-3TPCS, https://doi.org/10.5067/SMP40-3SPCS, http://dx.doi.org/10.20350/digitalCSIC/12601 and https://doi.org/10.17882/52804.

3. Results and Discussion

3.1. In-Situ Data Results in the Sub-Antarctic Zone

More than half of the ship's path was through the sub-Antarctic zone, between the subtropical and polar fronts. The in-situ filtered temperature, salinity, density and velocity of the boat along the trajectory southern than 40° S are shown in Figure 5. When the boat reached 40° S for the first time, heading east, appears to be likely under a stormy weather helping the skipper to navigate downwind to make significant progress. The boat traveled through the Agulhas retroflection zone and then it traversed the whole Southern Indian Ocean. At roughly 0° E, the ship crossed the subtropical front (both temperature and salinity decreased) and then again at around 50° E (increase both in temperature and salinity), and finally in 90° E (temperature and salinity decrease). The subantarctic front was crossed for the first time after Tasmania (a clear temperature and salinity drop), then crossed again at 140° W (increase in temperature and salinity) and at 130° W (temperature and salinity decrease). After traversing the whole Pacific Ocean, the Vendée Globe trajectory reaches the south of Patagonia and Cape Horn, an area susceptible to have relevant freshwater inputs due to ice melt, and again returns north through the Atlantic Ocean to finish the race.

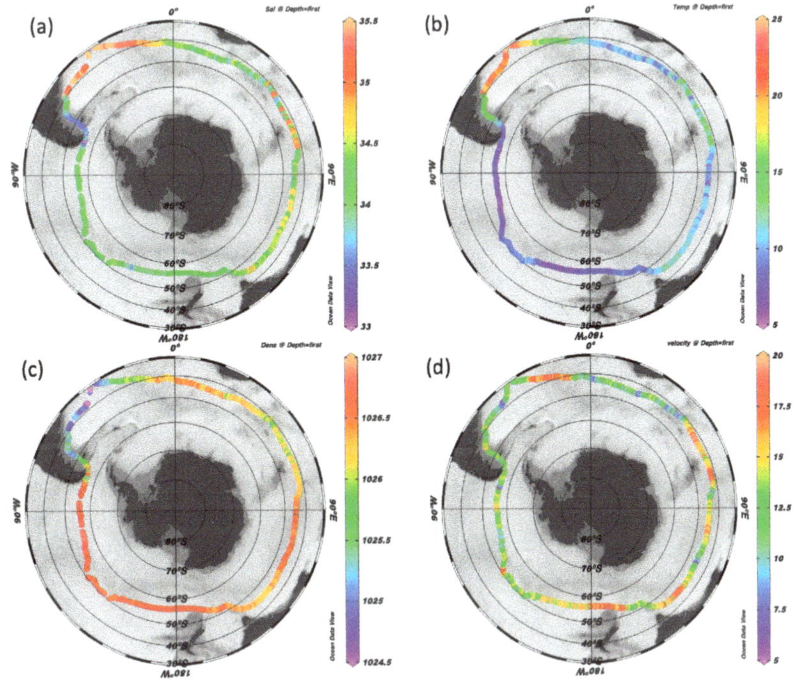

Figure 5. (**a**) Salinity (psu), (**b**) temperature (°C), (**c**) density (kg/m^3) and (**d**) velocity (knots) along the trajectory.

Temperature–Salinity (θS) diagram of in-situ data obtained south of 40° S (Figure 6) allow us to differentiate the saltier (>34.5 psu) and hotter (>13 °C) SubAntarctic waters (SAW), the subtropical front, Southern Indian Ocean waters (SIW) (10–15 °C, and 34–35 psu), the sub-antarctic front, and the fresher (33.5–34.5 psu) and colder (5–10 °C) Southern Pacific waters (SPW). Isopycnals refer to the density anomaly (σ_θ) at 0 m (surface), since all the θS points lie at surface, computed as $\sigma_\theta = \sigma_0 = \rho - 1000$ (kg/m^3).

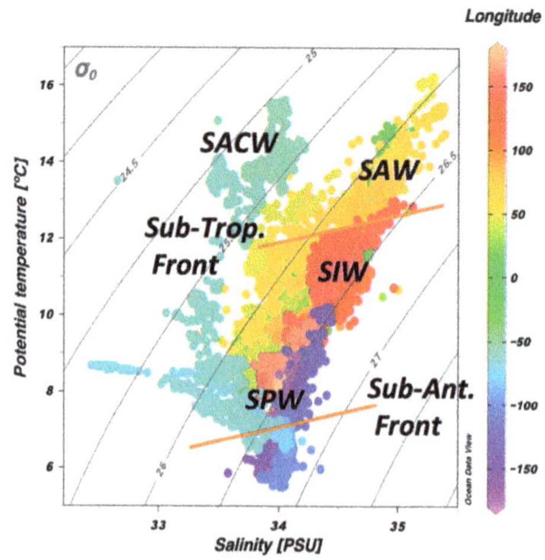

Figure 6. Temperature−Salinity diagram of surface in-situ filtered data showing the water masses: Southern Indian Waters (SIW), Southern Pacific Waters (SPW), SubAntarctic Waters (SAW) and South Atlantic Central Waters (SAW), and fronts found along the trajectory. Isopycnals (σ_θ) are shown and colours correspond to the longitude (°) as shown in the right side colourbar.

If we focus only in the Pacific Ocean (Figure 7), we can detect the boat crosses of the Sub-Antarcic front from south to north (longitude 140° W) gaining 2 °C and 0.5 psu and again from north to south at longitude 130° W.

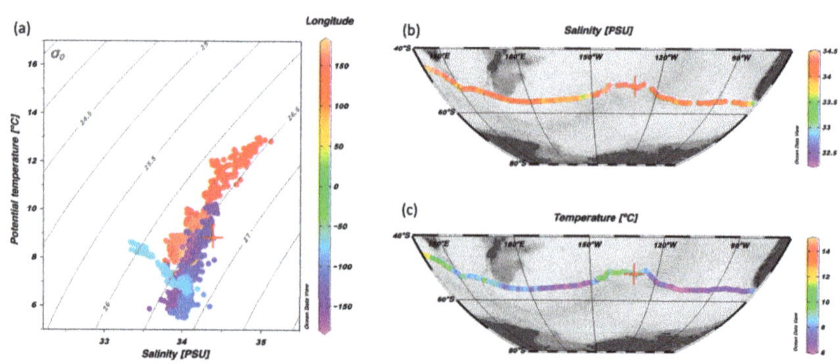

Figure 7. (**a**) Temperature−Salinity diagram using the filtered in-situ data in the Pacific Ocean (colors indicate the longitude (°) and contours represent the ispycnals σ_θ). (**b**) Salinity (psu) and (**c**) temperature (°C) recorded in the Pacific Ocean.

When the vessel arrived to the South Patagonia-Cape Horn, encounters continental-influenced waters (Figure 8), recording a significant drop in salinity of 1.5 psu due to glacier melting. Those waters have low salinity (32–33 psu), low temperature (8–9 °C) and relatively low density (25–25.5 kg/m^3). Later as it navigated northwards, waters temperature increased arriving at the Brazil-Malvinas confluence and further north reaching to South Atlantic Central Waters (14–15 °C, 33.5–34 psu).

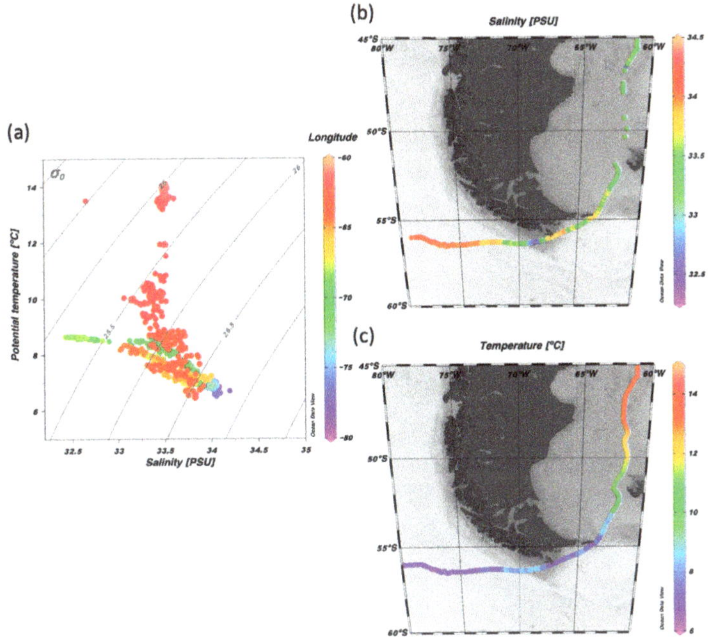

Figure 8. (**a**) Temperature−Salinity diagram using the filtered in-situ data along the South Patagonia-Cape Horn (colors indicate the longitude (°) and contours represent the isopycnals σ_θ). (**b**) Salinity (psu) and (**c**) temperature (°C) recorded in the South Patagonia-Cape Horn area.

3.2. Validation of Satellite SSS

We used the Vendée Globe SSS in-situ measurements southern than 40° S to validate the four satellite SSS products in the sub-Antartic zone. We compared four time-space composite satellite products (so-called Level 3): SMAP JPL, SMAP RSS, SMOS CATDS and SMOS BEC. Each SSS product presents some differences in missing data close to ice edge close to Antarctica, and different levels of noise (changes of navy blue and cyan colors in Figure 9). It seems that the SMOS BEC product is less affected by noise but has less data coverage. All satellite SSS products are able to represent the saltiest Atlantic ocean compared to the Pacific and the Indian Ocean in the subtropical zone.

The Antarctic coast extends further North between 0 and 180° East (between Good Hope cape and Leeuwin cape) bringing colder waters to the Indian Ocean than to the Pacific ocean. Because of that, colder and fresher water rises northward in the Indian Ocean, as ice edge also extend further North. SMOS BEC and SMAP RSS products represents well this characteristic, even if the later seems to be more affected by noise. All four products depicted the subtropical front. Circumpolar current interacts with the continent landmass only in Patagonia, where all the products represent the Humboldt current transporting freshwater to the equator along the Chilean coast, and the Falklands current to the north, while the Brazil current brings warmer and saltier water to the south along Argentinian coast.

The match-up was performed after filtering the input in-situ data and by co-localising the in-situ SSS measurements into the spatio-temporal grid of satellite products. To avoid error representation due to differences in the spatial and temporal sampling, VG2020 SSS data have been averaged into the spatio-temporal grid of the satellite products (daily and 0.25 degree grid). For high-spatial resolution in-situ SSS measurements such as the Vendée Globe SSS data, this match-up procedure minimizes the spatial representation uncertainty when comparing to the lower spatial resolution of the satellite SSS product. The internal variability of filtered VG2020 data inside the satellite grid is computed as the STD of VG2020 data inside the spatio-temporal grid of the satellite products. The STD is in average 0.06 psu with maximum values of 0.44 psu in frontal and continental affected areas.

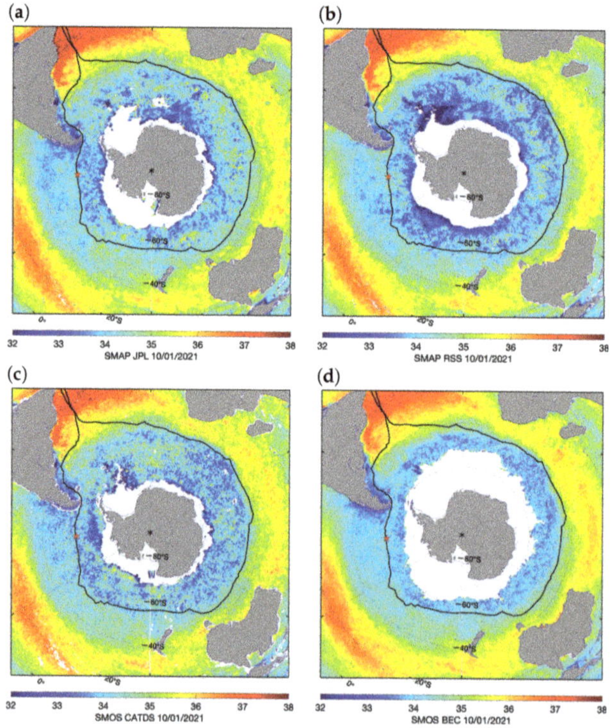

Figure 9. Example of SSS remotely sensed products for the 10 January 2021. (**a**) SMAP JPL; (**b**) SMAP RSS; (**c**) SMOS CATDS and; (**d**) SMOS BEC. The black line indicates the trajectory of One Ocean One Planet ship and the red star its position on 10 January 2021.

When analysing the time series south of 40° S of the VG2020 SSS filtered data and the averaged VG2020 SSS data (Figure 10), we can observe that the original high-frequency sampling provides information on the sub mesoscale variability not resolved at the remote sensed SSS grids. However, the typical SSS variability in the different oceans and the significant SSS changes when crossing fronts are still present in the averaged VG2020 SSS data.

Once the co-localization is performed, we analyse the difference between the satellite and co-localized in-situ SSS ($\Delta SSS = SSS_{SAT} - SSS_{In-situ}$). Time series of daily mean of in-situ and satellite SSS and of their difference (ΔSSS) are shown in Figure 11. The time series of the daily SSS mean estimated southern of 40° S for SMAP JPL (in yellow), SMAP RSS (in red), SMOS-CATDS (in blue), SMOS BEC (in green) products and the co-localized Vendée Globe in-situ dataset (in black) are shown in Figure 11a. Their anomaly with respect to the averaged VG2020 SSS data (ΔSSS) is showed in Figure 11b. We observe that the match-up pairs over the Sub-Antarctic region exhibit typical anomalies from −0.5 to 0.8 psu, with lower anomalies for the SMOS BEC product. Typical in-situ SSS variability south of 40° S (sub-Antarctic and sub-tropical fronts, glacier melting) is in general detected by all satellite products.

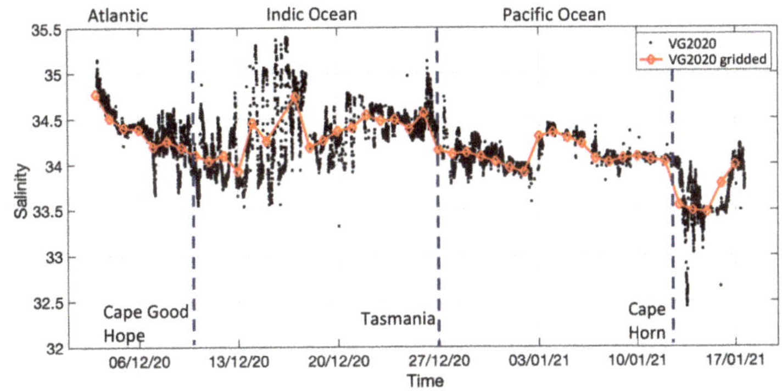

Figure 10. Original high-frequency VG2020 salinity measurements (black dots) and the collocated VG2020 SSS data to satellite 0.25, daily grid (red diamonds) southern than 40° S.

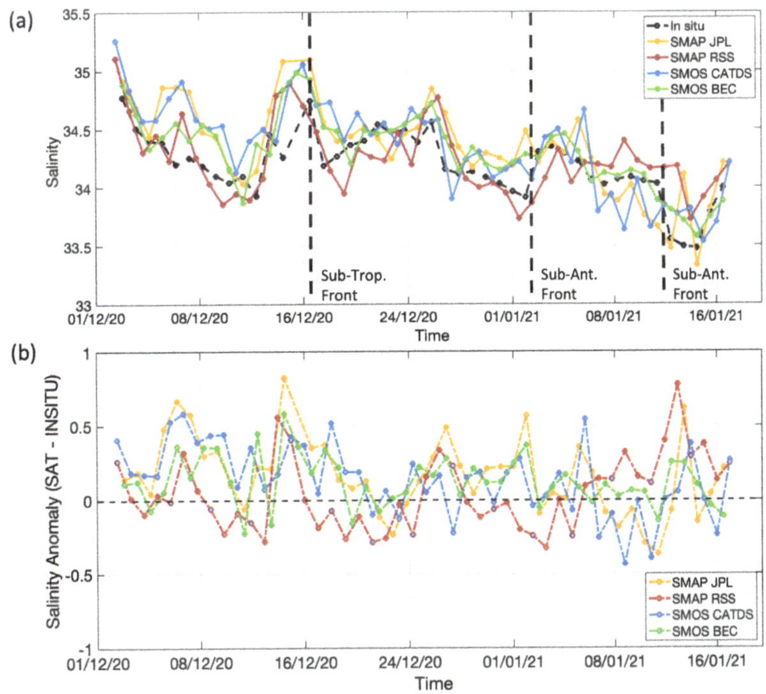

Figure 11. (**a**) Temporal evolution of daily-mean co-localized VG2020 measurements and the four satellite products along the vessel trajectory; (**b**) SSS anomaly between satellite products and VG2020 SSS data.

Figure 12 shows the scatterplots of satellite SSS (y-axis) versus VG2020 SSS (x-axis) below 40° S. For each plot, the red dashed line shows the x = y. The black dashed line indicates the linear fit through the data cloud. The number match-up pairs n, the slope and R2 coefficients of the linear fit are indicated in each plots. Better correlations with VG2020 data are found for SMOS BEC (R = 0.64), followed by SMAP JPL (R = 0.60), SMOS CATDS (R = 0.53) and SMAP RSS (R = 0.45).

In Figure 13, we represent the histograms of the SSS anomalies match-ups that correspond to the ΔSSS distributions for the four satellite products. Each plot includes among other statistics the standard deviation (STD), the mean, the median, the minimum and maximum bias between satellite products and averaged VG2020 SSS data. Better accuracy is found when using SMOS BEC product (STD = 0.30 psu), and lower accuracy using SMOS CATDS product (STD = 0.40 psu). SMAP SSS products presented STD = 0.37 and 0.39 for JPL and RSS products respectively. The lower bias is found for SMAP RSS procuct (median = 0.01 psu), followed by SMOS BEC (median = 0.12 psu), SMOS CATDS (median = 0.13 psu), and SMAP JPL (median = 0.15 psu).

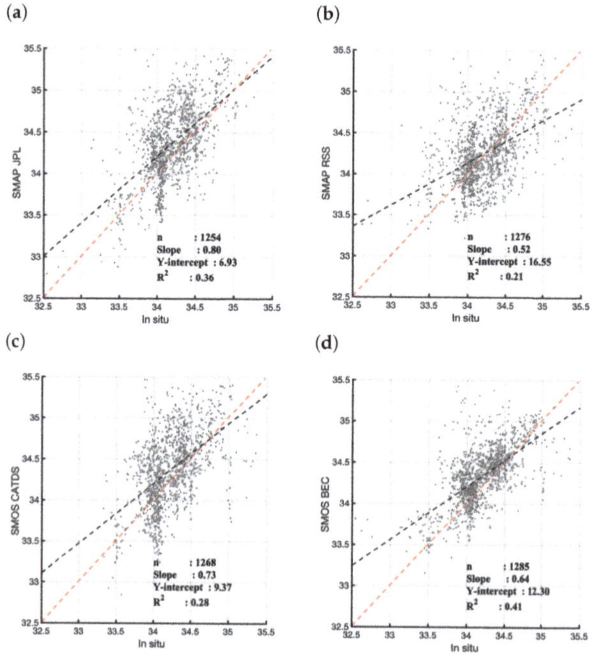

Figure 12. Scatter plots between VG2020 SSS and satellite SSS data for (**a**) SMAP JPL, (**b**) SMAP RSS, (**c**) SMOS CATDS, (**d**) SMOS BEC. The number of match-up pairs (n), slope, y−intercept and R2 coefficients of the linear fit are indicated in each plot.

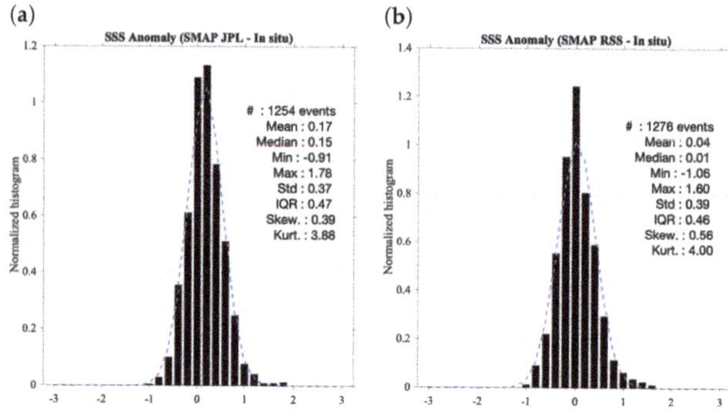

Figure 13. *Cont.*

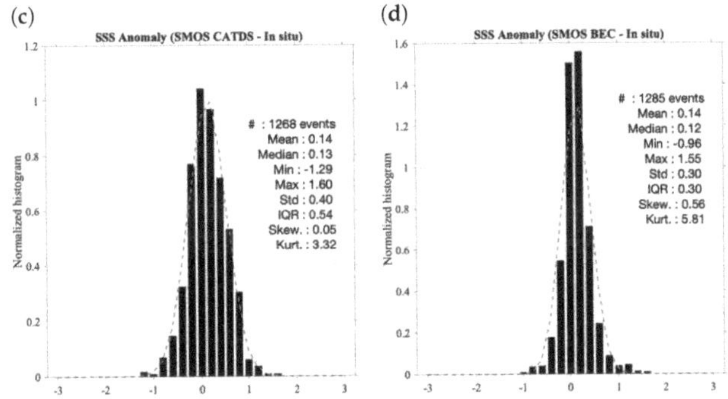

Figure 13. Histograms of SSS anomalies (satellite−in-situ) using (**a**) SMAP JPL, (**b**) SMAP RSS, (**c**) SMOS CATDS, (**d**) SMOS BEC.

4. Conclusions

The in-situ data obtained with the One Ocean One Planet boat competing in the Vendée Globe race are presented in this study. Results show that reliable data can be obtained for scientific purposes using vessels competing in round-the-world regattas. The data were thoroughly filtered and used to identify the typical sub-Antarctic oceanic features. Signatures of melting-affected regions, as well as sub-tropical and sub-polar fronts, were clearly recognized. Even if some parts of the race vessel measurements exhibit foam and air particle effects that must be filtered, they still represent a source of real surface information that is extremely beneficial as a sea-truth reference as well as for climatological studies.

For instance, this new data collection proves to be of great use for the validation of L-band SSS products in the Southern Ocean using ESA and NASA SSS missions. Because the race route includes infrequently sampled locations, the data gathered by One Ocean One Planet are very relevant for the validation of remotely sensed products. This provides significant insight into the assessment of the satellite SSS products since satellite data are typically validated using Argo data, despite the fact that their shallowest observations are between 2 and 10 m depth, which may overlook surface features observed by satellites. The values obtained with the One Ocean One Planet boat are much shallower (60 cm depth) which is closer to satellite observations than the Argo data.

The present contribution to ocean surface salinity data gathered within the Vendée Globe has been successfully used to validate satellite SSS products for the Southern Ocean, provided by ESA and NASA SSS missions. During the period that the regatta sailed south of 40S (3 December 2020–18 January 2021), the satellite products revealed average differences of 0.01–0.15 psu and a range of 0.3–0.4 psu standard deviation with respect to satellite observations.

These results are a step forward to analyzing the evolution of ocean salinity in sub-Antarctic zones and the impact of changes in the ice extent around Antarctica. Having periodic in-situ measurements in the sub Antarctic zone should allow us to analyze the impacts of the ice extent changes in the Antarctica and in the sub Antarctic zone. The study reveals the interest in data recording efforts through competing boats in the periodic round world races to contribute to ocean scientific knowledge. This novel research proves that the regatta's various vessels competing in Vendée Globe race every four years can be equipped for research in the future.

Author Contributions: Conceptualization, M.U., N.H., J.S. (Joaquin Salvador), J.S. (Jordi Salat); methodology, J.S. (Jordi Salat), M.U., N.H.; software, M.U., J.S. (Jordi Salat), N.H., S.G.; validation, M.U., S.G. and N.H.; investigation, M.U., J.S. (Jordi Salat), and N.H.; data curation, J.S. (Jordi Salat), M.U., S.G., J.S. (Joaquin Salvador); writing—original draft preparation, M.U., J.S. (Jordi Salat), J.S. (Joaquin Salvador); writing—review and editing, N.H., J.S. (Jordi Salat), E.O., C.G., S.G.; funding acquisition, M.U., J.S. (Jordi Salat), E.O. and C.G. All authors have read and agreed to the published version of the manuscript.

Funding: This work has been carried out thanks to European Union's Horizon 2020 research and innovation programme under the Marie Skłodowska-Curie Grant Agreement No. 840374. We also received funding from the Spanish government through the "Severo Ochoa Centre of Excellence" accreditation (CEX2019-000928-S).

Institutional Review Board Statement: Not applicable.

Informed Consent Statement: Not applicable.

Data Availability Statement: In-situ filtered data are available on demand by contacting the BEC Support Desk: smos-bec@icm.csic.es.

Acknowledgments: This work represents a contribution to the CSIC Thematic Interdisciplinary Platform PTI POLARCSIC and PTI TELEDETECT.

Conflicts of Interest: The authors declare no conflict of interest.

References

1. Tang, W.; Yueh, S.; Yang, D.; Fore, A.; Hayashi, A.; Lee, T.; Fournier, S.; Holt, B. The potential and challenges of using Soil Moisture Active Passive (SMAP) sea surface salinity to monitor Arctic Ocean freshwater changes. *Remote Sens.* **2018**, *10*, 869. [CrossRef]
2. Olmedo, E.; Gabarró, C.; González-Gambau, V.; Martínez, J.; Ballabrera-Poy, J.; Turiel, A.; Portabella, M.; Fournier, S.; Lee, T. Seven years of SMOS sea surface salinity at high latitudes: Variability in Arctic and Sub-Arctic regions. *Remote Sens.* **2018**, *10*, 1772. [CrossRef]
3. Umbert, M.; Gabarro, C.; Olmedo, E.; Gonçalves-Araujo, R.; Guimbard, S.; Martinez, J. Using Remotely Sensed Sea Surface Salinity and Colored Detrital Matter to Characterize Freshened Surface Layers in the Kara and Laptev Seas during the Ice-Free Season. *Remote Sens.* **2021**, *13*, 3828. [CrossRef]
4. Fournier, S.; Lee, T.; Wang, X.; Armitage, T.W.; Wang, O.; Fukumori, I.; Kwok, R. Sea surface salinity as a proxy for Arctic Ocean freshwater changes. *J. Geophys. Res. Ocean.* **2020**, *125*, e2020JC016110. [CrossRef]
5. Martínez, J.; Gabarró, C.; Turiel, A.; González-Gambau, V.; Umbert, M.; Hoareau, N.; González-Haro, C.; Olmedo, E.; Arias, M.; Catany, R.; et al. Improved BEC SMOS Arctic Sea surface salinity product v3. 1. *Earth Syst. Sci. Data* **2022**, *14*, 307–323. [CrossRef]
6. Banks, C.J.; Gommenginger, C.P.; Srokosz, M.A.; Snaith, H.M. Validating SMOS ocean surface salinity in the Atlantic with Argo and operational ocean model data. *IEEE Trans. Geosci. Remote Sens.* **2012**, *50*, 1688–1702. [CrossRef]
7. Tang, W.; Yueh, S.H.; Fore, A.G.; Hayashi, A. Validation of A quarius sea surface salinity with in situ measurements from A rgo floats and moored buoys. *J. Geophys. Res. Ocean.* **2014**, *119*, 6171–6189. [CrossRef]
8. Salat, J.; Umbert, M.; Ballabrera-Poy, J.; Fernández, P.; Salvador, K.; Martínez, J. The contribution of the Barcelona World Race to improved ocean surface information. A validation of the SMOS remotely sensed salinity. *Contrib. Sci.* **2013**, *9*, 89–100.
9. Gourrion, J.; Szekely, T.; Killick, R.; Owens, B.; Reverdin, G.; Chapron, B. Improved statistical method for quality control of hydrographic observations. *J. Atmos. Ocean. Technol.* **2020**, *37*, 789–806. [CrossRef]
10. Cunningham, S.; Alderson, S.; King, B.; Brandon, M. Transport and variability of the Antarctic circumpolar current in drake passage. *J. Geophys. Res. Ocean.* **2003**, *108*. [CrossRef]
11. Donohue, K.; Tracey, K.; Watts, D.; Chidichimo, M.P.; Chereskin, T. Mean antarctic circumpolar current transport measured in drake passage. *Geophys. Res. Lett.* **2016**, *43*, 11760–11767. [CrossRef]
12. Orsi, A.H.; Whitworth, T.; Nowling, W. On the meridional extent and fronts of the Antarctic Circumpolar Current. *Deep Sea Res. Ser. I* **1995**, *42*, 641–673. [CrossRef]
13. Sokolov, S.; Rintoul, S.R. Circumpolar structure and distribution of the Antarctic Circumpolar Current fronts: 2. Variability and relationship to sea surface height. *J. Geophys. Res. Ocean.* **2009**, *114*. [CrossRef]
14. Giglio, D.; Johnson, G.C. Subantarctic and polar fronts of the Antarctic Circumpolar Current and Southern Ocean heat and freshwater content variability: A view from Argo. *J. Phys. Oceanogr.* **2016**, *46*, 749–768. [CrossRef]
15. Fore, A.G.; Yueh, S.H.; Tang, W.; Stiles, B.W.; Hayashi, A.K. Combined active/passive retrievals of ocean vector wind and sea surface salinity with SMAP. *IEEE Trans. Geosci. Remote Sens.* **2016**, *54*, 7396–7404. [CrossRef]
16. Meissner, T.; Wentz, F.; Manaster, A.; Lindsley, R. *Remote Sensing Systems SMAP Ocean Surface Salinities Level 3 Running 8-Day, Version 4.0 Validated Release*; Remote Sensing Systems: Santa Rosa, CA, USA, 2019.

17. Olmedo, E.; González-Haro, C.; Hoareau, N.; Umbert, M.; González-Gambau, V.; Martínez, J.; Gabarró, C.; Turiel, A. Nine years of SMOS sea surface salinity global maps at the Barcelona Expert Center. *Earth Syst. Sci. Data* **2021**, *13*, 857–888. [CrossRef]
18. Boutin, J.; Vergely, J.L.; Marchand, S.; d'Amico, F.; Hasson, A.; Kolodziejczyk, N.; Reul, N.; Reverdin, G.; Vialard, J. New SMOS Sea Surface Salinity with reduced systematic errors and improved variability. *Remote Sens. Environ.* **2018**, *214*, 115–134. [CrossRef]

Journal of
Marine Science and Engineering

MDPI

Article

An Efficient Method for Nested High-Resolution Ocean Modelling Incorporating a Data Assimilation Technique

Georgy I. Shapiro [1,*] and Jose M. Gonzalez-Ondina [2]

1 School of Biological and Marine Sciences, University of Plymouth, Plymouth PL4 8AA, UK
2 University of Plymouth Enterprise Ltd., Plymouth PL4 8AA, UK; jose.ondina@plymouth.ac.uk
* Correspondence: gshapiro@plymouth.ac.uk

Abstract: A simple and computationally efficient method is presented for creating a high-resolution regional (child) model nested within a coarse-resolution, good-quality data-assimilating (parent) model. The method, named Nesting with Downscaling and Data Assimilation (NDA), reduces bias and root mean square errors (RMSE) of the child model and does not allow the child model to drift from reality. Usually coarser resolution models, e.g., global scale, are used to provide boundary conditions for the nested child model. The basic idea of the NDA method is to use a complete 3D set of output data from the parent model using a process which is similar to data assimilation of observations into an ocean model. In this way, the child model is physically aware of observations via the parent model. The method allows for avoiding a complex process of assimilating the same observations which were already assimilated into the parent model. The NDA method is illustrated in several simple 2D synthetic cases where the true solution is known. The NDA method reduces the child model bias to the same level as in the parent model and reduces the RMSE, typically by a factor of two to five, occasionally more.

Keywords: ocean modelling; data assimilation; oceanography

Citation: Shapiro, G.I.; Gonzalez-Ondina, J.M. An Efficient Method for Nested High-Resolution Ocean Modelling Incorporating a Data Assimilation Technique. *J. Mar. Sci. Eng.* **2022**, *10*, 432. https://doi.org/10.3390/jmse10030432

Academic Editor: Marcello Di Risio

Received: 16 February 2022
Accepted: 11 March 2022
Published: 16 March 2022

1. Introduction

Fine-resolution ocean modelling is becoming a ubiquitous practice to resolve important mesoscale and submesoscale features such as eddies, fronts, boundary currents, and localised upwellings, which play important roles in ocean dynamics (e.g., [1–4]). These localised models can be run by relatively small groups due to the availability of good-quality ocean models such as ROMS or NEMO to the wider oceanographic community [5,6]. Local fine-resolution models require initial and boundary conditions that can be obtained from good-quality but coarser resolution (parent) models run by major ocean modelling centres such as Mercator Ocean International (France) or Met Office (UK) via Copernicus Marine Service [7]. However, if only the boundary data are used for running the nested model (hereafter called the child model), the bulk of the data produced the parent model within the child model domain is not used for improvement of the child model. Therefore, most of the parent outputs which are constantly validated and kept on track by assimilating observations are lost for the child model. Philosophically, it seems promising to use more data from the good-quality parent to improve the child model. The method designed in this study and named Nesting with Downscaling and Data Assimilation (NDA) is focussed on extending the link between the parent and child models.

Because of inevitable approximations in the equations, numerical schemes, parameterisation, and uncertainties in input data, ocean models tend to drift from reality. A process called data assimilation (hereafter DA) is often regarded as a way of keeping a model 'on the tracks' by constantly correcting it with fresh observations [8–10]. Many modern DA methods originated from the methods of objective analysis [9,11]. While the DA problem can be formulated precisely, finding a solution to it is challenging [12–14].

Complex software systems have been developed for operational data assimilation, such as in [15–17].

A critical element of modern variational DA is the estimate of the model and observations error covariance matrices (ECMs). The true state of the ocean is unknown; therefore, ECMs are estimated by some proxy [18]. Various methods use different assumptions and simplifications in order to make the process feasible, e.g., [19,20]. Despite data assimilation being currently ubiquitous in geosciences, it has so far remained a topic mostly reserved to experts [21,22]

The purpose of this paper is to develop a simple and computationally efficient method of implementing DA for a high-resolution child model that is nested into a data-assimilating parent model. The method is mainly intended for smaller academic and operational centres that do not have resources of the same scale as the major ocean forecasting institutions. An example is a regional forecasting group that relies on downloading the outputs from global/basin-scale models available from a reputable ocean forecasting service such as EU CMEMS but also prefers to tune their local high-resolution models.

The basic idea suggested in this study is to assimilate data from a good-quality but coarser resolution parent model instead of using observations. The parent model is assumed to be data-assimilating itself, ensuring that the observations, however indirectly, will not allow the fine-resolution child model to deviate significantly from the true state of the sea. In other words, the coarse model can be regarded as a mean for incorporating a sparse number of observations into a fine-resolution model in a way that honours the laws of physics.

While in principle several existing DA methods could be adapted to assimilate data from the parent into the child model, this study uses the new NDA method. This method combines, from time to time, the outputs from the child model forecast with the forecast from the data-assimilating parent model. This combination creates an improved representation of the ocean state called the analysis, which is used, together with boundary conditions from the parent model, to generate the next high-resolution forecast.

The paper is arranged as follows. The NDA algorithm is described in the Data and Methods section. This section also describes the synthetic idealised case where the true solution is known and is used to illustrate the method. The Results section presents three configurations of synthetic cases for demonstrating the performance of the NDA method when the parent model is 'perfect', i.e., its only error is the representativeness error [23] due to insufficient resolution. The three configurations are a front, a single eddy, and multiple eddies. Each of these examples was tested for different sizes of the field structures and error parameters of the child model. The section also presents results when the parent model is not 'perfect' but produces some spatially correlated and uncorrelated errors. The strengths and limitations of the NDA method are discussed in the final section of the paper in comparison with the commonly used combination of Hollingsworth–Lönnberg and variational DA (VAR) methods [24,25].

2. Materials and Methods

This section presents the algorithm used for Nesting with Downscaling and Data Assimilation and describes a number of simple synthetic illustrative examples which are processed using this method in the subsequent section. The NDA method consists of two steps. The first step uses stochastic–deterministic downscaling [26] of the parent model forecast onto the child model grid. The second step combines, using the same philosophy of variational data assimilation, this downscaled forecast with the forecast generated by the child model.

In general, the algorithm detailed below is applicable when the vectors representing the condition of the ocean are composed of a number of state variables (e.g., temperature, salinity, pollution concentration, etc.) and are discretised in a 3D domain. For example, the vector of true state of the ocean x can take the form $x = \{x_i\} = \{T_1, \dots T_N, S_1, \dots S_N, C_1, \dots C_N\}$ where T_i, S_i, C_i are the temperature, salinity, and concentration of a chemical, and $i = 1, \ldots N$ are indices for the N nodes in the child model domain. In the case when x represents multiple variables, the method can take into account cross-correlations between different state variables. When the state vectors represent a 3D field, the covariances along the vertical axis are also taken into account. In a simpler approach, the method can be applied to only one variable at a time and layer-by-layer. The method itself is not restricted by an assumption of isotropy; however, this assumption is used in the illustrative examples. For illustration purposes, the simple synthetic cases are all 2D and are related to one state variable. In these illustrative examples, a layer-by-layer data assimilation is presented as the simplest and most practical variant.

2.1. The NDA Algorithm

The proposed algorithm is designed for assimilating data from an ocean model of a coarser resolution parent model into a finer resolution child model. The algorithm should be used at each time point when the analysis state for the child model is generated as an initial state for the next forecast.

A common approach to data assimilation into an ocean (or atmospheric) model is based on a variational principle, i.e., on minimising the cost function J given in Equation (1):

$$J(x) = \left(x - x^b\right)^T \mathbf{B}^{-1}\left(x - x^b\right) + (y - H(x))^T \mathbf{R}^{-1}(y - H(x)), \quad (1)$$

where x is the (unknown) vector which estimates the true state of the ocean, x^b is the model forecast before data assimilation, y is the vector of observed values, H is the operator which projects data from the model grid onto the locations of observations, and $\mathbf{B}$ and $\mathbf{R}$ are the forecast and observations error covariance matrices, respectively. The superscripts 'T' and '-1' denote the transposed and inverted matrices, respectively (see [19,22,23,25] for more detail).

The value of $x = x^a$ that minimises the cost function J is called the 'analysis' and it is closest in an RMS (root mean square) sense to the true state x^t [23]. This approach works well for a relatively small number of data. When the number of observations and model grid nodes is large, the matrices used in Equation (1) become very large and their inversion causes significant computational problems. For example, Bouttier and Courtier [23] state that 'except in analysis problems of very small dimension (like one-dimensional retrievals), it is impossible to compute exactly the least-squares analysis.' Therefore, a number of approximate methods are used to calculate the analysis state, and even this process is computationally expensive.

This study is related to assimilation of large numbers of data (which take the place of observations used in standard DA methods) from one model to another. For this purpose, an alternative cost function J_S is proposed, namely

$$J_S(x') = \left(x' - x'^b\right)^T \mathbf{B}^{-1}\left(x' - x'^b\right) + \left(S'(y) - x'\right)^T \mathbf{R}^{-1}\left(S'(y) - x'\right), \quad (2)$$

where the primed variables denote deviations of the respective variable from their statistical mean, which is defined in probability theory as the average of a theoretically infinite number of measurements. It is denoted by the angle brackets in the following equations:

$$x' = x - \langle x \rangle x'^b = x^b - \langle x^b \rangle S'(y) = S(y) - \langle S(y) \rangle$$

Here, y is the vector of data from the parent model (instead of observations) and S is the operator that projects (downscales) the data from the parent model onto the fine grid of the child model.

The best estimate x' for the true value is denoted x'^a and represents the fluctuation of the analysis vector around its statistical mean. It is obtained by minimising the cost function J_S given in Equation (2), as follows:

$$\nabla J_S = 0.$$

Taking into account that matrices **B** and **R** are symmetric so that $\mathbf{B}^T = \mathbf{B}$ and $\mathbf{R}^T = \mathbf{R}$, the following equation is obtained:

$$\begin{gathered}\nabla J_S(x'^a) = \left(x'^a - x'^b\right)^T\left(\mathbf{B}^{-1} + (\mathbf{B}^{-1})^T\right) - (S'(y) - x'^a)^T\left(\mathbf{R}^{-1} + (\mathbf{R}^{-1})^T\right) = \\ 2\left(x'^a - x'^b\right)^T\mathbf{B}^{-1} - 2(S'(y) - x'^a)^T\mathbf{R}^{-1} = 0,\end{gathered} \tag{3}$$

Taking the transpose of Equation (3) and dividing by two gives

$$\mathbf{B}^{-1}\left(x'^a - x'^b\right) = \mathbf{R}^{-1}\left(S'(y) - x'^a\right). \tag{4}$$

Let us introduce the error correlation matrices $\mathbf{C}_\mathrm{B}$ and $\mathbf{C}_\mathrm{R}$ for the child and downscaled parent models:

$$\mathbf{B} = \mathbf{V}_\mathrm{B}\mathbf{C}_\mathrm{B}, \quad \mathbf{R} = \mathbf{V}_\mathrm{R}\mathbf{C}_\mathrm{R}$$

where $\mathbf{V}_\mathrm{B}$, $\mathbf{V}_R$ are diagonal matrices containing the respective error variances at each fine grid node. The inverted diagonal matrices are also diagonal. Equation (4) then becomes

$$\mathbf{C}_\mathrm{B}^{-1}\mathbf{V}_\mathrm{B}^{-1}\left(x'^a - x'^b\right) = \mathbf{C}_\mathrm{R}^{-1}\mathbf{V}_\mathrm{R}^{-1}\left(S'(y) - x'^a\right). \tag{5}$$

Most assimilation systems assume **R** to be diagonal (no spatial observation error correlations) so that its inverse can be calculated quickly and easily. This approach is obviously not appropriate in the case where the 'observation errors' coming from the parent model have significant spatial correlations and the NDA method does not make this assumption. Correlation matrices for both the child and downscaled parent models relate to the same resolution and the same area of the ocean; hence, it is reasonable to use the same correlation function, e.g., a Gaussian of a certain length scale and therefore the same correlation matrix for both models, i.e., $\mathbf{C}_\mathrm{B} = \mathbf{C}_\mathrm{R} = \mathbf{C}$. After premultiplying Equation (5) by **C** and rearranging the terms:

$$\left(\mathbf{V}_\mathrm{B}^{-1} + \mathbf{V}_\mathrm{R}^{-1}\right)x'^a = \mathbf{V}_\mathrm{B}^{-1}x'^b + \mathbf{V}_\mathrm{R}^{-1}S'(y) \tag{6}$$

or

$$x'^a = \left(\mathbf{V}_\mathrm{B}^{-1} + \mathbf{V}_\mathrm{R}^{-1}\right)^{-1}(\mathbf{V}_\mathrm{B}^{-1}x'^b + \mathbf{V}_\mathrm{R}^{-1}S'(y)). \tag{7}$$

Using the commutative properties of diagonal matrices and the following identity

$$\left(\mathbf{V}_\mathrm{B}^{-1} + \mathbf{V}_\mathrm{R}^{-1}\right)^{-1} = \mathbf{V}_\mathrm{R}(\mathbf{V}_\mathrm{B} + \mathbf{V}_\mathrm{R})^{-1}\mathbf{V}_\mathrm{B}, \tag{8}$$

the solution for the analysis state given by Equation (7) can be rewritten as

$$\begin{gathered}x'^a = \left(\mathbf{V}_\mathrm{B}^{-1} + \mathbf{V}_\mathrm{R}^{-1}\right)^{-1}(\mathbf{V}_\mathrm{B}^{-1}x'^b + \mathbf{V}_\mathrm{R}^{-1}S'(y)) = \\ \mathbf{V}_\mathrm{R}(\mathbf{V}_\mathrm{B} + \mathbf{V}_\mathrm{R})^{-1}x'^b + \mathbf{V}_\mathrm{B}(\mathbf{V}_\mathrm{B} + \mathbf{V}_\mathrm{R})^{-1}S'(y).\end{gathered} \tag{9}$$

Equation (9) can be interpreted as a zero-dimensional Kalman gain formula applied to the fluctuations of state variables in the parent and child models at each fine grid node independently. The term 'zero-dimensional' reflects the fact that in this case, the matrices used in the Kalman gain formula have the size 1×1, i.e., are reduced to a scalar. For a single $x_i'^a$ element of the state vector $\boldsymbol{x}'$, Equation (9) gives

$$x_i'^a = \frac{V_{Rii}}{V_{Rii} + V_{Bii}} x_i'^b + \frac{V_{Bii}}{V_{Rii} + V_{Bii}} S'(y)_i. \quad (10)$$

where $x_i'^b$ and $S(y)_i'$ are the corresponding values of the vectors $\boldsymbol{x}'^b$ and $S'(y)$ at the same grid node i.

For the S operator, which downscales the fluctuations of the field variable from the coarse to the fine grid, we use the stochastic–deterministic downscaling (SDD) method developed in [26]. The spatial error correlation matrix $\mathbf{C}$ is not explicitly included in Equations (9) and (10), but it is used at the downscaling stage of data assimilation, hence the error covariances between data at the parent model grid are implicitly present in the operator S. This step represents a major difference from standard methodology where the observation errors are usually considered to be uncorrelated (e.g., [23,25]), and therefore all non-diagonal elements in the matrix $\mathbf{C}_\mathrm{R}$ are ignored.

The diagonal values $V_{B,ii}$ and $V_{R,ii}$ represent the variance of forecasting errors in child and parent models, respectively, and in principle they may be computed in any way designed to compute matrix $\mathbf{B}$ in practical implementations applied for assimilation of observational data, such as the NMC [27] or the Canadian [28]. For consistency, the implementation of the NDA method presented in this paper calculates values $V_{B,ii}$ and $V_{R,ii}$ for the DA step in a similar way as was used for the first step, i.e., SDD downscaling, see [26] for details. The SDD and therefore the implementation of NDA given below requires the knowledge of the ensemble statistical means and variances at each grid node. By the usual assumption of local (i.e., in a small 2D or 3D domain around each grid node) statistical homogeneity, the means $\langle \boldsymbol{x}^b \rangle_i$ and $\langle S(y) \rangle_i$ as well as the variances $V_{B,ii}$ and $V_{R,ii}$ are estimated using the ergodic hypothesis, i.e., by spatial averaging over a small trial domain around the node at the same time point. Generally, the NDA method can be applied for both isotropic and anisotropic situations. In the latter case, the covariances are calculated taking into account directions between the grid nodes in question. The trial domain can be 3D, in which case both horizonal and vertical covariances are taken into account, or 2D so that covariances are calculated layer-by-layer, such as in [29]. In the examples below we use the trial domain in the form of a 2D area on of size $4L \times 4L$, where L is the correlation length of $\mathbf{C} = \mathbf{C}_\mathrm{B} = \mathbf{C}_\mathrm{R}$. As the values $V_{R,ii}$ and $V_{B,ii}$ are computed at each grid node i separately, their values vary across the domain. The nodes that are masked out due to coastline or bathymetry, are excluded from calculation of averages and variances within the trial domain.

Equation (10) performs data assimilation on the fluctuations; however, the statistical means for the parent and child model are, in a general case, different. The NDA method should be applied to a good-quality parent model that has been debiased during its own data assimilation cycle and therefore its statistical mean represents the true value as closely as possible. Therefore, it is reasonable to replace the (potentially biased) statistical mean from the child model with the statistical mean of the parent model.

This gives the final equation for calculation of the analysis state:

$$\boldsymbol{x}^a = \mathbf{V}_\mathrm{R}(\mathbf{V}_\mathrm{B} + \mathbf{V}_\mathrm{R})^{-1}\boldsymbol{x}'^b + \mathbf{V}_\mathrm{B}(\mathbf{V}_\mathrm{B} + \mathbf{V}_\mathrm{R})^{-1}S'(y) + \langle S(y) \rangle. \quad (11)$$

Or, for an element of the state vector at a particular fine grid node i ($i = 1 \ldots N$), where N is the number of grid nodes in the child model:

$$x_i^a = \frac{V_{Rii}}{V_{Rii} + V_{Bii}} \boldsymbol{x}'^b{}_i + \frac{V_{Bii}}{V_{Rii} + V_{Bii}} S'(y)_i + \langle S(y) \rangle_i. \quad (12)$$

To summarise, the NDA model-to-model data assimilation procedure includes two steps. First, the parent model output is downscaled from the coarse grid onto the child model fine grid using the SDD method. The result is that at each fine grid node there are two values of the same state variable. The best estimate of the true field is obtained at the second step by combining these two values using a zero-dimensional Kalman filter. The workflow diagram of the algorithm is shown in Figure 1.

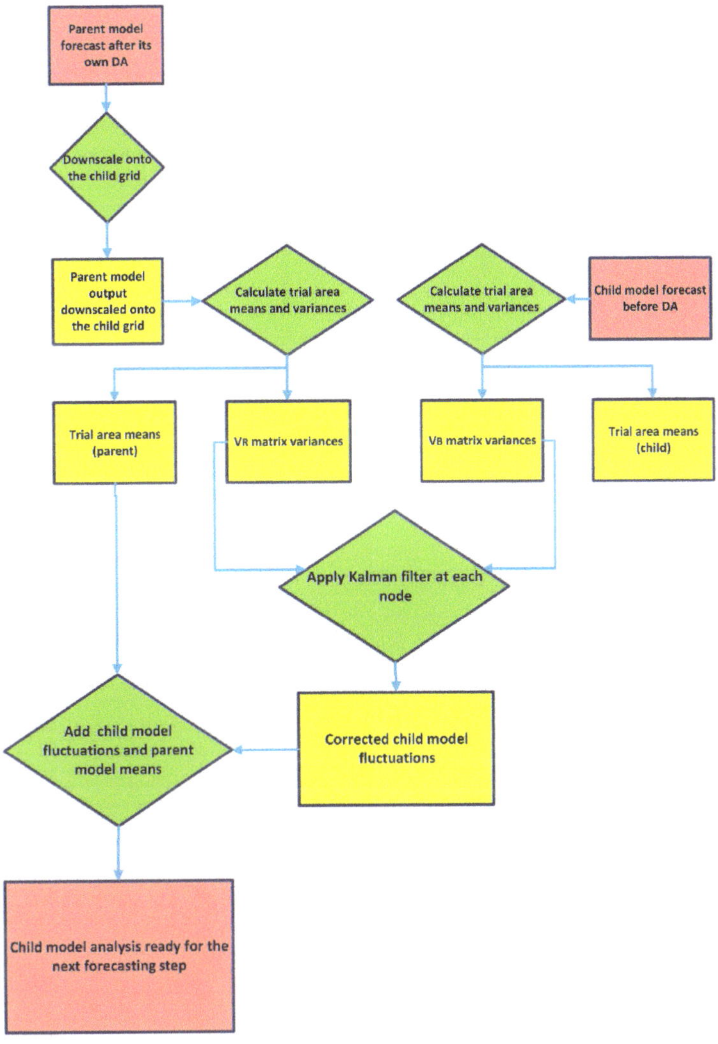

Figure 1. Workflow diagram for the NDA method showing one data assimilation cycle.

This algorithm is computationally efficient, as it does not require inversion of large matrices or solving a very large system of algebraic equations at the second step, which is required if using the variational methods based on Equation (1). The inversion of correlation matrices to obtain the weight coefficients for the SDD step can be done only once at the beginning of the model run, as these coefficients do not depend on time. Another benefit of the described method is that the correlation matrices for the downscaling of the parent model have a relatively small rank and condition number, and their inverse counterparts can be calculated without the need for any type of matrix regularisation, just using double precision for the computer representation of the variables. This is because the SDD method assumes local homogeneity of statistical properties of the field variable. More details on the philosophy and the algorithm of the SDD method can be found in Shapiro et al. [26].

2.2. Data for the Synthetic Idealised Case

In this subsection, the NDA algorithm is illustrated in a synthetic idealised case where the true solution is known. The task is to generate an analysis state using the fine-resolution (child) model forecast and the output from a good-quality data-assimilating parent coarse model, which is used instead of actual observations. It should be noted that, even if both coarse and fine models were perfect, there would be some unavoidable discrepancies, or representativeness errors, between the two models due to the different meshes they use [23]. One of the problems in data assimilation is that the true solution is unknown and therefore the error covariance matrices required to minimise the cost function must be estimated based on various assumptions, e.g., see the review paper by Bannister [18]. Comparison with observations has its own drawbacks as observations are not perfect. The testing of a DA method in a simplified synthetic setting when the true solution is known was used in previous studies and showed its advantages, e.g., [22].

In this study, the parent model output is simulated by sampling the true solution (a prescribed function) on the parent model coarse grid, to which either correlated or uncorrelated error might be added. The fine-resolution forecast is simulated by sampling the same true solution on the fine grid, adding random noise, bias, and some spatial shift. Whilst generally the NDA method calculates the bias locally so it can vary across the domain (Equation (12)). In the following examples, the bias was applied uniformly to keep the illustrative example simple. The spatial shift, a type of spatially correlated error, is added to simulate the 'double penalty effect', which is common to fine-resolution models, e.g., [29].

2.3. Case Studies

Three distinct examples are considered, all relate to data assimilation of one field variable at a single computational surface, whether it be a geopotential ('horizontal') level, such as used in Bell et al. [10], or a curved level such as used in sigma [30] or multienvelope vertical coordinate system [31]. It is common practice to perform the data assimilation level-by-level using a 2D approach at each level, where matrix **B** only includes spatial correlations along the level surface [32]. Occasionally, vertical correlations are used in a preprocessing step for the DA, where observations are interpolated into the model levels [10]. Whilst the NDA method can be applied in 3D, here it is applied to 2D fields for simplicity.

The properties of the NDA method are analysed for (i) an ocean front, (ii) an isolated isotropic eddy, and (iii) a system of densely packed anisotropic mesoscale eddies. In all examples the parent model has resolution of $\Delta x_p = \Delta y_p = 10$ km, and the child model has resolution of $\Delta x_c = \Delta y_c = 2.5$ km and a correlation length of $L = 17$ km. The trial domain used to calculate local spatial averages and variances of fluctuations was a square of $4L \times 4L$ km^2 centred at each node. The correlation Gaussian function has a value of less than 2% beyond the distance of $2L$ and its contribution can be safely neglected for all the examples below. These dimensions were chosen to simulate typical ocean structures. However, it must be pointed out that classification of ocean features as mesoscale or submesoscale are relative to the local Rossby radius. For example, the same eddy of a radius of 12 km could be classed as mesoscale if the Rossby radius is, for example, 10 km, as in some shelf or high latitude seas, and as submesoscale if the Rossby radius is, for example, 40 km, as in the middle of open ocean. The NDA method itself is not specific to whether the features are mesoscale, large-scale, or submesoscale; however, the quality of the first step (downscaling) depends on the ratios between Δx_p, Δy_p, Δx_c, and Δy_c, and the Rossby radius as discussed in [26].

2.3.1. Ocean Front

An ocean front is a narrow area separating two water masses, and it has significant impact on horizontal and vertical exchanges, e.g., [33]. This example is set in a square domain of 200 × 200 km^2, and the front extends in the meridional (y) direction at the centre of the domain.

The true solution is set to be in the form

$$F = A \tanh\left(\frac{x}{L_x}\right), \tag{13}$$

where A is the amplitude of the front and L_x is its half-width. For this exercise, $A = 1$ and L_x ranges between 6 and 40 km. The coordinates x and y are referenced to the centre of the domain in the zonal and meridional directions, respectively.

2.3.2. Isolated Mesoscale Eddy

Mesoscale eddies are a ubiquitous feature of oceans worldwide. Originally, they were thought to exist only next to jet currents such as the Gulf Stream; however, since the 1960s–1970s, it became clear that mesoscale eddies exist nearly anywhere in the ocean. Most of the kinetic energy of the ocean is contained in mesoscale eddies [34]. In this example, an isotropic eddy is placed in the centre of a domain of 200 × 200 km^2 and the true solution is set to be in the form

$$F = A \exp\left\{-\left[\left(\frac{x}{L_e}\right)^2 + \left(\frac{y}{L_e}\right)^2\right]\right\}, \tag{14}$$

where A is the amplitude of the eddy and L_e is its radius. For this exercise $A = 1$ and L_e ranges between 6 and 46 km.

2.3.3. Multiple Mesoscale Eddies

For this example, let us consider a square domain 1000 × 1000 km^2 and take the true solution for a variable F in the form of multiple anisotropic mesoscale eddies

$$F = \sin\left(\frac{\pi x}{L_x}\right) sin\left(\frac{\pi y}{L_y}\right), \tag{15}$$

where L_x is the eddy half-size in the x-direction, and $L_y = 105$ km is the eddy half-size in the y-direction. Let us consider the range of eddy sizes L_x between 12 and 24 km. At $L_x = 24$ km the parent model can be classed as eddy-permitting, as it has 2.4 grid points over the smaller 'diameter' of the eddy. At $L_x = 12$ km, the parent model is not even eddy-permitting but only showing an embryonic representation of the eddy. The child fine-resolution model is eddy resolving for any L_x within the chosen range of eddy sizes.

3. Results

The NDA data assimilation method detailed in the subsection algorithm above was applied to the simulated child model forecast in order to create analysis which is used as an initial condition for the next forecasting cycle. The results for three examples are presented in this section when the parent model is error-free: an ocean front, an isolated single eddy, and a set of multiple eddies. An additional example is considered when the parent model has some errors (either correlated or uncorrelated, extra to the representativeness errors) that are not eliminated by its own DA procedure.

3.1. Ocean Front

This example uses Equation (13) for the representation of an ocean front in the meridional direction. Figure 2 shows a map of the true solution F for a sharp ocean front at L_x = 6 km (Figure 2a), its representation by the parent model (Figure 2b), by the child model forecast before data assimilation (Figure 2c), and as transects across the front. The following parameters of added errors are used to simulate the child model forecast before data assimilation: normally distributed noise with standard deviation 0.15, positive bias of 0.3, and shift of the field by 4 km to the west.

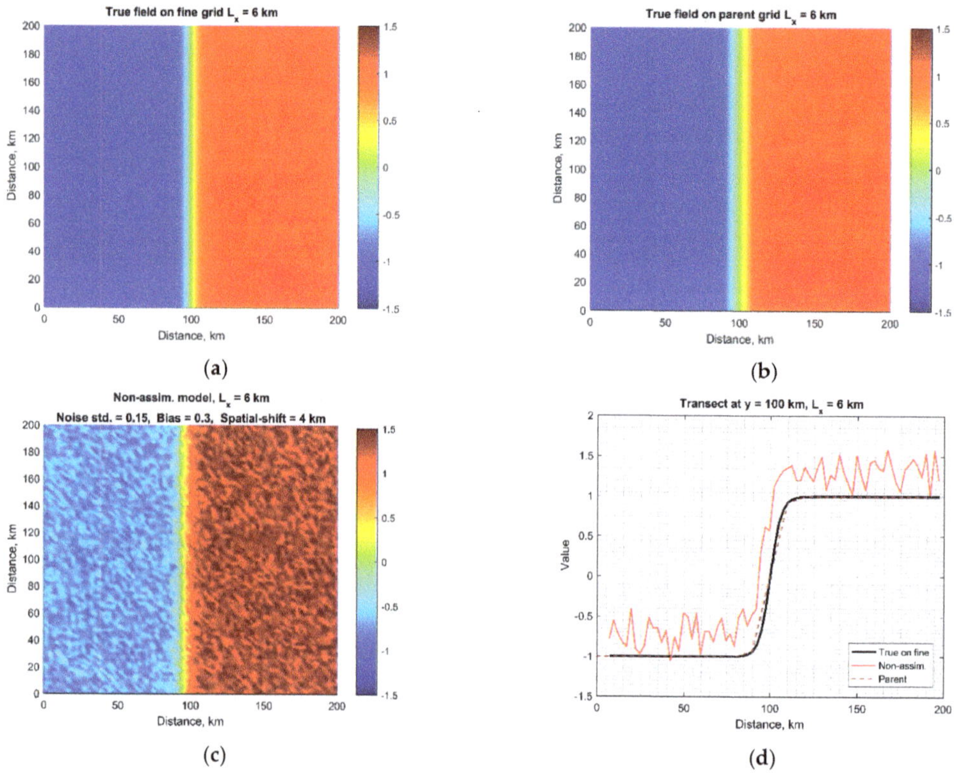

Figure 2. Ocean front of width $L_x = 6$ km. (**a**) True field on the fine mesh, (**b**) parent model ouput, (**c**) simulated child model before data assimilation, (**d**) transects of the previous fields along the line at $y = 100$ km.

Even at a resolution of 10 km that does not resolve the structure of the front (half-width of 6 km), the parent model gives a reasonable representation of areas outside the front where the changes in the state variable are smooth (Figure 2b). However, as expected, the width of the front is exaggerated due to insufficient resolution. The child model forecast shown in Figure 2c is noisy and clearly shifted westward relative to the true state. Figure 3 shows the results of assimilation of output from the parent model using the NDA method. The analysis state (after assimilation) removes the spatial shift and bias and reduces the noise even for such a sharp front where the resolution of the parent model is inadequate. For comparison, Figure 4 shows the improvement provided by the NDA assimilation for fronts of different sharpness, with L_x = 10, 20, 30, and 40 km. In all examples, the errors in non-assimilating child model forecasts are simulated by adding bias = 0.3, random normally distributed noise with standard deviation STD = 0.15, and spatial shift of 4 km to the west.

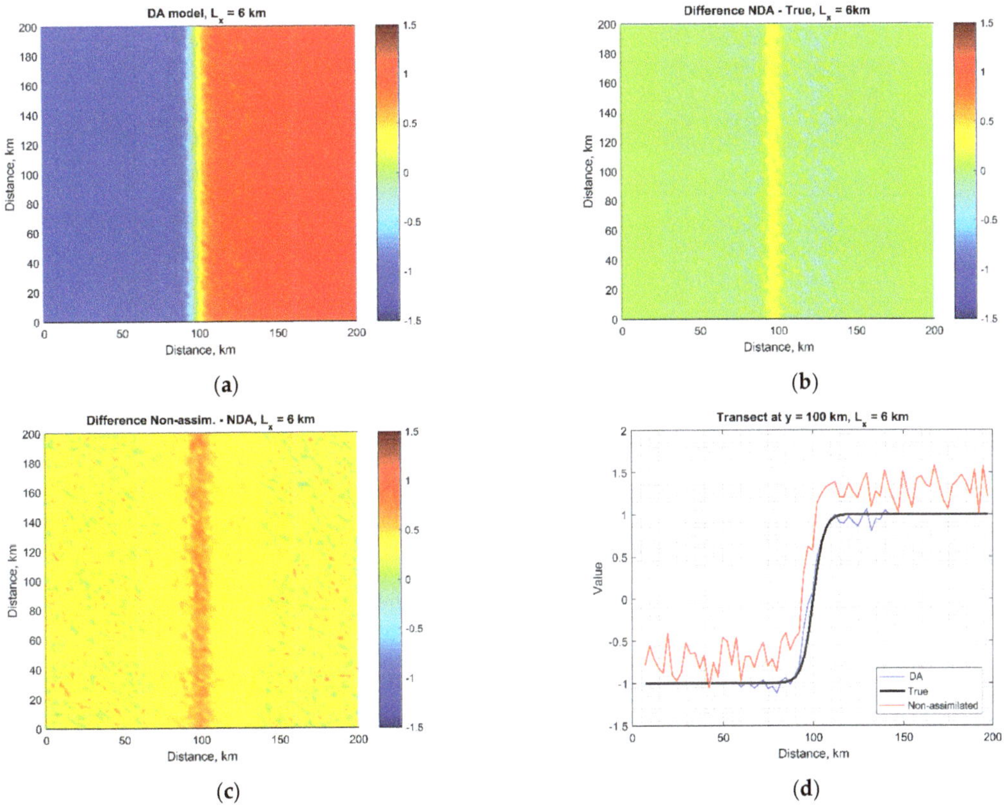

Figure 3. Assimilation results for the ocean front shown in Figure 2: (**a**) child model after NDA assimilation, (**b**) difference between the assimilated model and the true field, (**c**) difference between the non-assimilated and assimilated child models, (**d**) zonal transects of the previous fields along a line at $y = 100$ km.

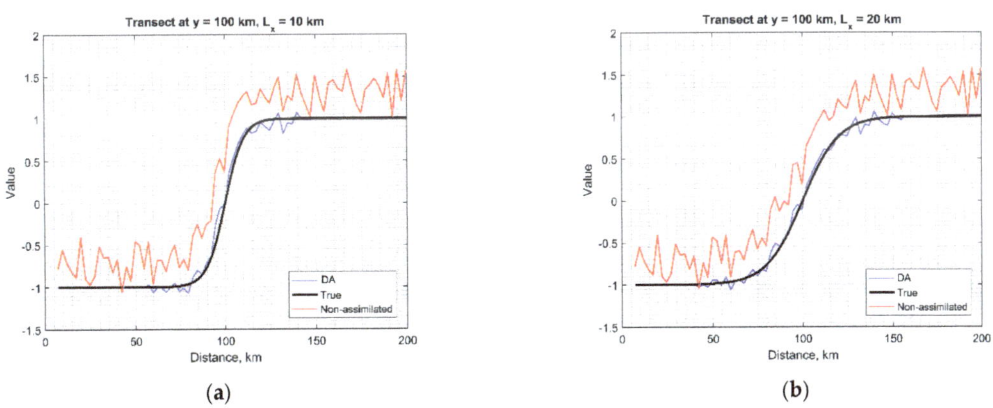

Figure 4. *Cont.*

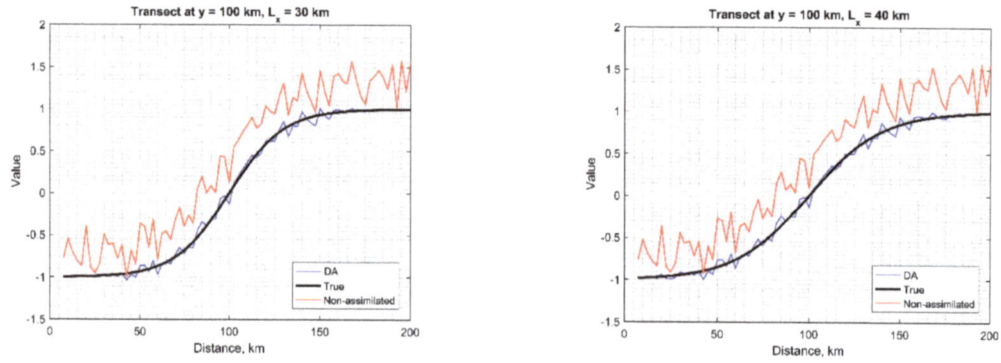

(c) **(d)**

Figure 4. Results of NDA assimilation for fronts of different sharpness: (**a**) $L_X = 10$ km, (**b**) km, (**c**) $L_X = 30$ km, (**d**) $L_X = 40$ km. The curves present data for the true field (black), non-assimilated noisy child model (red) and child model after NDA assimilation (blue).

Comparison of panels in Figures 4a–d and 3d shows that the improvement due to data assimilation from the parent model is achieved both for sharp fronts not resolved by the parent model and for smooth fronts.

Figure 5 shows how the bias and RMSE of the child model against the true state change with the sharpness of the front before and after data assimilation.

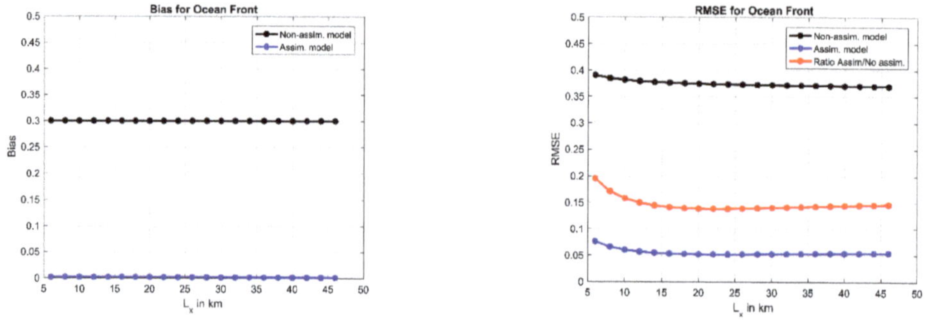

(a) **(b)**

Figure 5. Plots of bias (**a**) and RMSE (**b**) for the ocean front case with different front sharpness for non-assimilated (black) and assimilated (blue) child models. The red line shows the ratio of assimilated to non-assimilated RMSE.

After applying the NDA method, the bias is practically removed and the remaining values are on the order of 5×10^{-3} or less. The RMSE calculated against the true state is more than five times lower than before data assimilation, as shown by the red line in Figure 5b.

Similar properties are demonstrated in the example of a single eddy.

3.2. Single Eddy

Figure 6 shows a map of the true solution F for an axisymmetric mesoscale eddy with radius $L_e = 16$ km, its representation by the parent coarse model and by the child model before data assimilation. It also shows transects across the eddy centre. The parameters of added errors are the same as in the previous example: normally distributed noise with a standard deviation of 0.15, positive bias of 0.3, and shift of the field by 4 km to the west.

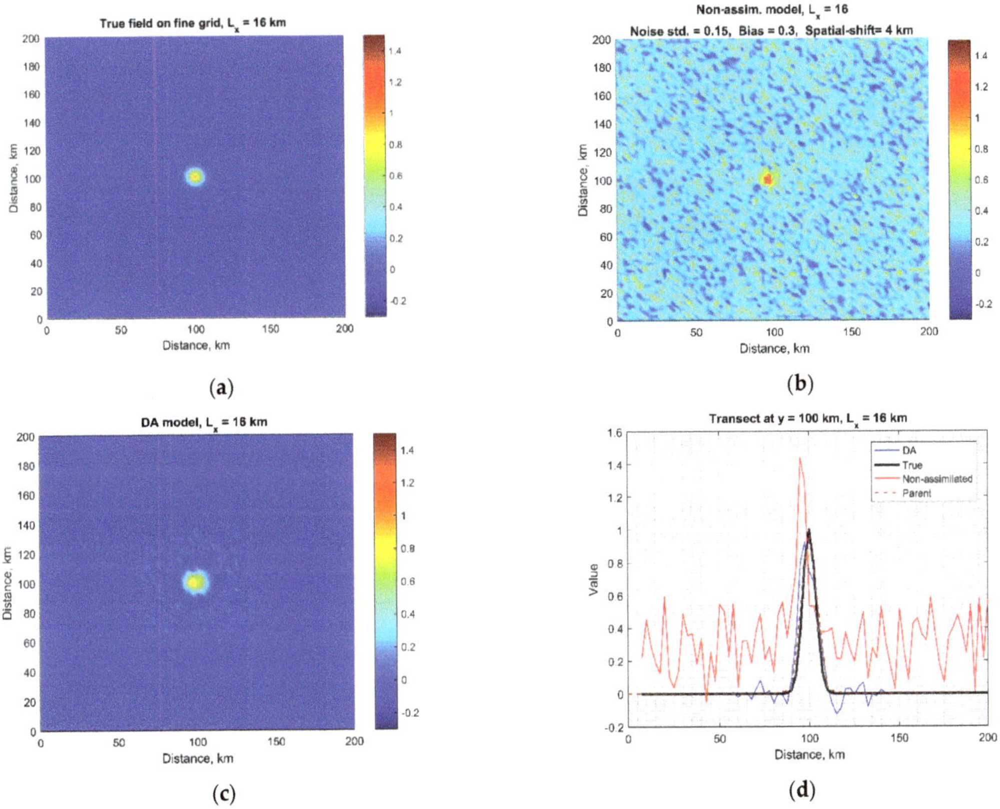

Figure 6. Results of NDA assimilation for a single isolated eddy: (**a**) true field sampled on the fine grid, (**b**) child model before data assimilation, (**c**) child model after NDA data assimilation, (**d**) transects showing the same fields as in (**a–c**) in comparison with the parent model.

The NDA method reduces errors nearly to zero outside of the eddy and greatly reduces them inside the eddy (Figure 6b,d). The bias is reduced from 0.3 before data assimilation to 0.002 or less after assimilation. The reduction in RMSE for various values of eddy radius for the range 6 to 46 km is shown in Figure 7.

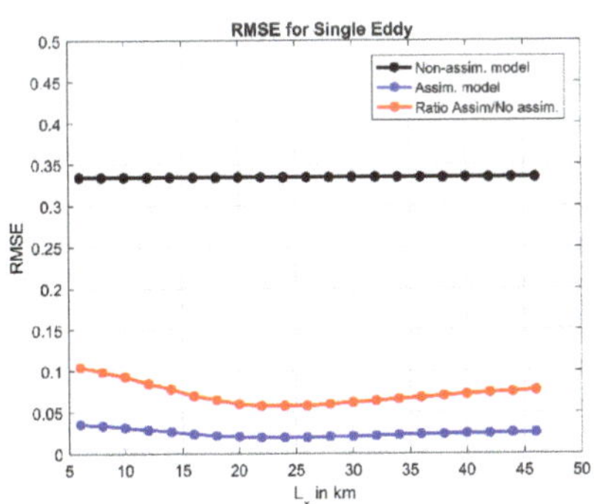

Figure 7. Dependence of RMSE for different eddy sizes: before data assimilation (black), after NDA assimilation (blue), the ratio of RMSE after and before data assimilation (red).

For illustration purposes, the spatial distributions of the variances $V_{B,ii}$ and $V_{R,ii}$ for the case of a single eddy with $L_x = 16$ km are presented in Figure 8. The following parameters are used for the child model: STD of noise = 0.15, bias = 0.3, and spatial shift = 4 km. The non-uniform colour in Figure 8b is a representation of noise in the child model output before data assimilation.

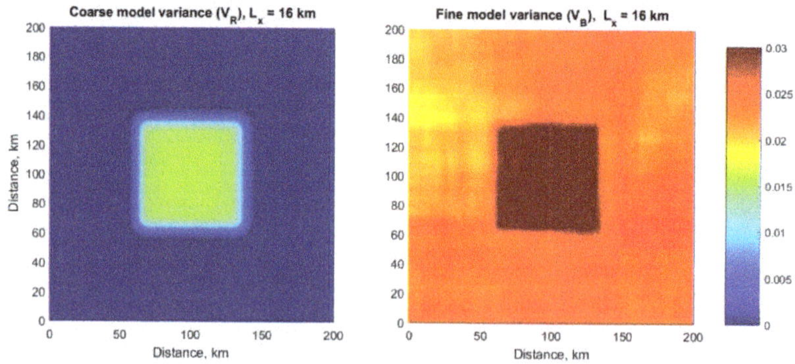

(**a**) (**b**)

Figure 8. Spatial distribution of the diagonal elements $V_{B,ii}$ (**a**) and $V_{R,ii}$ (**b**) in the example of a single eddy of a radius of 16 km.

The next example illustrates the qualities of the NDA method for a domain 1000 × 1000 km^2 packed with anisotropic eddies.

3.3. Multiple Eddies

Figure 9 shows maps of the field variable F from the parent and child models with the following parameters of added errors to simulate child model forecast before data assimilation: normally distributed noise with a standard deviation of 0.15, positive bias of 0.3, and spatial shift of the field by 4 km to the west.

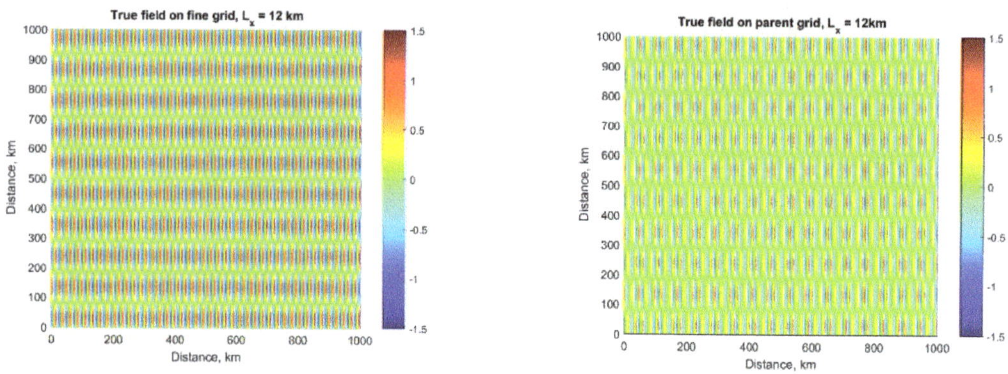

(**a**) (**b**)

Figure 9. *Cont.*

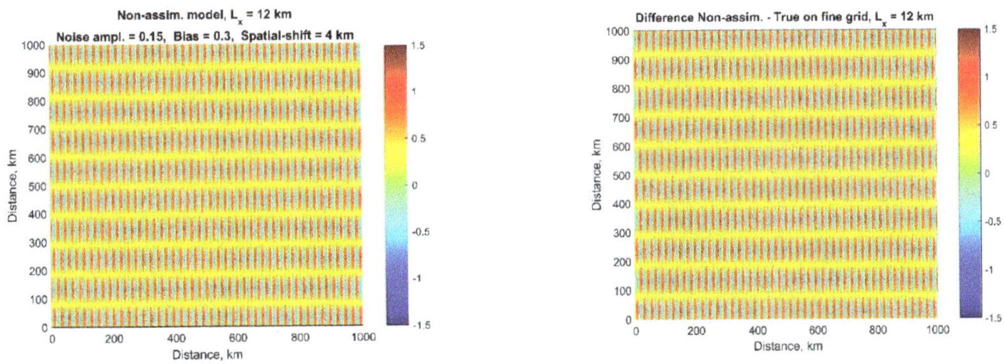

(c) **(d)**

Figure 9. Maps of the field variable before data assimilation for multiple eddy case with $L_x = 12$ km: (**a**) true field on fine grid, (**b**) true field on parent grid, (**c**) non-assimilated model, (**d**) difference between non-assimilated model and true field.

Figure 9b shows that the parent model generally underestimates the true field shown in Figure 9a due to representativity errors. The difference between the non-assimilated child model forecast and true state shown in Figure 9d is substantial, showing an RMSE = 0.61, which is about 30% of the range of the true field. The anatomy of the differences between the true state, and outputs from parent (coarse), and non-assimilated child models on a zoomed-in section of the zonal transect are shown in Figure 10. The largest errors produced by the non-assimilated model are due to spatial shift and bias. The errors produced by the parent model are exclusively due its insufficient resolution (i.e., are representativeness errors), as we assume that otherwise the parent model is perfect.

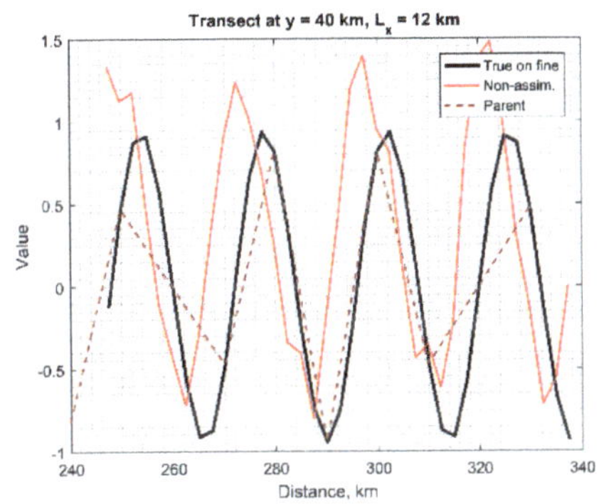

Figure 10. Transects of true field on the fine grid (black) and outputs from the parent (dashed brown) and non-assimilated child (red) models.

The NDA method presented in the previous section is then applied to assimilate the data from the parent model into child model forecast.

Figure 11 shows the results of data assimilation (analysis) and a map of differences between the assimilated child model and true field.

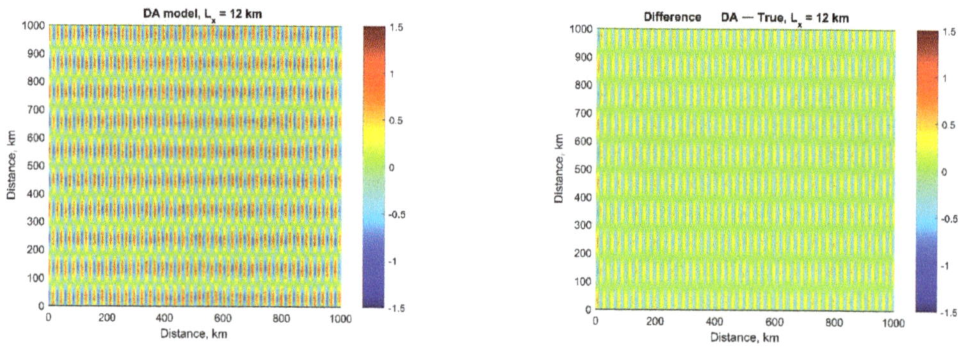

(a) (b)

Figure 11. Maps for the multiple eddies case with width $L_x = 12$ km: (**a**) NDA assimilated model, (**b**) difference between NDA assimilated model and true field. Parameters of the non-assimilated model are the same as in Figures 9 and 10.

The details of the improvement achieved by NDA data assimilation on a zoomed-in zonal transect are shown in Figure 12.

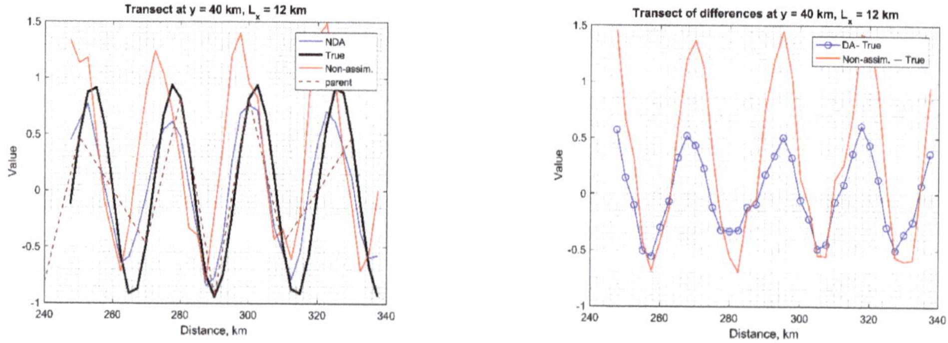

(a) (b)

Figure 12. Transects at $y = 40$ km for the multiple eddies case with $L_x = 12$ km. (**a**) Transects of NDA model output (blue), true solution (black), parent model (dashed dark red), and the non-assimilated child model (red); (**b**) differences between assimilated (blue) and non-assimilated (red) models and the true solution.

Data assimilation partially reduces the spatial shift which simulated the double penalty effect common to fine-resolution models. The reduction in spatial shift is due to the properties of the SDD component of the NDA method, see [26] for details. The significant reduction in the bias is due to the assimilation part of the NDA algorithm; however, it also causes some reduction in the amplitudes of the eddies (Figure 12a). The remaining errors shown in Figure 12b are mainly due to incompletely corrected spatial shift and random noise in the assimilated model. The RMSE of the assimilated model is lower at 0.25 vs. a non-assimilated value of 0.61, and the bias is reduced by orders of magnitude from 0.3 to $-1.3\ 10^{-5}$.

Figure 13 shows the results of the sensitivity analysis with different sizes of eddies and levels of various sources of errors in the non-assimilated child model forecast noise, bias, and shift.

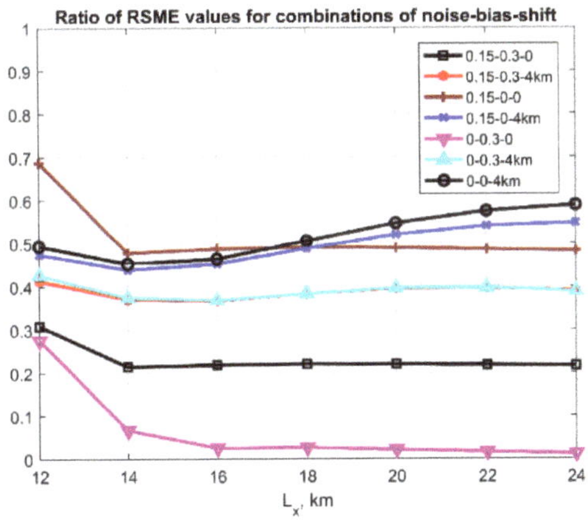

Figure 13. Ratio of RSME values of the child model relative to the true solution after and before NDA data assimilation as a function of eddy radius in the zonal direction x. Ratios are calculated for different combinations of errors generated by the child model: random noise (STD equal to either zero or 0.15), spatial shift (zero or 4 km), and bias (zero or 0.3). Percentages are shown with respect to the amplitude of the true signal. The first number in the legend shows the level of noise, the second shows the bias, and the third one shows the spatial shift.

In this example, the NDA method works best when the only source of error is the bias, which is removed nearly completely. The second-best results are achieved when only random noise and bias are present but not the spatial shift. The RMSE ratio for cases containing errors due to spatial shift grows slightly at larger eddy sizes. This is because a relatively small shift of 4 km does not distort large eddies to the same extent as small eddies even in the non-assimilated model. Generally, the curves in Figure 13 show that after application of NDA data assimilation algorithm the RMSE is reduced by half or better. Bias between the child and parent models is reduced by orders of magnitude. Assuming that the parent data-assimilating model is unbiased, it means that the fine resolution model becomes unbiased after application of NDA data assimilation process.

3.4. *Effect of Errors in the Parent Model*

The next example shows the quality of the NDA method when the parent model still contains errors, even after its own data assimilation cycle. To investigate how these errors affect the reduction in RMSE in the child model when the NDA data assimilation is applied, we tested two settings. In the first example, a quite large random noise was added to the parent model output with a standard deviation of 0.1, i.e., 10% of the signal. The results for the ocean front and multiple eddies examples are shown in Figures 14 and 15.

The transect in Figure 14 shows that after NDA, the front is well represented with small random noise even when the parent model is quite noisy. The improvement is up to five-fold and is consistent across the range of front widths from 6 to 46 km. The second example for the ocean front illustrates the efficiency of the NDA method when the parent model has spatially correlated errors, namely a lateral shift of 1 km to the true parent model (no random noise in this case). The simulated errors to the child model are the same as specified in the Figure 14 caption. The improvements in the child model outputs after DA follow: bias is reduced from 0.3365 to 0.0118 (approximately by a factor of 28) and RMSE is reduced from 0.3778 to 0.0616 (approximately 6-fold).

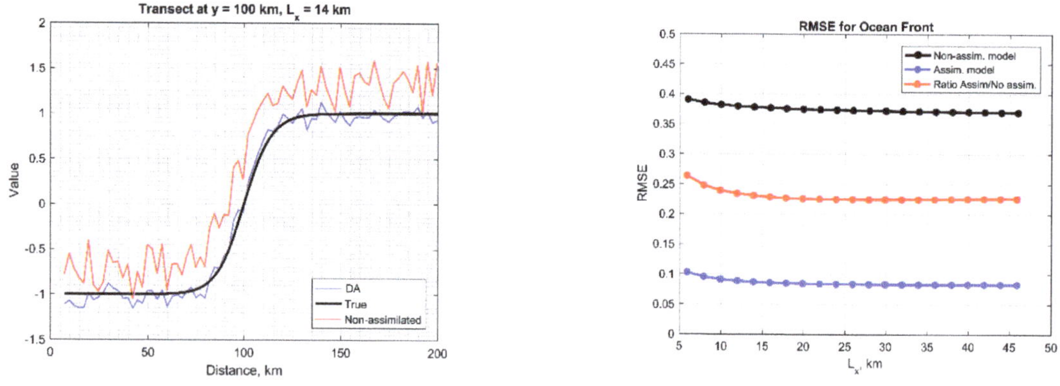

(a) (b)

Figure 14. Ocean front case of $L_x = 14$ km, with noisy parent model (added noise STD = 0.1) and noisy non-assimilated child model (added noise STD = 0.15, bias = 0.3 and shift of 4 km to the west). (**a**) Transects at $y = 100$ km of assimilated (blue), non-assimilated (red) child models, and the true field (black), (**b**) RMSE calculated for non-assimilated (black) and assimilated (blue) child models and the ratio between them at different front steepness.

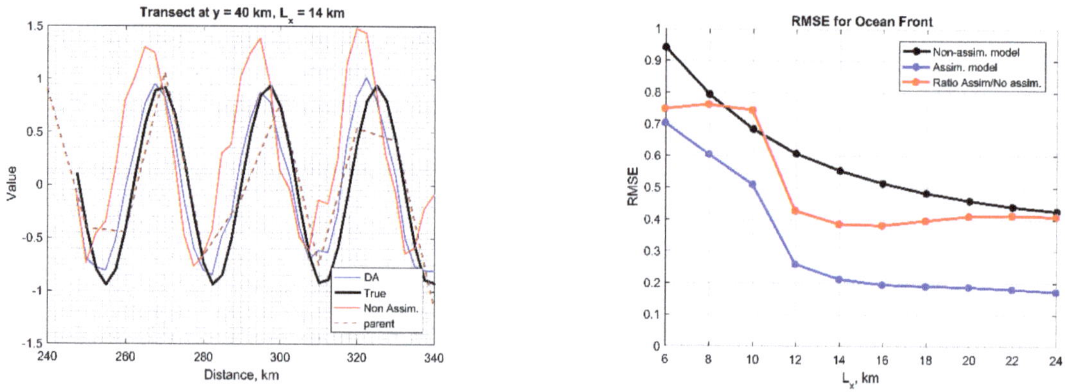

(a) (b)

Figure 15. Multiple eddy case of $L_x = 14$ km with noisy parent model (added noise STD = 0.1) and noisy non-assimilated child model (added noise STD = 0.15, bias = 0.3, and shift of 4 km to the west). (**a**) Transects at $y = 100$ km of true field (black), parent (dashed brown), assimilated (blue), and non-assimilated (red) child models, (**b**) RMSE for different eddy sizes and the ratio between them.

In case of multiple eddies, the improvement is less dramatic; however, for eddy diameters larger than 12 km, the RMSE is improved approximately by half (Figure 14). For eddy sizes 10 km and less the improvement due to assimilation of the parent model is limited, which is expected as such eddies are strongly distorted by the parent model of only 10 km resolution.

3.5. Spectral Characteristics

The mechanism of improvement of fine model outputs by the NDA method can be seen from the analysis of spectral characteristics of the downscaled parent model, fine model forecast (before data assimilation) and analysis (after data assimilation) in Figure 16. The bias represented by the peak at zero wavenumber is removed at the data assimilation step of NDA. Noise in the range below Nyquist wavenumber is reduced by melding the noisy child model forecast with the clean parent model data. It should be noted that the SDD downscaling honours the data of the parent model on the child nodes that coincide with a coarse grid node. A small peak between the coarse and fine grid Nyquist wavenumbers (outside of the scale in this figure) is an artefact created at the SDD step [26], as seen from the spectrum of the downscaled field. Other than this peak, the downscaled field has much lower level of noise than the child model forecast. Reduction in noise at wavenumbers higher than the parent model Nyquist frequency is due to the low level of noise generated at the downscaling step of the process. Figure 16 shows an efficient removal of bias in the Fourier space. The amplitude of the signal peak is also slightly reduced after DA.

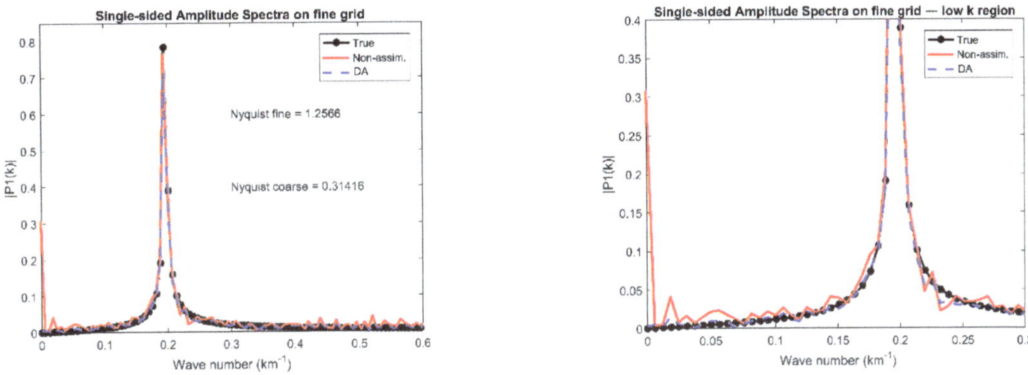

(**a**) (**b**)

Figure 16. Amplitude spectra on a zonal transect at $y = 40$ km for the example of multiple eddies of $L_x = 16$ km. The STD of noise in the non-assimilated child model is 0.15, bias is 0.3, and the spatial shift is 4 km to the west. The spectra are calculated for the true field, the child model forecast (before assimilation), and analysis (after assimilation). (**a**) Spectra in the full range of wave numbers and (**b**) zoomed in area for low wave numbers.

In real applications, there could be a potential mismatch between the bathymetry and coastlines in the fine and coarse models. For instance, some details such as small islands might be absent in the coarse mesh but present in the fine model, causing a mismatch between the land masks in fine and coarse models. This issue relates only to the first step of the NDA method, which performs the downscaling of the coarse model data onto the fine model grid, and this mismatch is natively resolved during the SDD downscaling [26]. As shown there, in the example of the coastal areas of the Red Sea, better resolution of the fine model results in higher vorticity/enstrophy values in the downscaled field compared to the parent coarse model, as shown in Figures 12–16 of that paper. The potential mismatch of the land mask is not specific to the NDA method; the same issue may appear when assimilation is carried out using satellite imagery (i.e., observations) of different resolution than the fine model. After downscaling, at the DA stage the data from the parent and child model are available exactly on the same (fine) grid with the same land mask, and the grid mismatch issue does not appear.

3.6. Comparison with a 'Standard' Method

The quality of the NDA method can be illustrated by comparison with existing data assimilation techniques applied to model-to-model data assimilation. Let us consider the combination of H–L [24] and variational method, which is considered to be the most commonly used technique for the estimation of the covariance matrices (e.g., [22,35]). There are several variants of this method; here, we applied the practical algorithm described in the textbook by Kalnay [25], which is named hereafter as the 'standard' method. The 'standard' method consists of the following stages: calculate innovations (differences between observations and model) at each observational point, estimate the covariances between innovations at different locations, fit the best-fit curve (usually Gaussian) using all covariances except at zero distance, and estimate the model and observation error variances using the value where the best-fit curve intersects the $r = 0$ vertical line on the covariance plot. In the practical implementations of this algorithm, the model and observation error variances are assumed spatially homogenous [25]. The model variances and the best-fit curve are used to calculate the background error covariance **B**-matrix, and the diagonal observational error covariance **R**-matrix. Because of spatial homogeneity, all diagonal elements in **B** and **R** matrices are equal but different between matrices. The H–L method requires combination of innovations into spatial bins, so that all innovations within a certain bin are allocated to the same distance from the central point. In practice, the covariances are calculated by averaging individual products of innovations over a period of time instead of statistical averaging assuming the ergodic hypothesis [36]. In the model-to-model DA approach, the role of observations is played by the output from the parent data-assimilating model.

The cost functions for the 'standard', Equation (1), and the NDA, Equation (2) methods are different; in other words, both methods minimize their own cost functions. The procedures for calculation of **B** and **R** matrices in the NDA and H–L methods are different and therefore they produce different matrices. For NDA, both matrices represent the covariances of the background error of the downscaled parent and child models, respectively, and are calculated in a similar and consistent way as both data sets are from models. Therefore, whilst both matrices have different error variances (diagonal elements), they have the same spatial correlations. Conversely, in the 'standard' method, the matrices are calculated using the H–L method that assumes that the observation errors have no spatial correlation. This assumption can be valid for some types of observations, but it is not valid for model-to-model DA.

For comparison between the NDA and 'standard' methods, we selected the multiple eddies example (as in Results, Section 3.3). The covariances required to build the **B** and **R** matrices for the standard method were estimated using H–L as follows. We generated 150 random realisations of the child model output, where each realisation was the sum of the true field given by Equation (15) ($L_x = 40$ km and $L_y = 105$ km) and the following errors: (i) random noise with standard deviation of 0.15, (ii) random spatial shift with a standard deviation of 4 km, and (iii) constant positive bias of 0.3. Random components were normally distributed with zero mean. Because of significantly larger computing resources required by the 'standard' method compared to NDA, the 'standard' DA used the reduced domain 300 × 300 km instead of 1000 × 1000 km for the NDA method. The bins required for the H–L method were 1 km in size.

The innovations were computed as $d = \mathbf{H}x^b - y$, where $\mathbf{H}$ is a subsampling matrix operator composed only of ones and zeroes that select from the coarse model the node that coincides in location with a child node. The 'observations' are located at all parent model grid nodes. Following the 'standard' method, we used a simple isotropic Gaussian function for the covariance of innovations at every parent model grid point x:

$$C_x(r) = a\, exp\left(-r^2/D^2\right), \tag{16}$$

where a and D are fitting parameters. The binned covariances and the fitting curve are shown in Figure 17.

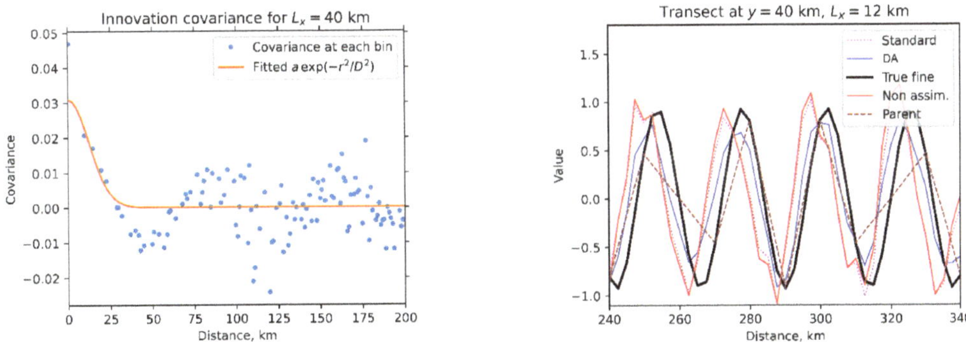

(a) (b)

Figure 17. (**a**) Estimates of background and observational error covariances for multiple eddy field with $L_x = 40$ km and $L_y = 105$ km and the corresponding fitting curves. The error variances estimated at $r = 0$ are used as diagonal elements of the background matrix **B** (equal to 0.031) and the 'observational' matrix **R** (equal to 0.016); (**b**) the zonal transect at $y = 40$ km showing the true field (black line), parent model (dashed brown), child model noisy forecast before DA (red), analysis state after 'standard' DA (dotted magenta), and analysis after NDA (blue).

Despite the parent model is assumed 'perfect', i.e., it has no errors relative to the true field (apart from representativeness errors due to insufficient resolution), the H–L method gives it an error variance (diagonal elements of **R**) of 0.016. This is approximately 50% of the noisy child model error variance values (diagonal elements of **B**) of 0.031. The **B** matrix is then constructed using a variance equal to 0.031 (Figure 17a) and the Gaussian formula Equation (16), and the diagonal matrix **R** is built using a value for its variance of 0.016.

The **B** and **R** matrices were used to carry out the variational DA cycle to the simulated child model forecast with the following parameters: multiple eddy field with $L_x = 12$ km, $L_y = 105$ km, random noise with STD = 0.15, bias = 0.3, and spatial shift = 4 km to the west. The analysis state is estimated using the following equations of the 'standard' method ([25], p. 155)

$$x^a = x^b + \mathbf{W}\left[y - H\left(x^b\right)\right] = x^b + \mathbf{W}d\mathbf{W} = \left(\mathbf{B}^{-1} + \mathbf{H}^T\mathbf{R}^{-1}\mathbf{H}\right)^{-1}\mathbf{H}^T\mathbf{R}^{-1}.$$

The transect in Figure 17b shows the analysis state for eddies with $L_x = 12$ km and $L_y = 105$ km obtained by the 'standard' method together with the true solution, parent model output, noisy child model forecast (before DA), and, for comparison, the analysis state obtained by the NDA method.

The bias and RMSE relative to the true solution for the child model forecast and the analysis state after 'standard' and NDA data assimilation process are shown in Table 1.

Table 1. Errors in the child model outputs before and after data assimilation.

	Forecast (before DA)	**'Standard' DA**	**NDA**
Bias	0.300	−0.0066	0.000
RMSE	0.608	0.5268	0.250

Figure 18 shows the map of the analysis state produced by the NDA and the standard methods. The errors are removed more efficiently in the areas of low values of field variable F.

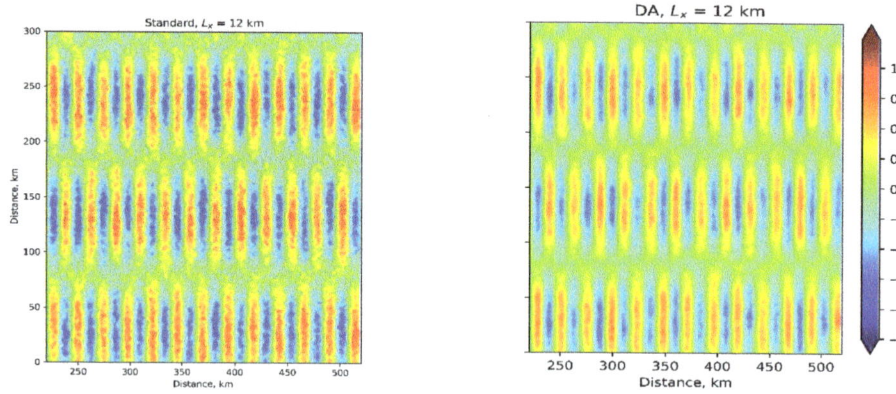

(a) (b)

Figure 18. The child model output after DA using (**a**) the 'standard' method and (**b**) the data assimilation with stochastic–deterministic downscaling (NDA). The NDA produces less noise.

4. Discussion

The examples presented show the data assimilation cycle using the NDA method when the data used for assimilation are not coming from observation as in the common assimilation methods but from a coarser parent model. The examples relate to a synthetic situation when the true values of the state variable are known. The forecast by the fine-resolution child model is simulated by sampling the true field on the model's fine grid and adding various sources of correlated and uncorrelated errors—random noise, bias, and spatial shift. The output from the coarser parent model is regarded as computer generated 'observations' and is simulated by sampling the true field on the coarser grid to which random noise can be added. In practice, the parent model is assumed to be a good-quality, dynamic ocean model that properly assimilates observational data. Therefore, the actual field measurements are used in the NDA method indirectly; instead of assimilating observations, the child model assimilates data from the coarse model, which in turn assimilates data from observations. Therefore, the child and parent models are double-linked, the parent model repeatedly provides to the child model: (i) 2D boundary conditions for creating a forecast and (ii) 3D data to improve the forecast and create the starting point for the next forecast.

The NDA is a two-step process. The first step of the NDA method is the application of stochastic–deterministic downscaling presented in [26]. As a result, at each fine grid point there are two generally different values of the field variable. The second step is to combine these values separately at each grid point using a suitable data assimilation method. In this paper, a modified cost function is used to obtain the best possible estimate of the analysis state. This approach leads to the use of a zero-dimensional Kalman, similar to what has been done in atmospheric chemistry [32].

The contribution of the first (SDD) and second (DA) steps to the improvement of RMSE and bias depends on a number of factors, such as the ratio of model resolution to the size of the feature in question, level of noise, and general quality of parent and child models. As the parent model in the examples considered above is assumed to be unbiased after its own DA procedure, it does not generate any bias after the downscaling step. By design of the DA step, the zero bias from the first step is carried forward to the final analysis state. The SDD downscaling can produce good results on its own, even without the need for running a fine-scale numerical model, examples of this are presented in [26]. Therefore, in case the child model has large errors, a combination of good-quality data from the first step (downscaling) with not so good-quality output from the child model using the Kalman gain formula (Equation (10)) will worsen the RMSE. However, if the errors of the fine model are relatively small or if the fine model contains features that are not well represented in the coarse model, the two-step NDA RMSE can be significantly smaller than the first-step SDD RMSE. The gain formula used in the Kalman filter requires the knowledge of error variances. In this paper, the variances are assessed using the ergodic hypothesis, e.g., [36], and the ensemble statistical mean at each grid node is estimated by spatial averaging in a small trial domain around the node at the same time point. This scheme is fast and does not consume significant computational resources. If necessary, it can be extended by including data from preceding time points; however, these data may not be statistically independent and hence the potential improvement requires further study.

The study by [37] used a spectral nudging method in order to restrict the drift of large-scale spectral components in the fine model from the global model. The main difference with the NDA is that the nudging technique uses weighting coefficients that are tuning parameters and are prescribed in advance, while the NDA variational method uses weights computed from the variance of the errors by minimising the cost function and therefore the weights cannot be changed at will. In contrast to the spectral nudging, the NDA method corrects both large- and small-scale components of the child model as seen in the amplitude spectra shown in Figure 16. The NDA automatically adjust the bias of the child model to the level of parent model which can be interpreted as an aggressive nudging of a single long-wave component of the field.

The forecast error covariance matrix **B** (for the child model) may be different from the error covariance matrix **R** (for the parent model) at very small distances not resolved by the parent model. It has to be noted that in any DA method, forecast error covariance matrices are not exact and are estimated based on using various approximations. Bannister [18] states: 'Owing partly to its very large rank, the B-matrix is impossible to use in an explicit fashion in an operational setting and so methods have been sought to model its important properties in a practical way'.

Whatever the approximation for the **B** (and R) matrix is used, the final judgement of the DA quality should be made on how well the DA reduces the model errors. The NDA method is shown to reduce the errors in the child model (Figures 3–16). It also demonstrates that the assumption of a similar spatial correlation structure between **B** and **R** matrices used in NDA for model-to-model data assimilation produces better results than the assumption of a non-correlated **R** matrix as applied in the widely used H–L method (Table 1). Therefore, the NDA method demonstrates the appropriateness of underlying assumptions and algorithms.

The analysis presented above shows that the NDA method is not the only one that can be applied to model-to-model data assimilation; the 'standard' method could also give reasonable results. However, when compared with the 'standard' DA method, the NDA gives better accuracy, stronger reduction in the RMS errors, and a complete removal of bias. The NDA is also more computationally efficient. We had no restrictions or limitations in computing the analysis for 1000×1000 km^2 domain at 2.5 km resolution when using the NDA on our office PC. However, we were unable to carry out the 'standard' DA for a domain greater than 300×300 km^2 on the same PC due to computing resource restrictions. In terms of speed of calculations, it required about 2 min to complete one full DA cycle, including calculation of the covariance matrix and weighting coefficients using the NDA method in the multiple eddy example. On the other hand, the standard method for the reduced domain took 24 min, including the creation of **B** and **R** matrices and calculation the analysis state.

As an illustration of the use of the NDA method for operational oceanography, we present below a snapshot of SST in the Lakshadweep Sea (Indian Ocean) from the parent global model at 1/12 degree of resolution available from Copernicus Marine Service, product GLOBAL_MULTIYEAR_PHY_001_030 [7,38] and two nested NEMO models: LD20 at 1/20 degree resolution and LD60 at 1/60 degree resolution, see Figure 19. The parent model assimilates sea surface temperature, sea surface level, and in situ TS profiles, and is described in detail by their authors [38]. Both LD20 and LD60 child models were developed by the Plymouth Ocean Forecasting Centre team. The child models were run with and without NDA from 1 January 2015. The LD20 model used 3D temperature, salinity, and velocities data from the global model for NDA. The LD60 model was implemented with two versions of the NDA: (i) using the global model as the parent and (2) using the intermediate nested LD20 model as the parent. In all cases, the NDA was applied to 3D temperature, salinity, and velocities data from the parent models. The details of the LDA20 and LDA60 models will be published elsewhere [39].

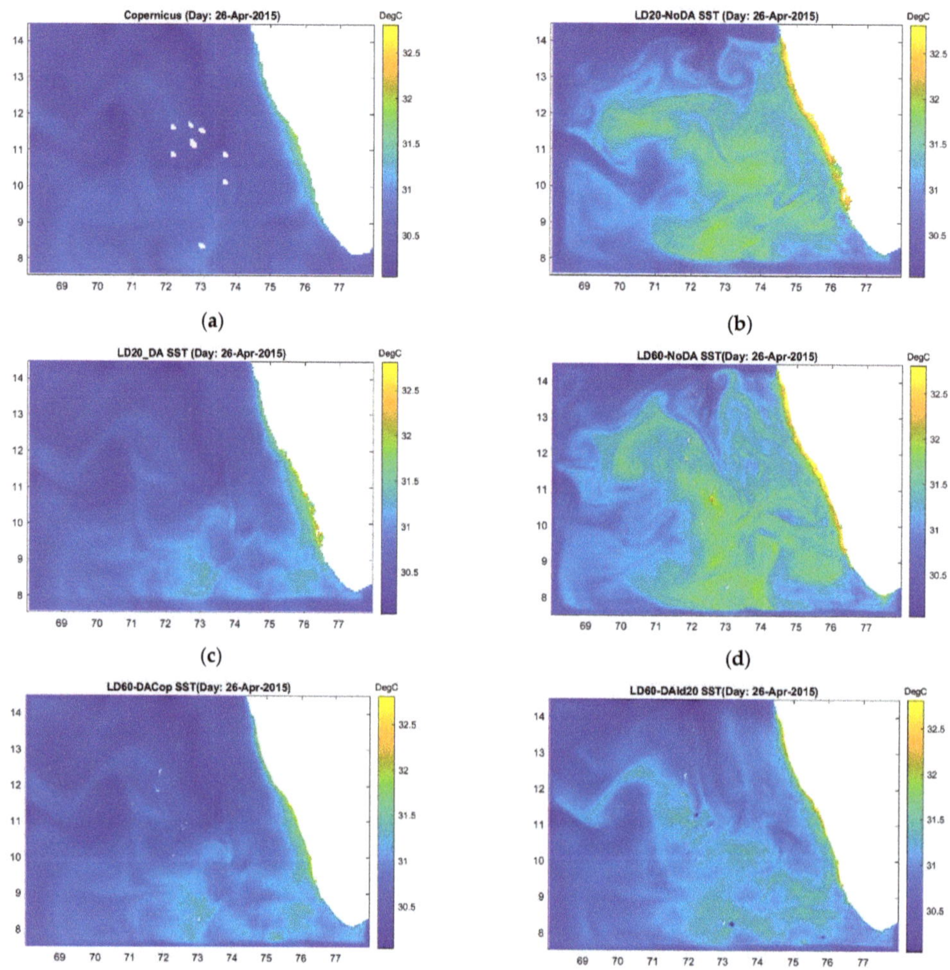

Figure 19. Daily averaged SST for 26/04/2015 from (**a**) the operational data-assimilating global model at 1/120 resolution, generated using EU Copernicus Marine Service Information [38], (**b**) nested model LD20 at $1/20^0$ without NDA, (**c**) nested model LD20 at $1/20^0$ with NDA using 3D temperature, salinity, and velocities data from the global model, (**d**) nested model LD60 at $1/60^0$ without NDA, (**e**) nested model LD60 at $1/60^0$ with NDA using 3D temperature, salinity, and velocities data from the global model, (**f**) nested model LD60 at $1/60^0$ with NDA using 3D temperature, salinity, and velocities data from the nested LD20 model.

5. Conclusions

This paper introduces a data assimilation method where the data are assimilated into a high-resolution model from a coarser good-quality data-assimilating model, not directly from observations. An efficient and simple algorithm for model-to-model variational data assimilation method named Nesting with Downscaling and Data Assimilation (NDA) is developed. The theoretical background behind the NDA algorithm is discussed, and its application is illustrated in a number of idealised synthetic situations that resemble real-world practice in fine-resolution ocean modelling. Here, we followed the philosophy used by Stommel in his famous work [40], which explained the existence of the Gulf Stream using a synthetic idealised case—square ocean, flat bottom, and sinusoidal wind distribution. Our results demonstrate that the model-to-model data assimilation is an efficient way for improving the accuracy of a fine-resolution model. This approach allows for avoiding a repetition of a complex and resource-hungry assimilation of actual observations, which has already been done in the parent model. It is likely that the same basic idea of model-to-model data assimilation would also work for other methods currently used in observational data assimilation. In this paper, the NDA is compared with a commonly used variational method and shown to be more accurate and significantly less computationally expensive.

Author Contributions: Conceptualization, G.I.S.; methodology, G.I.S.; software, G.I.S. and J.M.G.-O.; formal analysis, G.I.S.; writing—original draft preparation, G.I.S.; writing—review and editing, G.I.S. and J.M.G.-O. All authors have read and agreed to the published version of the manuscript.

Funding: This research received no external funding.

Data Availability Statement: For code availability, please contact the corresponding author.

Acknowledgments: The authors are thankful to the members of Plymouth Ocean Forecasting Centre for their support. We are also thankful to M. S. Poovadiyil for providing Figure 19.

Conflicts of Interest: The authors declare no conflict of interest.

References

1. GFDL. High-Resolution Modeling. 2020. Available online: https://www.gfdl.noaa.gov/high-resolution-modeling (accessed on 6 July 2020).
2. Dufour, C.O.; Griffies, S.; de Souza, G.; Frenger, I.; Morrison, A.; Palter, J.; Sarmiento, J.L.; Galbraith, E.; Dunne, J.P.; Anderson, W.G.; et al. Role of mesoscale eddies in cross-frontal transport of heat and biogeochemical tracers in the Southern Ocean. *J. Phys. Oceanogr.* **2015**, *45*, 3057–3081. [CrossRef]
3. Shapiro, G.I.; Stanichny, S.; Stanychna, R.R. Anatomy of shelf–deep sea exchanges by a mesoscale eddy in the North West Black Sea as derived from remotely sensed data. *Remote Sens. Environ.* **2010**, *114*, 867–875. [CrossRef]
4. Meunier, T.; Barton, E.D.; Barreiro, B.; Torres, R. Upwelling filaments off Cap Blanc: Interaction of the NW African upwelling current and the Cape Verde frontal zone eddy field? *J. Geophys. Res. Atmos.* **2012**, *117*, C08031. [CrossRef]
5. NEMO. Annual Mean of Sea Surface Salinity in 1/12° (NEMO-WRF Coupling). Available online: https://www.nemo-ocean.eu (accessed on 16 June 2021).
6. ROMS. Regional Ocean Modeling System (ROMS). Available online: https://www.myroms.org (accessed on 16 June 2021).
7. CMEMS. Ocean Explainers: Oceanography Educational Webpages. Available online: https://marine.copernicus.eu (accessed on 20 May 2021).
8. DARC. What Is Data Assimilation? Available online: https://research.reading.ac.uk/met-darc/aboutus/what-is-data-assimilation (accessed on 5 August 2021).
9. Lorenc, A.C. Analysis methods for numerical weather prediction. *Q. J. R. Meteorol. Soc.* **1986**, *112*, 1177–1194. [CrossRef]
10. Bell, M.J.; Forbes, R.M.; Hines, A. Assessment of the FOAM global data assimilation system for real-time operational ocean forecasting. *J. Mar. Syst.* **2000**, *25*, 1–22. [CrossRef]
11. Gandin, L.S. The problem of optimal interpolation, Scientific papers. *Main Geophys. Obs.* **1959**, *99*, 67–75.
12. Moore, A.M.; Martin, M.J.; Akella, S.; Arango, H.G.; Balmaseda, M.A.; Bertino, L.; Ciavatta, S.; Cornuelle, B.; Cummings, J.; Frolov, S.; et al. Synthesis of Ocean Observations Using Data Assimilation for Operational, Real-Time and Reanalysis Systems: A More Complete Picture of the State of the Ocean. *Front. Mar. Sci.* **2019**, *6*, 90. [CrossRef]
13. Dobricic, S.; Pinardi, N. An oceanographic three-dimensional variational data assimilation scheme. *Ocean. Model* **2008**, *22*, 89–105. [CrossRef]
14. Ghil, M.; Malanotte-Rizzoli, P. Data Assimilation in Meteorology and Oceanography. *Adv. Geophys.* **1991**, *33*, 141–266. [CrossRef]

15. NEMOVAR. The NEMOVAR Ocean Data Assimilation System as Implemented in the ECMWF Ocean Analysis for System 4. Available online: https://www.ecmwf.int/en/elibrary/11174-nemovar-ocean-data-assimilation-system-implemented-ecmwf-ocean-analysis-system-4 (accessed on 1 October 2021).
16. DART, Data Assimilation Research Testbed. Available online: https://dart.ucar.edu (accessed on 1 October 2021).
17. PDAF, Parallel Data Assimilation Network. Available online: https://pdaf.awi.de/trac/wiki (accessed on 1 October 2021).
18. Bannister, R.N. A review of forecast error covariance statistics in atmospheric variational data assimilation. II: Modelling the forecast error covariance statistics. *Q. J. R. Meteorol. Soc.* **2008**, *134*, 1971–1996. [CrossRef]
19. Waters, J.; Lea, D.J.; Martin, M.J.; Mirouze, I.; Weaver, A.; While, J. Implementing a variational data assimilation system in an operational 1/4 degree global ocean model. *Q. J. R. Meteorol. Soc.* **2015**, *141*, 333–349. [CrossRef]
20. Haben, S.; Lawless, A.; Nichols, N. Conditioning and preconditioning of the variational data assimilation problem. *Comput. Fluids* **2011**, *46*, 252–256. [CrossRef]
21. Kubryakov, A.I.; Korotaev, G.K.; Dorofeev, V.L.; Ratner, Y.B.; Palazov, A.; Valchev, N.; Malciu, V.; Mateescu, R.; Oguz, T. Black Sea coastal forecasting system. *Ocean Sci.* **2012**, *8*, 183–196. [CrossRef]
22. Carrassi, A.; Bocquet, M.; Bertino, L.; Evensen, G. Data assimilation in the geosciences: An overview of methods, issues, and perspectives. *WIREs Clim. Chang.* **2018**, *9*, e535. [CrossRef]
23. Bouttier, F.; Courtier, P. Data Assimilation Concepts and Methods. 1999. Available online: https://www.ecmwf.int/en/elibrary/16928-data-assimilation-concepts-and-methods (accessed on 5 August 2021).
24. Hollingsworth, A.; Lönnberg, P. The statistical structure of short-range forecast errors as determined from radiosonde data, Part 1: The wind field. *Tellus A* **1986**, *38*, 111–136. [CrossRef]
25. Kalnay, E. *Atmospheric Modeling, Data Assimilation and Predictability*; Cambridge University Press (CUP): Cambridge, UK, 2003; 369p.
26. Shapiro, G.I.; Gonzalez-Ondina, J.M.; Belokopytov, V.N. High-resolution stochastic downscaling method for ocean forecasting models and its application to the Red Sea dynamics. *Ocean Sci.* **2021**, *17*, 891–907. [CrossRef]
27. Parrish, D.F.; Derber, J.C. The National Meteorological Center's spectral statistical interpolation analysis system. *Mon. Weather Rev.* **1992**, *120*, 1747–1763. [CrossRef]
28. Polavarapu, S.; Ren, S.; Rochon, Y.; Sankey, D.; Ek, N.; Koshyk, J.; Tarasick, D. Data assimilation with the Canadian middle atmosphere model. *Atmosphere-Ocean* **2005**, *43*, 77–100. [CrossRef]
29. Zingerlea, C.; Nurmib, P. Monitoring and verifying cloud forecasts originating from operational numerical models. *Meteorol. Appl.* **2008**, *15*, 325–330. [CrossRef]
30. Bruciaferri, D.; Shapiro, G.; Wobus, F. A multi-envelope vertical coordinate system for numerical ocean modelling. *Ocean Dyn.* **2018**, *68*, 1239–1258. [CrossRef]
31. Mellor, G.L.; Blumberg, A.F. Modeling Vertical and Horizontal Diffusivities with the Sigma Coordinate System. *Mon. Weather Rev.* **1985**, *113*, 1379–1383. [CrossRef]
32. Adhikary, B.; Kulkarni, S.; Dallura, A.; Tang, Y.; Chai, T.; Leung, L.; Qian, Y.; Chung, C.; Ramanathan, V.; Carmichael, G. A regional scale chemical transport modeling of Asian aerosols with data assimilation of AOD observations using optimal interpolation technique. *Atmos. Environ.* **2008**, *42*, 8600–8615. [CrossRef]
33. Fedorov, K.N. *The Physical Nature and Structure of Oceanic Fronts*; Springer: Berlin/Heidelberg, Germany, 1986; 333p, ISBN 978-0-387-96445-4.
34. Robinson, A.R. (Ed.) . *Eddies in Marine Science*; Springer: Berlin/Heidelberg, Germany, 1983; 612p, ISBN 978-3-642-69003-7.
35. Stewart, L.M.; Dance, S.L.; Nichols, N.K.; Eyre, J.R.; Cameron, J. Estimating interchannel observation-error correlations for IASI radiance data in the Met Office system. *Q. J. R. Meteorol. Soc.* **2014**, *140*, 1236–1244. [CrossRef]
36. Stull, R.B. An introduction to boundary layer meteorology. *Kluwer* **2003**, 670. [CrossRef]
37. Katavouta, A.; Thompson, K.R. Downscaling ocean conditions with application to the Gulf of Maine, Scotian Shelf and adjacent deep ocean. *Ocean Model.* **2016**, *104*, 54–72. [CrossRef]
38. Global Ocean Physics Reanalysis (GLOBAL_MULTIYEAR_PHY_001_030). Available online: https://resources.marine.copernicus.eu/product-detail/GLOBAL_MULTIYEAR_PHY_001_030/INFORMATION (accessed on 7 March 2022).
39. Stommel, H. The westward intensification of wind-driven ocean currents. *Trans. Am. Geophys. Union* **1948**, *29*, 202–206. [CrossRef]
40. Shapiro, G.I.; Ondina, J.M.; Poovadiyil, M.S.; Tu, J.; Asif, M. *The Operational Ocean Modelling System of the Lakshadweep Sea*; University of Plymouth: Plymouth, UK, 2022; *(Manuscript in Preparation)*.

Journal of
Marine Science and Engineering

MDPI

Article

A Projection Method for the Estimation of Error Covariance Matrices for Variational Data Assimilation in Ocean Modelling

Jose M. Gonzalez-Ondina [1,*], Lewis Sampson [2] and Georgy I. Shapiro [3]

1 University of Plymouth Enterprise Ltd. (UoPEL), Drake Circus, Plymouth PL4 8AA, UK
2 Met Office, FitzRoy Road, Exeter EX1 3PB, UK; lewis.sampson@metoffice.gov.uk
3 School of Biological and Marine Sciences, University of Plymouth, Drake Circus, Plymouth PL4 8AA, UK; g.shapiro@plymouth.ac.uk
* Correspondence: jose.ondina@plymouth.ac.uk

Abstract: Data assimilation methods are an invaluable tool for operational ocean models. These methods are often based on a variational approach and require the knowledge of the spatial covariances of the background errors (differences between the numerical model and the true values) and the observation errors (differences between true and measured values). Since the true values are never known in practice, the error covariance matrices containing values of the covariance functions at different locations, are estimated approximately. Several methods have been devised to compute these matrices, one of the most widely used is the one developed by Hollingsworth and Lönnberg (H-L). This method requires to bin (combine) the data points separated by similar distances, compute covariances in each bin and then to find a best fit covariance function. While being a helpful tool, the H-L method has its limitations. We have developed a new mathematical method for computing the background and observation error covariance functions and therefore the error covariance matrices. The method uses functional analysis which allows to overcome some shortcomings of the H-L method, for example, the assumption of statistical isotropy. It also eliminates the intermediate steps used in the H-L method such as binning the innovations (differences between observations and the model), and the computation of innovation covariances for each bin, before the best-fit curve can be found. We show that the new method works in situations where the standard H-L method experiences difficulties, especially when observations are scarce. It gives a better estimate than the H-L in a synthetic idealised case where the true covariance function is known. We also demonstrate that in many cases the new method allows to use the separable convolution mathematical algorithm to increase the computational speed significantly, up to an order of magnitude. The Projection Method (PROM) also allows computing 2D and 3D covariance functions in addition to the standard 1D case.

Keywords: data assimilation; variational methods; analysis of innovations; ocean modelling; operational forecast

Citation: Gonzalez-Ondina, J.M.; Sampson, L.; Shapiro, G.I. A Projection Method for the Estimation of Error Covariance Matrices for Variational Data Assimilation in Ocean Modelling. *J. Mar. Sci. Eng.* **2021**, *9*, 1461. https://doi.org/10.3390/jmse9121461

Academic Editor: Mikhail Emelianov

Received: 19 November 2021
Accepted: 14 December 2021
Published: 20 December 2021

1. Introduction

Due to intrinsic inaccuracies in the model equations, numerical schemes and quality of input data streams, even the best ocean models gradually deviate from reality and can only be considered an estimate of the true ocean state [1]. The introduction of Data Assimilation (DA) techniques allowed to reduce the deviation of models from the true state, vastly improving the accuracy of ocean forecasting [2].

There are a number of DA methods currently in use [1,3]. Many modern operational Variational Data Assimilation schemes have originated from the method of optimal interpolation Gandin [4,5], and are based on minimising the so-called cost function, see, e.g., Lorenc [6], Waters et al. [7] and Carrassi et al. [1]. The cost function includes a combination of the model forecasts and the observational data, weighted by the relative correctness of each component as represented by their error covariance matrices (ECM). This DA technique allows to compute a more accurate state called the analysis, which then is used as an initial condition for a new forecasting cycle.

The cost function used to obtain the analysis vector with the maximum likelihood of concordance with the truth [2] is given below (see, e.g., [8])

$$J[\delta x] = \delta x^T \mathbf{B}^{-1} \delta x + (H(\delta x) - \boldsymbol{d})^T \mathbf{R}^{-1} (H(\delta x) - \boldsymbol{d}), \tag{1}$$

$$\delta x = x^a - x^b, \quad \boldsymbol{d} = y^o - H\left(x^b\right), \tag{2}$$

where observational and background data are represented by vectors y^o and x^b, respectively. Vector $\boldsymbol{d}$ contains the so-called innovations (differences between observation values and model values interpolated at the observation location), δx the analysis increment, and x^a the analysis vector. The operator H, referred to as the observation operator, takes model data into observational space. The matrices **B** and **R** are the ECMs for background and observational data, respectively (see Ide et al. [9], for a description of the notation).

The calculation of the background (or forecast) and observation error covariance matrices (ECM) is a key component of variational data assimilation systems. The more precisely they are estimated, the smaller difference there is between the analysis fields and the true state of the ocean [10]. In practice, the estimation of the ECMs is not an easy task. Firstly, observations are usually scarce. Even today, when satellite imagery is at one's disposal, only some data are available in large quantities and only at the ocean surface. Data within the water column are limited to places where buoys or drifters exist. Secondly, the true state of the ocean is never known, so the methods rely on statistics on the available observations to estimate it. Errors has to be estimated by some proxy, typically either from innovations or model differences [10]. Thirdly, ECM estimation requires to process large amounts of data coming from the model and the observations. The computation of all correlations from a set of N observations requires to carry out $\mathcal{O}(N^2)$ floating point operations. This is a mammoth task even for modern high-performance computers.

Since there are such computational difficulties with calculating the ECMs, simplifications have been developed in order to make the process feasible. These include parametrising the covariance matrices [7] and the assumption of isotropic error correlation [11,12]. The isotropic assumption means that the direction is ignored when calculating correlation, and distance becomes the only determining variable for the correlation. A widely used method for computing the error covariances under assumptions of statistical isotropy and lack of spatial correlation between observational errors was designed by Hollingsworth and Lönnberg [13] and is hereafter called the H-L method, which is used in many operational systems (see, e.g., [1]). Due to the nature observations it is common to assume that they are spatially uncorrelated [14], and therefore allow **R** to be a diagonal ECM. Using this assumption, the H-L method is able to produce a joint estimate for both the observation and background EMCs.

The H-L method produces the covariance function for each model grid node using the following steps [13]. First, all possible products of pairs of innovations relevant to each model grid node, are calculated and binned by their distances. Second, the products related to a certain bin are averaged (usually using time averaging) to give an approximate value of covariance. Third, a curve is fitted through the covariance points (not taking into account the first bin) to produce a covariance function which then is used to calculate the diagonal element of **R** related to that grid node, and a row of elements for the **B** matrix, both diagonal and non-diagonal. Although the binning step was not considered an essential part of the H-L method by the authors, it was recommended when the amount of data is large. In modern practice, the number of innovations can be in the order of millions and, more importantly, observations are rarely taken at the same locations, therefore, binning is unavoidable (see, e.g., [15,16]). Despite being widely used, the H-L method has some weaknesses such as inability to include anisotropy and underperformance in sparsely observed areas.

The proposed Projection Method (PROM) follows the general variational approach based on minimisation of the cost function and aims to improve upon the H-L method. Instead of calculating the covariances in bins and then fitting a curve to them, PROM uses exact locations of the observations and fits the covariance curve using all individual products of innovations. The PROM method does not require the assumption of isotropy as it always finds the best fitting surface instead of a curve.

Section 2 presents a description of the method and the data used for the numerical experiments, Section 3 presents the results of numerical experiments using PROM for two cases from an operational model of the North Indian Sea. One case considers a slightly idealised situation when the true covariance function is known, and the second relates to the actual outputs from the North Indian Sea model. Performance of PROM in comparison to the H-L method is considered in the Section 4. The Section 5 gives a final summary of the main points presented in the paper.

2. Materials and Methods

We have developed a method to jointly estimate covariance matrices **B** and **R** based on the analysis of innovations. This method is similar in concept to H-L, but it removes the intermediate step of binning the innovations. As we shall see, removing this step requires to look at the analysis of innovations and function fitting from a different point of view. The following subsections describe the PROM method in detail.

2.1. *Analysis of Innovations*

The core concept of the analysis of innovations is that the model error covariances can be computed from the innovation covariances:

$$\eta = x^t - x^b, \quad \epsilon = x^t - y^o, \tag{3}$$

$$d = y^o - H\left(x^b\right) = \left(x^t + \epsilon\right) - \left(x^t + \eta\right) = \epsilon - \eta, \tag{4}$$

where η and ϵ are the background and observation errors, respectively. To calculate the covariance of the innovations, some assumptions are made. The first assumption is that the errors are not biased, which simplifies the equation of the covariance for innovations to:

$$f(r) = \overline{d(r_0)d(r)}, \tag{5}$$

where $d(r) = \epsilon(r) - \eta(r)$ is a random variable that represents the innovations at r, $f(r)$ is the spatial covariance of innovations and r is the 2- or 3-dimensional relative position vector from a grid point assumed to be located at r_0. The over-line means ensemble average but, in practice, the ergodic hypothesis is invoked allowing to replace the ensemble average with a time average over a period when the process can be considered statistically in a steady state [17].

By substituting the definition of d into Equation (5) we obtain:

$$\begin{aligned} f(r) &= \overline{(\epsilon(r_0) - \eta(r_0))(\epsilon(r) - \eta(r))} \\ &= \overline{\epsilon(r_0)\epsilon(r)} - \overline{\eta(r_0)\epsilon(r)} - \overline{\epsilon(r_0)\eta(r)} + \overline{\eta(r_0)\eta(r)}. \end{aligned} \tag{6}$$

Assuming that observations and forecast error are uncorrelated, and that for nonzero separation the observation errors are also uncorrelated, the innovation covariance can be written as follows:

$$f(r) = \begin{cases} \overline{\epsilon(r_0)\epsilon(r_0)} + \overline{\eta(r_0)\eta(r_0)}, & \text{for } r = r_0 \\ \overline{\eta(r_0)\eta(r)}, & \text{for } r \neq r_0. \end{cases} \tag{7}$$

As the number of observations increase the innovations statistics converge in Equation (5) to a function for forecast error covariance for $r \neq r_0$. At $r = r_0$, $f(r)$ is not continuous, because error variances include both forecast and observation errors [10,13] and for this reason, this point is treated differently in all methods based on innovations.

2.2. Fitting the Covariance Model to Innovations

The main idea of the PROM method is that the binning step used in H-L method is unnecessary, except for a "central bin" located at grid point $\boldsymbol{r}_0$. This is to cover the situation in which the actual observations near $\boldsymbol{r}_0$ are located at slightly different locations, but the covariance function will be computed at $\boldsymbol{r}_0$. For illustration, let us consider the 2D case shown in the sketch in Figure 1 on page 4, where the isotropy assumption is not used. The red/blue dots, located at $\Delta\boldsymbol{r} = \boldsymbol{r}_j - \boldsymbol{r}_i$, are all the products of pairs of innovations $d(\boldsymbol{r}_i)d(\boldsymbol{r}_j)$ for which $\boldsymbol{r}_i$ is inside the central bin of the target point $\boldsymbol{r}_0$, defined as all the points at a small distance L_c from $\boldsymbol{r}_0$. Interestingly, if a surface (e.g., a 2D Gaussian) is fitted to these products, it will be a good approximation to the true covariance function. In what follows we will prove why fitting to the products produces a good estimate of the covariance and we will show how to compute this fit efficiently.

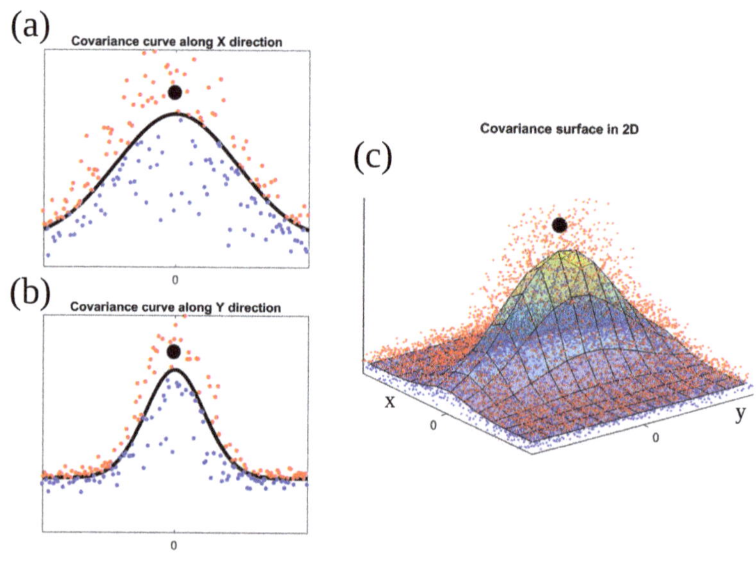

Figure 1. Sketch of the PROM method for a 2D case (i.e., anisotropic). The dots are products of pairs innovations (where one of the elements of the pair being close to the point where the covariance is computed) located at position $\Delta\boldsymbol{r}$ computed as the vector difference between the locations of the "far" innovation and the one in the central bin. The model covariance (an anisotropic 2D Gaussian surface in this case) is fitted to all innovation products. For clarity we have plotted in red colour the products above the fitted Gaussian surface and in blue those below, representing the fact that the fitted covariance function tends to be somewhere in the middle. The plots in panels (**a**,**b**) represent profiles of the covariance functions along the X and Y axis to show the anisotropy of the function and data while the plot in panel (**c**) shows the full covariance function fitted to all the data. The black dot represents the variance of the innovations at the centre $\boldsymbol{r}_0$.

From a mathematical point of view, the PROM method defines a subspace of functions (a parametric model of the covariance) and then finds the function with minimum distance to the data *projected* into that subspace. The method also assumes that the background model and observations are unbiased and the local homogeneity of the covariance, as it is done for the H-L method. The subspace can be defined by any basis of functions, including anisotropic ones.

Let us assume that, for certain physical variable, we have a set of N innovations d_i, $i = 1, \ldots N$ computed as differences of observations performed at locations $\boldsymbol{r}_i$ and times t_i and a background model interpolated at $(\boldsymbol{r}_i, t_i)$. All innovations are contained in the spatial domain and time range of interest. The times are assumed to be multiple of some interval: $t_i = k\Delta t$ (for example, Δt is equal to one day) for $k = 1 \ldots T$. To simplify the following derivation, we will define $f_{\boldsymbol{r}}$ to be the covariance function centred at $\boldsymbol{r}$ as (see Equation (7)):

$$f_{\boldsymbol{r}}(\Delta\boldsymbol{r}) = \overline{d(\boldsymbol{r}, t)d(\boldsymbol{r} + \Delta\boldsymbol{r}, t)}, \quad \text{for } \boldsymbol{r} \neq \boldsymbol{0}, \tag{8}$$

where $\boldsymbol{r} \in \Omega$ is any location in the domain, Ω, $\Delta\boldsymbol{r}$ is a relative location to $\boldsymbol{r}$, t is time and the over-line stands for the time average, which by invoking the ergodic hypothesis we use in place of the ensemble average. In practice, $f_{\boldsymbol{r}}$ is never computed exactly, instead it is approximated by a simple model $\tilde{f}_{\boldsymbol{r}}(\Delta\boldsymbol{r}, a_0, a_1, \ldots)$, where a_i are parameters that, in general, depend on $\boldsymbol{r}$.

In order to obtain the values of the model parameters a_i that better approximate $\tilde{f}_{\boldsymbol{r}}$ to $f_{\boldsymbol{r}}$, the first step is to define what we mean by "approximate", that is, a measure of how close a function is to another. For this we will define a $||\bullet||$ that allows to compute a distance between two functions $||f_{\boldsymbol{r},t} - \tilde{f}_{\boldsymbol{r},t}||$.The definition of this norm is the most important aspect of the method because it determines how faithful the model will be to the true covariance, and how difficult the computation of the model will be. The next subsection will be devoted to defining this norm. Once this is done, the fitting parameters a_i will be computed using a norm minimisation approach similar to the least squares method.

2.3. Distance between Functions

The only points were we have information about innovations are the spatial locations where the observations were made. For this reason we chose the following norm defined for functions $f_{\boldsymbol{r}} \in V$, where V is the space of all real functions on the domain Ω with origin at $\boldsymbol{r}$:

$$||f_{\boldsymbol{r}}|| \equiv \sqrt{\sum_{i=1}^{N}[f_{\boldsymbol{r}}(\Delta\boldsymbol{r}_i)]^2}, \tag{9}$$

where $\Delta\boldsymbol{r}_i = \boldsymbol{r}_i - \boldsymbol{r}$. This discrete norm takes into account the values of the function at all locations where the observations were made. It can be thought as a "sampling" norm, where the values of the function are only picked at certain locations. This type of norm is a good choice if the locations $\boldsymbol{r}_i$ are scattered more or less uniformly in the domain, which is true in most practical cases.

This norm can be derived from an inner product as the canonical norm for said inner product: $||f_{\boldsymbol{r}}|| = \sqrt{\langle f_{\boldsymbol{r}}, f_{\boldsymbol{r}}\rangle}$, where:

$$\langle f, g\rangle = \sum_{i=1}^{N} f(\boldsymbol{r}_i)g(\boldsymbol{r}_i). \tag{10}$$

Assuming that functions are identified if and only if they coincide on all the points $\boldsymbol{r}_i$, it is trivial to check that the definitions above satisfy the conditions for norms and inner products for the space of functions V.

However, there is still one difficulty. The covariance functions $f_{\boldsymbol{r}}$ are not known, not even at points $\boldsymbol{r}_i$, only the innovations $d(\boldsymbol{r}_i, t_i)$ are. This means that norms and inner products including $f_{\boldsymbol{r}}$ cannot be computed exactly, they have to be approximated. This will be described in Section 2.5, meanwhile, we will assume that we can compute any norm or inner product.

2.4. Finding the Projection

We search for a function $\tilde{f}_{\boldsymbol{r}}$ as close as possible to the true $f_{\boldsymbol{r}}$. In other words, we search for:

$$\min_{a_1,a_2,\dots a_m} \left\| \tilde{f}_r(\Delta \boldsymbol{r}, a_1, a_2, \dots a_m) - f_r \right\| \tag{11}$$

Let us assume that all $\tilde{f}_r$ conform a finite dimensional function subspace $V^m \subset V$, so we can write:

$$\tilde{f}_r(\Delta \boldsymbol{r}, a_1, a_2, \dots) = a_1\phi_1(\Delta \boldsymbol{r}) + a_2\phi_2(\Delta \boldsymbol{r}) + \dots + a_m\phi_m(\Delta \boldsymbol{r}), \tag{12}$$

where $\phi_0, \phi_1, \dots$ are a basis of V^m, the m-dimensional space of all the Equations $\tilde{f}_r$. (11) is equivalent to:

$$\min_{a_1,a_2,\dots a_m} \langle \tilde{f}_r(\Delta \boldsymbol{r}, a_1, a_2, \dots a_m) - f_r, \tilde{f}_r(\Delta \boldsymbol{r}, a_1, a_2, \dots a_m) - f_r \rangle, \tag{13}$$

which can be computed by solving the system resulting from differentiating the inner product with respect to each coefficient a_i and then equalling to zero. After simplification this system can be written as:

$$\langle \tilde{f}_r(\Delta \boldsymbol{r}, a_1, a_2, \dots a_m), \phi_j \rangle = \langle f_r, \phi_j \rangle, \quad \forall j = 1, 2, \dots m, \tag{14}$$

which is equivalent to finding the projection of f_r onto V^m.

For simplicity, in what follows we will consider $m = 2$, so there are only two basis functions ϕ_1 and ϕ_2, but the process is exactly the same for the case of more functions. Substituting $\tilde{f}_r$ by $a_1\phi_1(\Delta \boldsymbol{r}) + a_2\phi_2(\Delta \boldsymbol{r})$:

$$\begin{aligned} \langle a_1\phi_1 + a_2\phi_2, \phi_1 \rangle &= \langle f_r, \phi_1 \rangle \\ \langle a_1\phi_1 + a_2\phi_2, \phi_2 \rangle &= \langle f_r, \phi_2 \rangle. \end{aligned} \tag{15}$$

Simplifying by using the linearity of the inner product:

$$\begin{aligned} a_1\langle \phi_1, \phi_1 \rangle + a_2\langle \phi_2, \phi_1 \rangle &= \langle f_r, \phi_1 \rangle \\ a_1\langle \phi_1, \phi_2 \rangle + a_2\langle \phi_2, \phi_2 \rangle &= \langle f_r, \phi_2 \rangle. \end{aligned} \tag{16}$$

Now, under the assumption that all the inner products in the system of equations above can be computed, the system can be solved for the coefficients a_1 and a_2. The resulting function $\tilde{f}_r$ will be the one in V^2 closest to f_r in terms of the discrete norm.

The system of Equation (16) must be solved for all points for which we want to estimate the covariance. The coefficients a_1 and a_2 depend on $\boldsymbol{r}$ because the right-hand side of Equation (16) depends on $\boldsymbol{r}$. The inner products can be computed very quickly and efficiently in modern hardware and the system of two linear equations can be solved using any simple method as it is a very quick operation. Even in more general cases, the number of unknowns a_i is small and the typical system is well-conditioned. Only when the amount of available data for a particular node is very small, there may be cases where the system is ill-conditioned. These cases can be detected (for example, by looking at the number of available data or the condition number of the equation matrix) and removed from the final results.

Once the projection is found, matrix **B** can be parametrised using a_1 and a_2 [18]. Matrix **R** can then be computed by subtracting $\tilde{f}_r(\boldsymbol{0}) = a_1 + a_2$ from the variance at $\boldsymbol{r}$ computed in the central bin (see Equation (7)). This is completely analogous to what is done in the H-L method.

2.5. Approximating the Inner Product

There is still one difficulty—how to compute the inner products $\langle f_r, \phi \rangle$. Let us start by substituting the definition of the inner product and f_r:

$$\langle f_r, \phi \rangle = \sum_{i=1}^{N} f_r(\Delta \boldsymbol{r}_i)\phi(\Delta \boldsymbol{r}_i) = \sum_{i=1}^{N} \overline{d(\boldsymbol{r}, t)d(\boldsymbol{r} + \Delta \boldsymbol{r}, t)}\phi(\Delta \boldsymbol{r}_i). \tag{17}$$

This expression is not computable because the innovations are known only at the points and times of observations, and not at any point $d(\boldsymbol{r}, t)$; however, this expression can be approximated by using the Central Limit Theorem. This theorem states that we can substitute each mean in a sum by one random value following the same distribution and the value of the sum will still converge to the original value. This can be done by assuming that choosing the particular time when the i-th observation was taken is a random choice.

$$\langle f_{\boldsymbol{r}}, \phi \rangle = \sum_{i=1}^{N} d(\boldsymbol{r}, t_i) d(\boldsymbol{r}_i, t_i) \phi(\Delta \boldsymbol{r}_i) + O\left(N^{-1/2}\right). \tag{18}$$

The value of the innovation $d(\boldsymbol{r}, t_i)$ is still not known. By using the hypothesis of local homogeneity we can approximate it as:

$$\langle f_{\boldsymbol{r}}, \phi \rangle = \sum_{i=1}^{N} d_0(t_i) d(\boldsymbol{r}_i, t_i) \phi(\Delta \boldsymbol{r}_i) + O\left(N^{-1/2}\right), \tag{19}$$

$$\text{for } D_0(t_i) = \frac{1}{n_i} \sum_{j=1}^{n_i} d\left(\boldsymbol{r}_{k(i,j)}, t_{k(i,j)}\right),$$

where $\Delta \boldsymbol{r}_i = \boldsymbol{r}_i - \boldsymbol{r}$; $k(i,j)$ the indices of the $j = 1, \ldots, n_i$ innovations inside the central bin at time t_i and $d\left(\boldsymbol{r}_{k(i,j)}, t_{k(i,j)}\right)$ are these innovations; in other words, those such that $\|\boldsymbol{r}_j - \boldsymbol{r}\| < L_c$, for some distance L_c. This distance L_c is usually chosen as half the size of the output grid cell. This approximation of the inner product converges fast when the number of innovations N is large enough. In contrast to the H-L method, which only uses one data point (the covariance) in each bin, the PROM method considers all the observations (and hence innovations) as if they were taken at the same time.

2.6. Data

To test the PROM method and perform comparisons with H-L we have used real innovations computed in an ocean operational model of the North Indian Sea [19]. This area was chosen because its circulation pattern is complex, as it is its bathymetry, containing areas of deep and shallow waters. This model uses NEMO v3.6 stable [20] for the dynamic simulation and NEMOVAR [21] as the DA engine. The computational mesh for NEMO uses hybrid enveloped sigma coordinate system [22] (see Figure 2 on page 9).

As it was done for the operational model, for this paper we have assumed that the covariances have seasonal variability, but that they can be considered statistically uniform during a season for the purpose of computing the error covariances. The innovation data we used were already preprocessed, as part of the operational system, to remove biases from observations and model and passed a quality control that removed the erroneous observations. We chose to use sea surface temperature (SST) because it had the largest amount of data available.

For some tests we used what we call "idealised" or "synthetic" data. The difference between "real" and "idealised" cases are explained in the following sections. These idealised data allow to test the PROM and H-L methods in controlled conditions of data spatial and time density, but knowing exactly what the true covariance function is. The way the synthetic innovation data are generated is described in Section 3.1. The "real" case uses the actual innovations supplied to the operational model for data assimilation.

3. Results

To test the method we have used data as realistic as possible to test the characteristics of our method. Firstly, we introduce an "idealised case" which uses the real location of real innovations for a domain in the North Indian Sea. The "idealised" part comes from the fact that, instead of using actual innovations for which the covariance function is unknown, we use a function with known correlation. This allows to compare the skills of both methods in a well-known scenario.

Secondly, we present a comparison with real temperature innovations. In this case the covariance is not known a priori, but the quality of the results obtained, combined with the knowledge obtained from the idealised case, allows to confirm the benefits of the proposed method.

3.1. Idealised Case

For this case we used the location and time of Sea Surface Temperature (SST) innovation data for one year's summer season (JJA) to mimic a seasonal time-average. Instead of the actual innovation data we associated to these locations synthetic data of known spatial correlation. This part of the data is not a perfect representation of real innovations, but it will be a good test of the skill of the methods. If a method does not perform well in this situation it is not likely to do better in a real situation. Additionally, the performance of the PROM method relative to the density of data can be analysed.

The synthetic innovation data used follow the expression:

$$d_s(\boldsymbol{r},t) = a_t\, e^{\|\boldsymbol{r}-\boldsymbol{r}_0\|^2/2b_0^2} + c\varepsilon_{\boldsymbol{r},t}, \tag{20}$$

where $\boldsymbol{r}_0$ is a test point in the domain, b_0 a length scale (0.88 km in the tests), c the amplitude of the random noise (0.5 in the tests) and a_t and $\varepsilon_{\boldsymbol{r},t}$ are random variables with normal distribution $N(0,1)$. It is assumed that idealised innovations are not correlated in time. For these data it is easy to prove that the covariance function at $\boldsymbol{r}_0$ is exactly the Gaussian curve $\Psi(\boldsymbol{r}) = e^{|\boldsymbol{r}-\boldsymbol{r}_0|^2/2b_0^2}$:

$$\overline{d_s(\boldsymbol{r}_0,t)d_s(\boldsymbol{r},t)} = \overline{a_t^2 e^{|\boldsymbol{r}-\boldsymbol{r}_0|^2/2b_0^2} + a_t\varepsilon_{\boldsymbol{r}_0,t}e^{\|\boldsymbol{r}-\boldsymbol{r}_0\|^2/2b_0^2} + a_t\varepsilon_{\boldsymbol{r},t} + c^2\varepsilon_{\boldsymbol{r},t}\varepsilon_{\boldsymbol{r}_0,t}}. \tag{21}$$

By using the facts that the average is linear, that the random variables are uncorrelated, and that $\overline{a_t^2} = 1$ we get:

$$\overline{d_s(\boldsymbol{r}_0,t)d_s(\boldsymbol{r},t)} = e^{\|\boldsymbol{r}-\boldsymbol{r}_0\|^2/2b_0^2}, \qquad \text{for } \boldsymbol{r} \in \Omega - \{\boldsymbol{r}_0\}. \tag{22}$$

For this exercise we consider only 2D "horizontal" isotropic covariance functions, which can be understood as a function along a constant z, σ, isopycnal or any other convenient computational surface. We tested these innovations at ten locations which cover areas with different characteristics (density of data and presence of land masses), as shown in Figure 2 on page 9. This type of innovation has only Gaussian covariance $\Psi(\boldsymbol{r})$ at $\boldsymbol{r}_0$, so the tests are repeated by making $\boldsymbol{r}_0$ equal to each location in the figure and constructing a different set of innovation data. The locations considered are a subset of nodes of a 2D grid of grid size 0.3° × 0.275° (lon × lat) covering the area between longitudes 45°–74° and latitudes 8°–32°. This grid was chosen for the simplicity of calculations and is different from the NEMO grid from which the innovations were computed. The benefit of using this synthetic innovation data is that the true covariance is perfectly known a priori, so it is possible to provide accurate measurements of the skills of the method. Since actual locations and times or real innovations were used, most of the problems associated to computing the covariance for real data were taken into account.

We used innovation data for three summer months (JJA), replicating an attempt of computing seasonal covariances. In total, we used nearly five million data points scattered over the whole domain, covering most of its surface with variable density of data. This amounts to an average of more than 50,000 data points per day in the domain.

For each location we also tested the sensitivity of having reduced amounts of data by randomly selecting different percentages of the innovation points. Tests were run for 100% (all the innovation points included), 75, 50, 20, 10, 5, 2 and 1%, where each test (or realisation) used different selection of innovation points. Most of the tests where performed 30 times, with different random innovations, but the last three of 5, 2 and 1% were performed 60 times accounting for the larger statistical variability of the results.

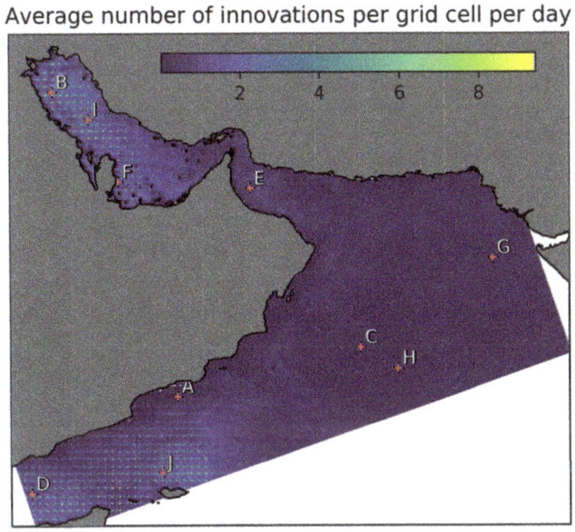

Figure 2. Domain of the operational system, showing the locations where the covariance functions were computed. The background colours show the density of innovations as the mean number of innovations per day on each grid cell.

In Figure 3 on page 10 we have plotted the results of the Gaussian curve $\Psi(r) = a\, e^{-(r-r_0)^2/2(88.8)^2}$ fitted to one realisation of innovation products using 100% of the data points. This is a 1D illustration of the actual 2D fitting process, but valid being a case of isotropic covariance. We have chosen locations B and H ($\boldsymbol{r}_0 = \boldsymbol{r}_B$ and $\boldsymbol{r}_0 = \boldsymbol{r}_H$) as being representative of how the fitting mechanism works. In the plots, the green dots are all pairs of products relevant to each location and the continuous black line is the Gaussian curve fitted to all these products using the PROM method. As it can be seen in Figure 2 on page 9, location B is in a region where there is high density of innovation data, while location H is in a region of relatively scarce amounts of data. This is mirrored in the density of innovation products in each plot of Figure 3 on page 10. It is worth noticing that the fitted line gives good results of the fitting parameter a, even if the green points seem to be scattered all over the place. For location B, where there is a large amount of data, the green dots cover densely a region that seems to be between two vaguely defined Gaussian curves, with the fitted function at an intermediate location. For H, located in an area with more scarce data, both in time and space, there are regions mostly empty of green points, but the fitting function still approximates quite well the covariance. The black dot at distance 0 is the variance of the innovations close to r_0 (i.e., in the central bin), which in this idealised case should tend to $1 + c^2 = 1.25$, where c is the amplitude of the random noise as given by Equation (20). In most cases we tested, the black dots are above the value of the Gaussian curve at r_0. The difference between the computed variance and the value of the fitted curve at r_0 is the variance of the observations, which corresponds to the diagonal element of $\mathbf{R}$.

Figure 4 on page 10 shows the errors of the fitting coefficient a (the difference between it and the true value of 1), for all the realisations for the same points B, H and different amounts of data. The crosses (+) are the average values of the variance and the lines represent the standard deviation based on many realisations of randomly generated innovations. As one would expect, the mean errors are closer to zero when the number of data points is higher, as it is the variation of these errors in different realisations of the experiment. When the number of data points is very low, there are cases where the method is not able to produce a solution for the fitting coefficient and even when it is able, the error can be quite high. However, there is a wide range of density of observations where the PROM method produces consistent results. This suggests that PROM is quite stable to changes in density of innovations.

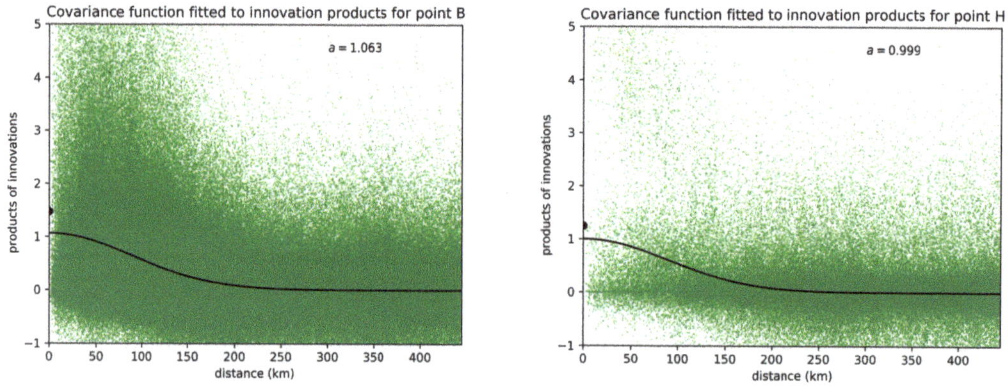

Figure 3. Gaussian function $\Psi(r) = a\, e^{-(r-r_0)^2/2(88.8)^2}$ fitted to all pairs of innovation products around points B and H for one particular realisation and 100% of the data. The green dots are products of innovations plotted against the distance between them and the black line is the Gaussian fitted to all points using the PROM method. Each plot has a label with the fitting coefficient a, the theoretical value for this coefficient is 1. The black dots represent the variances at r_0. The vertical axis limits have been chosen to make the figure more clear, about 3% of the green points lie outside of this range and they are not shown.

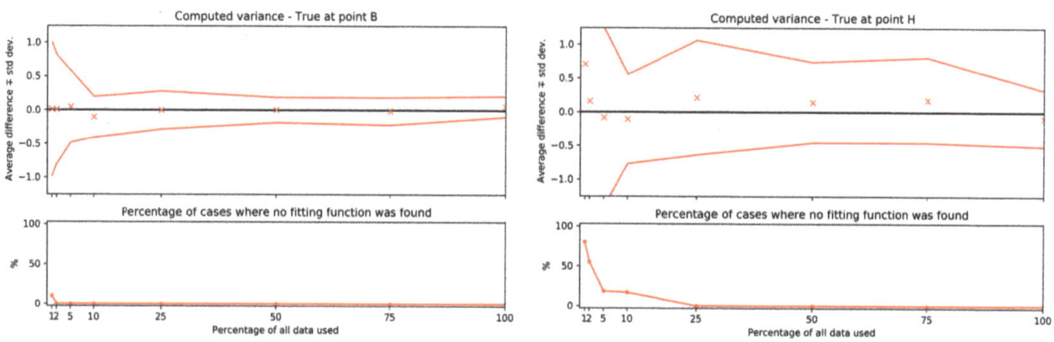

Figure 4. Sensitivity of PROM to the density of data. The top of each plot represents the error between the computed variance and the true value of 1 averaged for many realisations. The lines represent the standard deviation of these errors around the average. The bottom part of each plot shows how many of the realisations PROM method failed to obtain a result.

3.2. Real Case

The next set of experiments were designed to test the skill of the PROM method with real data. This is a more difficult task, since the covariance function is not known and hence, it is impossible to compute exactly the errors associated to each calculation of the covariance curve.

However, it is possible to test the validity of results in a different way that allow us to assess the skill of the method. For calculations we used the same summer season data from previous section, but in this case we used the actual innovations as computed by the NEMO operational model of the North Indian Sea. As pointed out before, we are assuming ergodicity of time averages and a seasonal uniformity of the statistical properties of the temperature field. Following standard practices [12], the covariance function is assumed to be isotropic and that it can be modelled by the sum of two horizontal Gaussian functions accounting for a small scale (of the order of the Rossby Radius) and a large scale (a fixed, predetermined length scale) [18]. In other words, the covariance function can be written as:

$$\tilde{f}_{r_0}(\Delta r) = a_0\, e^{\|\Delta r\|^2/2[L_r(r_0)]^2} + a_1\, e^{\|\Delta r\|^2/2L^2}, \tag{23}$$

where $L_r(r_0)$ is the Rossby Radius at r_0 (capped from below to 25 km), L is the large length scale fixed at 444 km and r_0 is the location of each node in the grid. The free parameters a_0 and a_1 were obtained by the PROM method. This choice of two Gaussian functions of different length scales is adopted from the literature [18] as it has shown good correspondence with real background error covariance functions, including the local effects due to ocean circulation and the regional effects of larger scale.

Figure 5 on page 11 shows the fitting of $\tilde{f}_{r_0}(\Delta r)$ to real innovation products at points B and H. As with the idealised case, there are products of innovation pairs scattered all over the place (green dots), but the PROM method was able to fit the covariance model to these data.

To study the sensitivity of PROM to data density, we also run many tests for different percentages of the total data. For each test, only a percentage of all innovations were randomly selected (100% meant that all available innovations were used). As for the ideal case, we repeated the test many times for each percentage, where each test (or realisation) differs from another only on the percentage of innovations and which innovations were selected. This contrasts with the idealised case, where the innovations also changed their values randomly.

Figure 6 on page 12 shows the coefficients a_0 and a_1 computed using PROM for different percentages of innovation data for the same two locations B and H. PROM produces results of the average variance that are stable in a wide range of data densities. For example, for location B, the estimated variance using all the innovation data was 0.058 °C^2 and for 25% of the data, 0.063 °C^2. For location H, the estimated variance for 100% was 0.028 °C^2 and for 25% of the data, 0.027 °C^2.

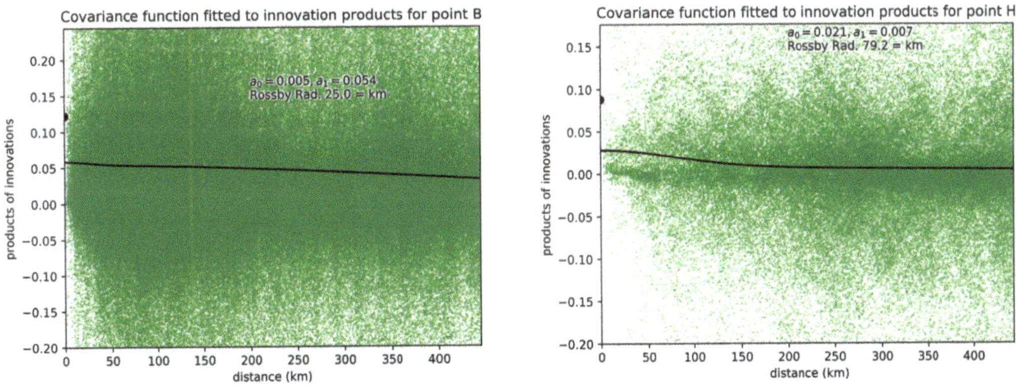

Figure 5. Linear combination of Gaussian functions, $\Psi_2(r) = a_0\, e^{\|r-r_0\|^2/2[L_r(r_0)]^2} + a_1\, e^{\|r-r_0\|^2/2L^2}$ fitted to all pairs of innovation products around points B and H for 100% of the data. The green dots are products of innovations located at the distance between them, and the black line is the function fitted to all points using the PROM method. The black dot represents the variance at r_0. The values of a_0 and a_1 and the Rossby Radius are shown in labels on each plot. The vertical axis limits were chosen to make the figure more clear, about 3% of the green points lie outside of this range and they are not shown.

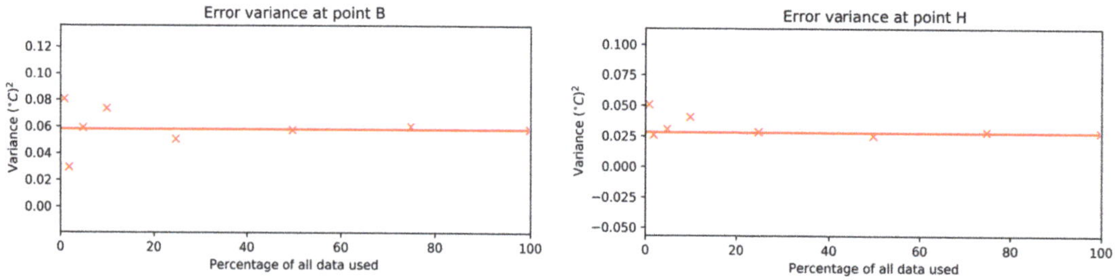

Figure 6. Sensitivity of PROM to the density of data. Each plot represents the computed error variance averaged for many realisations of randomly selected data points. The number of realisations for which PROM method failed to obtain a result are the same as in the idealised case (see Figure 4 on page 10).

Figure 9 on page 17 and Figure 10 show the variances and length scale ratios computed using PROM for 100% of the data. The method was able to produce results everywhere in the domain, even in the regions where the amount of data is scarce. Further analysis is given in the Discussion section.

As the true solution for the covariance function is not known, objective estimations on the quality of the results are obtained by indirect methods. The number of points where the method is not able to compute the covariance function is one of these objective measurements. Another test that can be applied to the grid points where is based on the Cauchy–Schwartz inequality applied to covariances. Let us consider two random variables α, β, the Cauchy–Schwartz inequality states:

$$\mathrm{Covar}(\alpha, \beta)^2 \leq \mathrm{Var}(\alpha)\mathrm{Var}(\beta). \tag{24}$$

This expression is always true, in particular, it can be applied to the innovations at any pair of locations in the domain $\boldsymbol{r}_1, \boldsymbol{r}_2$ as $\mathrm{Covar}(d(\boldsymbol{r}_1), d(\boldsymbol{r}_2))^2 \leq \mathrm{Var}(d(\boldsymbol{r}_1))\mathrm{Var}(d(\boldsymbol{r}_2))$. In reality, we do not know the covariances and variances exactly, only the approximate covariance functions $\tilde{f}_{\boldsymbol{r}_1}, \tilde{f}_{\boldsymbol{r}_2}$. For these functions, the inequality does not hold in general, but we can use it as a test. If $\tilde{f}_{\boldsymbol{r}_1}$ and $\tilde{f}_{\boldsymbol{r}_2}$ do not satisfy the inequality, they cannot be real covariance functions. One must be careful because many methods of calculating error covariances (PROM and H-L included) produce inconsistent covariances in the sense that, in general, for two different points $\boldsymbol{r}_1, \boldsymbol{r}_2$, $\tilde{f}_{\boldsymbol{r}_1}(\boldsymbol{r}_2 - \boldsymbol{r}_1) \neq \tilde{f}_{\boldsymbol{r}_2}(\boldsymbol{r}_1 - \boldsymbol{r}_2)$ (this problem is usually fixed at a later stage, see, for example, in Weaver and Courtier [23] how covariance matrices are made symmetric). To avoid this problem we compute the approximate covariance as the minimum value between the two values provided by $f_{\boldsymbol{r}_1}$ and $f_{\boldsymbol{r}_2}$:

$$\min\left(\tilde{f}_{\boldsymbol{r}_1}(\boldsymbol{r}_2 - \boldsymbol{r}_1), \tilde{f}_{\boldsymbol{r}_2}(\boldsymbol{r}_1 - \boldsymbol{r}_2)\right)^2 \leq \tilde{f}_{\boldsymbol{r}_1}(\boldsymbol{0})\tilde{f}_{\boldsymbol{r}_2}(\boldsymbol{0}) \tag{25}$$

For this test, we compared all nodes with the node to its immediate one the east using Equation (25) and plotted a yellow cross in the location of the nodes that do not satisfy it. These results can be seen in Figure 11 on page 18, where it can be seen that the concentration of "uncertain" grid points (i.e., those which do not satisfy the condition) is higher in regions of lower density of data or near the coast.

4. Discussion

The PROM method shares points in common with the H-L method and for this reason, most of the discussion consists of comparisons between both methods. For H-L we used a standard implementation of the method based on the initial description by Hollingsworth and Lönnberg [13]. Different groups may use different implementations which can differ in a few technical aspects. We have checked that the implementation of the H-L method used for comparison in this section reflects standard practices used by the MetOffice (private communication). Briefly, our implementation of the H-L method consists of the following steps. First, all possible products of pairs of innovations relevant to each model grid node, are calculated and binned by their distances. Second, the products related to a certain bin are averaged (usually using time averaging) to give an approximate value of covariance. Third, a curve is fitted through the covariance points (not including the first bin) to produce a covariance function which then is used to calculate the diagonal element of **R** related to that grid node, and a row of elements for the **B** matrix, both diagonal and non-diagonal. The central bin around the grid point in question in the H-L has the same size as in PROM method.

A bin is considered valid if it contains products of innovations coming from, at least, K different time intervals of length Δt (one day in our case). Otherwise, it is assumed that there are not enough data to compute the covariance in an statistically meaningful way and the bin is disregarded. This is done, instead of just putting a limit on the total amount of products of innovations per bin, to avoid cases where there are relative large amounts of data in a bin, but coming from just a few (or even just one) time intervals. In cases like this, all the innovations in the bin come from few different ocean states and the statistics cannot be trusted. In this section we present results for $K = 5, 10$ and 20. Additionally, as we did in PROM, if the central bin is empty for one time interval, the data for the whole interval are also disregarded. The results obtained by applying the H-L method are then compared with our PROM results.

Usually, the covariance matrices are not used directly as computed by methods like H-L. In practice, a post-processing stage deals with missing values and noise. We will not discuss these methods as they would be similar for any method, and will focus only on the raw results.

4.1. Idealised Case

For the idealised case we only considered the results at locations A–J (see Figure 2 on page 9). For the locations with a small number of available data, H-L often failed to produce results. Sometimes because there were not enough data for the fitting procedure, other times because the results obtained by H-L for the variance exceeded a prescribed threshold $|\text{variance} - 1| > 10$. In contrast, PROM produced results much more often, even in those cases where H-L fails. This is represented in the lower part of each plot in Figure 7 on page 15, where the blue lines correspond to the percentage of cases when the H-L method was not able to find a valid value for the variance and the dashed and the red line does the same for PROM. As it can be seen, the H-L method completely failed to produce results for locations A, C and H where data density is lower, and only produced results for high percentage of the data in the rest of the points. On the other hand, PROM produced results in all points when 100% of the data was included. In most cases, this happened even for percentages below 10%.

Finally, PROM usually produces better results to H-L. Not only in terms of the value of the variance averaged among all realisations, but also showing less variation of those values. Moreover, the method is more stable as the number of available data decreases, showing good skill in the whole range shown.

The results obtained are summarised in Figure 7 on page 15, where for all the locations and selected percentages we show the average error (difference between the computed variance and the true value of 1) and the range of the standard deviation of errors computed from all realisations.

4.2. Real Case

For the real case we produced results in the whole domain, however, firstly, we will consider only the results at points A–J to see how they compare. Additionally, we can use the knowledge obtained from the idealised case to estimate the quality of the results. In this case, the term "realisation" means the subset of innovations which are selected randomly from the full data set. For a particular percentage (for example 50%), different "realisations" contain 50% data of the full data set, but the set of innovations chosen is different in each case. The process of selection of innovations used the same random seed as in the idealised case and, for this reason, the number of cases where both methods failed is identical to the idealised case (see Figure 7 on page 15) being much more likely that H-L fails compared to PROM. Figure 8 on page 17 shows the variances obtained for all the percentages and locations in Figure 2 on page 9 averaged for all realisations. There are many more cases where PROM is able to produce a result compared to H-L. The variance results produced for 100% of the data are "similar" or, at least, not entirely dissimilar, for both methods, being them usually more similar for locations where the density of data is higher. In general, PROM stays close to its horizontal line (red) for a wider range of data amount percentages than H-L does to its line (dash blue). In practice this means that PROM method will produce valid covariance functions much more often than H-L.

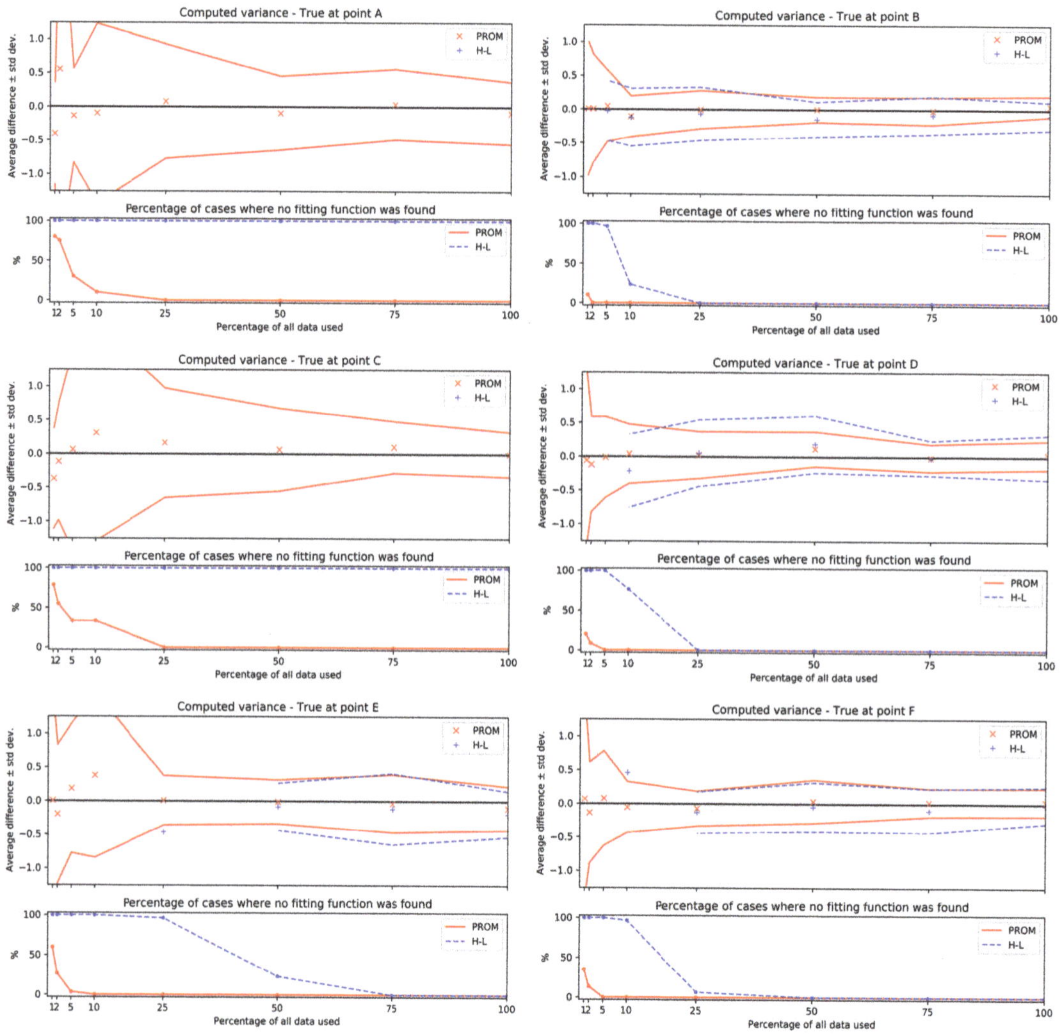

Figure 7. *Cont.*

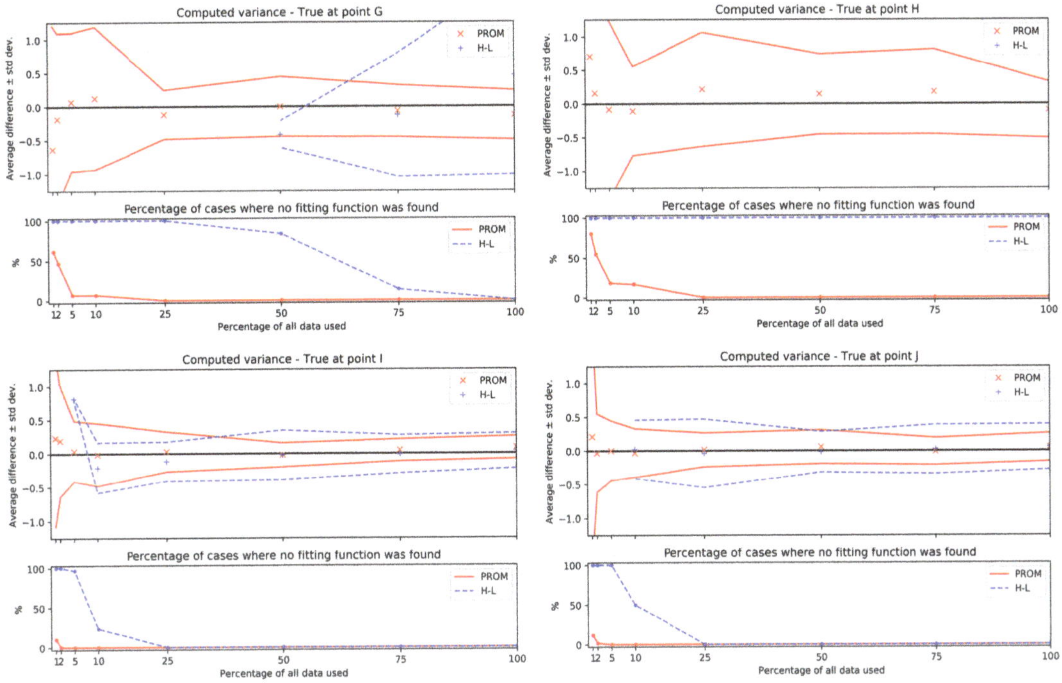

Figure 7. Comparison of variance values obtained by H-L and PROM methods. Each plot represents results obtained from at least 30 numerical experiments realised at each point (see Figure 2 on page 9). The top part of each plot shows comparisons of differences of the computed variance with respect to the true value of 1 (red $\times$ for PROM method, blue $+$ for H-L). The lines represent the standard deviation of computed values for all realisations of the same experiment with different random innovations. The bottom part of each plot shows the percentage of cases where each method failed to produce a value of the variance.

The results for the variance for the entire domain computed using PROM and H-L are shown in Figure 9 on page 17, where we have included results of H-L for different minimum number of days required per bin ($K = 5, 10, 20$).

Even for the rather small value of $K = 5$, H-L was unable to produce results in the some parts of the domain, while PROM was able to find a result of the covariance function for all grid nodes. This limitation of H-L can be mitigated somewhat by increasing the size of the central bin so there are more innovations and hence more product pairs, however, there are two things to consider. First, increasing the size of the central bin reduces the spatial resolution of the method and requires to stretch the condition of local homogeneity to larger scales; second, the results obtained this way should be compared with PROM results obtained with the same enlarged central points. In this case, PROM will also have more data to work with and the covariance results will improve as well.

In this real case, we do not know the true error variances, but by inspecting Figure 9 on page 17 and Figure 10 one can extract some conclusions about the quality of the results obtained. The modelled error variances and length scales obtained from H-L are slightly noisier than those obtained using PROM, even in the regions where data are less scarce. In these areas, the results of H-L and PROM are very similar, which means that when there are plenty of data, both methods converge to the same results. In the regions with smaller data density, H-L often fails to produce a result for the covariance.

The plots in Figure 11 on page 18 show that the number of "uncertain" points (those which do not satisfy Equation (25)) is much smaller for PROM (779) than for H-L (1622). Not surprisingly, the uncertain points are more common near the coast lines and in the regions where the density of innovations is low (see Figure 2 on page 9). Once again, PROM seems to be able to extract more information from the available data.

The PROM method can also be much faster than H-L. This is possible because PROM can be written in terms of convolutions of the basis functions ϕ_i with the covariance function. In the case when ϕ_i are separable (they can be written as products of functions depending only on each spatial coordinate) as Gaussian functions are, the convolution can be computed with reduced computational complexity. This is described in detail in Appendix A where computational complexity estimates are also given.

Our Python implementation of both methods was optimised using the JIT compilation library NUMBA [24] to increase the performance of two inner loop functions in the case of H-L and using the method described in Appendix A for PROM when computing the maps. The computing time for a typical case of 100% data was 27 s for PROM and 300 s for H-L. A very sizeable computational time reduction consistent with the time computational complexities is shown in Appendix A.

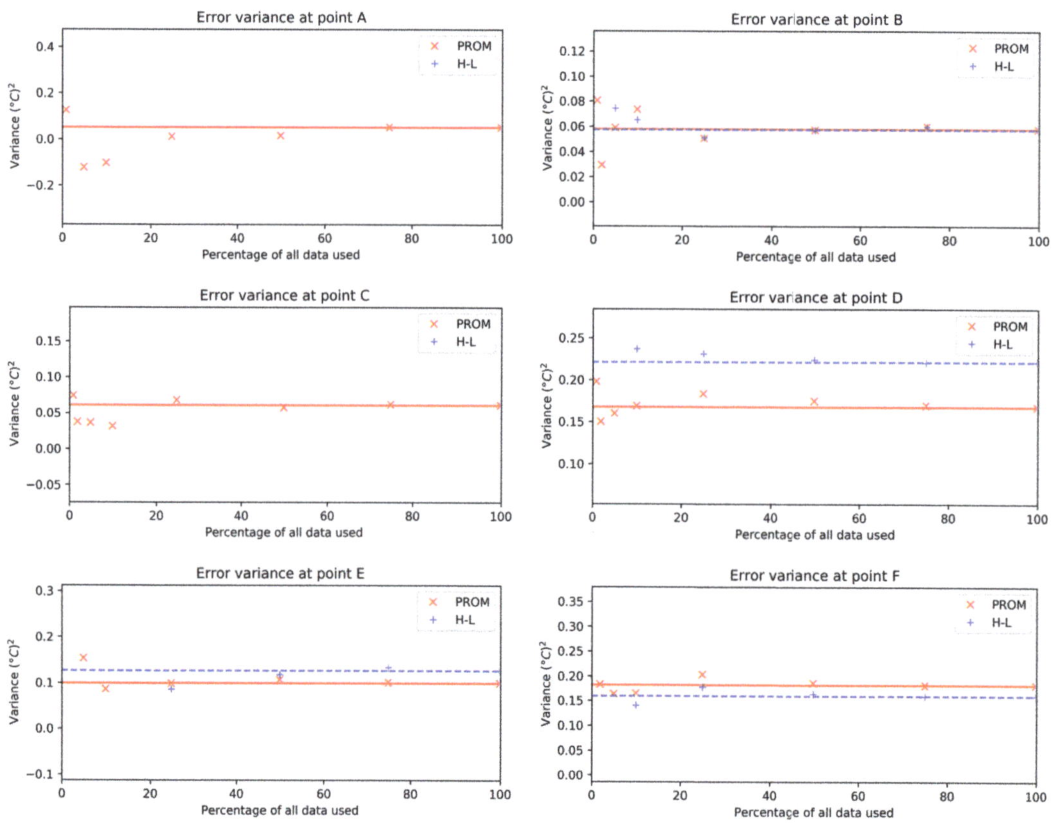

Figure 8. *Cont.*

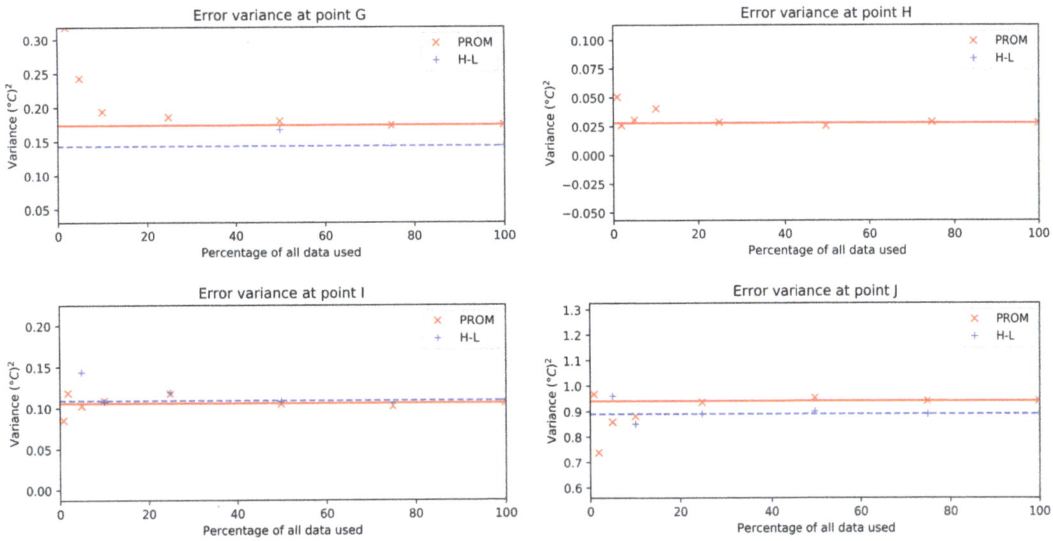

Figure 8. Comparison of H-L and PROM methods. Each plot represents results obtained from many numerical experiments realised at each location (see Figure 2 on page 9) showing comparisons of the computed variance (red × for PROM method, blue + for H-L). The respective horizontal lines are located at the value of the variance obtained for 100% in each method.

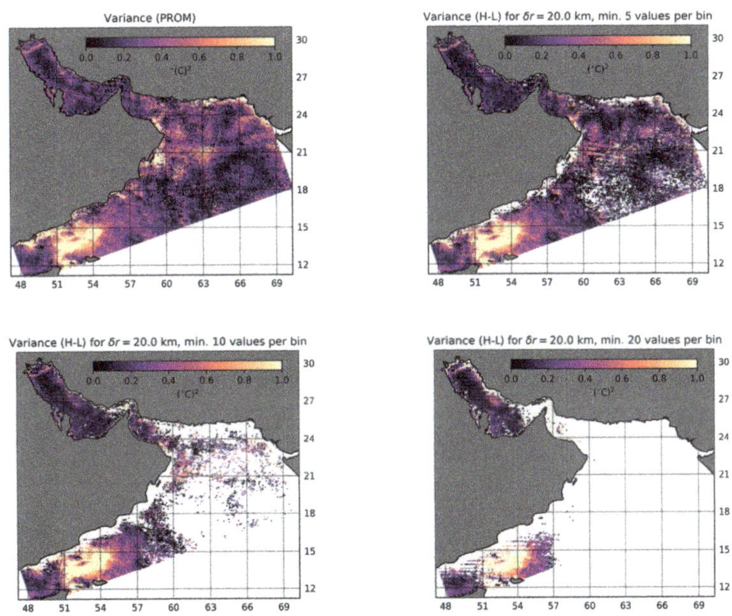

Figure 9. Background model error variances ($a_0 + a_1$) computed using both methods. For H-L we have included results for bin sizes $\delta r = 20$ m and different minimum number of days to consider a bin valid for computing covariances ($K = 5, 10, 20$). H-L method is very sensitive to K, and only produces results in most of the domain nodes for $K = 5$.

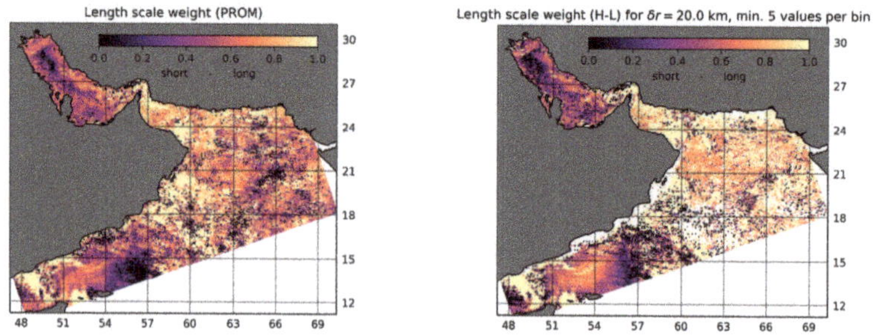

Figure 10. Weights of the length scales of the background model error ($a_1/(a_0+a_1)$) computed using both methods. For H-L we used $\delta r = 20$ m and a minimum of days per bin, $K = 5$.

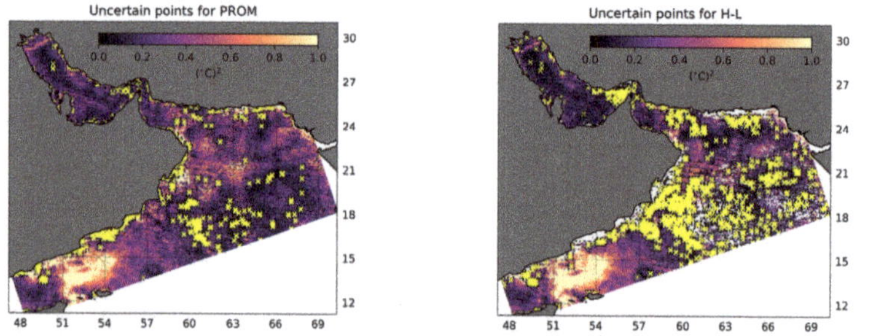

Figure 11. Points that do not satisfy the inequality in Equation (25) for horizontally adjacent valid grid nodes. In the left plot are the 779 uncertain points found for the PROM method, in the right, the 1622 uncertain points found for H-L with a bin size of 20 km and a minimum number of days per bin $K = 5$.

5. Conclusions

We have presented a novel method to compute the error covariance matrices based on analysis of innovations named Projection Method (PROM). Similarly to the widely used H-L method [13], PROM uses statistics of innovations, i.e., differences between model and observations. The main novelty is that PROM does not require spatial binning of the innovations and all the statistics are performed over the whole available data set. This means that no innovation value is discarded which allows to obtain better results by making better use of all the information. It also means that the method is simpler to implement, especially for anisotropic cases.

To test the new method we first applied it to an idealised case, where the location and timings of the innovations are real (coming from an operational model of the North Indian Sea), but the innovation values are substituted by a synthetic field based on random fluctuations following a known covariance function. Then, PROM was applied to real data from the same modelling system.

The idealised case demonstrates that the PROM method performs better than H-L when compared to a known, true covariance function. The results obtained in ten selected geographical locations and for different amounts of input data are used to test the sensitivity of each method to the data available. The PROM method is shown to produce reasonable results even when the H-L method fails due to insufficient amount of data. In areas with large number of observations both methods produce similar results.

In the case of real data for innovations, the true solution for the covariance function is not known, so the capabilities of both methods are assessed indirectly. In regions with low density of observations, the H-L method often fails to produce a valid result, especially for some choices of the bin sizes (δr) and minimum number time intervals per bin (K). On the other hand, PROM always produced a result. The test based on initial description inequality for covariances shows that the H-L method results fails the test in more than twice the number of points of grid nodes as compared with PROM.

Finally, we developed a computational optimisation scheme, which is based on the concept of "separable convolution", that can be applied to our method but not to H-L. This optimisation reduces the algorithmic complexity of the method and is able to reduce computation times greatly. Both methods were implemented in Python, and PROM computational time is seven or more times shorter than H-L using similar levels of code optimisation.

Author Contributions: Conceptualisation, J.M.G.-O., L.S. and G.I.S.; methodology, J.M.G.-O. and L.S.; software, J.M.G.-O. and L.S.; writing, J.M.G.-O., G.I.S. and L.S.; supervision, G.I.S. All authors have read and agreed to the published version of the manuscript.

Funding: This research received no external funding.

Institutional Review Board Statement: Not applicable.

Informed Consent Statement: Not applicable.

Conflicts of Interest: The authors declare no conflict of interest.

Appendix A. Optimisation of the Algorithm

PROM method is conceptually simpler than the H-L method because binning is not necessary nor is there the need to compute each bin's covariances. However, in terms of computational complexity both methods are equivalent when the trivial implementations are used. Let us assume for simplicity that we want to compute the covariance functions for a square mesh of $M \times M$ nodes and that, for each node, there are an average of n innovations inside its central bin, and N other innovations considered to form innovation pairs. With this assumptions the computational complexity will be $O(nNM^2)$ (number of nodes times the number of pairs) for both methods when using the trivial algorithm. Usually, N is a big number (typically, hundreds of thousands) so the time spent in the calculations can grow very large as M increases. Not requiring as many steps as H-L, PROM is simpler and the implementation can potentially be faster than H-L, but not by much.

However, PROM has one advantage that allows to reduce the algorithmic complexity to just $O(MT)$, which can result in huge time savings when computing large covariance matrices. Let us call $\boldsymbol{r}_{pq}$ the location of a grid node of the output grid with indices (p,q). The inner products that need to be computed for each node are (see Equation (19)):

$$\left\langle f_{\boldsymbol{r}_{pq}}, \phi \right\rangle \approx \sum_{i=1}^{N} d_0(t_i) d(\boldsymbol{r}_i, t_i) \phi(\Delta \boldsymbol{r}_i), \tag{A1}$$

where we have dropped the error term $O\left(N^{-1/2}\right)$ and opted for a more concise notation. Let us now suppose that instead of basis functions ϕ we use discretised versions of them ϕ', such that $\phi(\boldsymbol{r}_{pq}) = \phi'(\boldsymbol{r}_{pq})$, and that the values of ϕ' are constant inside each cell of the grid with centres at $\boldsymbol{r}_{pq}$. In other words, ϕ' are discretised, non-continuous, stair-like versions of ϕ functions. In this case, Equation (A1) can be rewritten as:

$$\left\langle f_{\boldsymbol{r}_{pq}}, \phi' \right\rangle \approx \sum_{l=1}^{M} \sum_{m=1}^{M} \sum_{t=1}^{T} d_0(t) D_{lm}(t) \phi(\Delta \boldsymbol{r}_{lm}), \tag{A2}$$

where $D_{lm}(t)$ are the sums of all the innovations inside each grid cell (l,m) at time index t. Equation (A2) is now formally a 2D spatial convolution that can be computed in $\mathcal{O}(MT)$ if ϕ' is separable, which is the case for 2D Gaussian functions and their discretised versions. Pre-computing the required $d_0(t)$ and $D_{lm}(t)$ takes a comparatively small amount of time and we will not take it into account here. In most cases, MT is much smaller than nNM^2 and hence PROM can be much faster than H-L. This was confirmed by the timings of our implementations for PROM and H-L methods.

One could argue that the discretisation of ϕ as ϕ' is another way of performing a spatial binning, and in some respects this is true. There are two main differences, however. First and more importantly, there is no statistical requirement on the minimum number of innovations inside each bin; second, the output grid can be the same mesh we are using in the ocean model, so function ϕ' is consistent with the rest of the variables (e.g., the SST) that are also discretised in a similar way.

At a first glance, the optimised code for PROM seems not well suited for non-homogeneous covariance functions like Equation (23). The reason is that the separable convolution method only works for homogeneous, separable functions. However, this limitation can be easily overcome by performing the inner product for Gaussian functions of several length scales in the range of values that take the Rossby radius in the area, and use these values to interpolate to the desired length scale.

In this case, we have performed convolutions with Gaussian curves of length scale equal to the large length scale (L) and for L_r taking 20 values between 25 and 133 km (the minimum and maximum Rossby radius values considered). Even with the added cost of computing the extra convolutions, PROM is almost an order of magnitude faster than H-L. In this paper, all the results computed for the real data have been computed using the method presented in this appendix after testing that they were almost identical to those obtained using the trivial algorithm.

References

1. Carrassi, A.; Bocquet, M.; Bertino, L.; Evensen, G. Data assimilation in the geosciences: An overview of methods, issues, and perspectives. *Wiley Interdiscip. Rev. Clim. Chang.* **2018**, *9*, e535.
2. Lahoz, W.A.; Schneider, P. Data assimilation: making sense of Earth Observation. *Front. Environ. Sci.* **2014**, *2*, 16.
3. Bannister, R. A review of operational methods of variational and ensemble-variational data assimilation. *Q. J. R. Meteorol. Soc.* **2017**, *143*, 607–633.
4. Gandin, L.S. The problem on optimal interpolation. *Trudy GGO* **1959**, *99*, 67–75.
5. Gandin, L.S. Objective analysis of meteorological fields. Translated from the Russian, Jerusalem program for scientific translations. *Q. J. R. Meteorol. Soc. Jerus. Isr.* **1965**, 286, 242.
6. Lorenc, A.C. Analysis methods for numerical weather prediction. *Q. J. R. Meteorol. Soc.* **1986**, *112*, 1177–1194.
7. Waters, J.; Lea, D.J.; Martin, M.J.; Mirouze, I.; Weaver, A.; While, J. Implementing a variational data assimilation system in an operational 1/4 degree global ocean model. *Q. J. R. Meteorol. Soc.* **2015**, *141*, 333–349.
8. Kalnay, E. *Atmospheric Modeling, Data Assimilation and Predictability*; Cambridge University Press: Cambridge, UK, 2003.
9. Ide, K.; Courtier, P.; Ghil, M.; Lorenc, A.C. Unified notation for data assimilation: Operational, sequential and variational (gtspecial issueltdata assimilation in meteology and oceanography: Theory and practice). *J. Meteorol. Soc. Jpn. Ser. II* **1997**, *75*, 181–189.
10. Bannister, R.N. A review of forecast error covariance statistics in atmospheric variational data assimilation. I: Characteristics and measurements of forecast error covariances. *Q. J. R. Meteorol. Soc. J. Atmos. Sci. Appl. Meteorol. Phys. Oceanogr.* **2008**, *134*, 1951–1970.
11. Haben, S.A.; Lawless, A.S.; Nichols, N.K. Conditioning and preconditioning of the variational data assimilation problem. *Comput. Fluids* **2011**, *46*, 252–256.
12. Bannister, R.N. A review of forecast error covariance statistics in atmospheric variational data assimilation. II: Modelling the forecast error covariance statistics. *Q. J. R. Meteorol. Soc. J. Atmos. Sci. Appl. Meteorol. Phys. Oceanogr.* **2008**, *134*, 1971–1996.
13. Hollingsworth, A.; Lönnberg, P. The statistical structure of short-range forecast errors as determined from radiosonde data. Part I: The wind field. *Tellus A* **1986**, *38*, 111–136.
14. Desroziers, G. Observation error specifications. In *Advanced Data Assimilation for Geosciences: Lecture Notes of the Les Houches School of Physics: Special Issue, June 2012*; OUP: Oxford, UK, 2012; p. 209.
15. Garand, L.; Heilliette, S.; Buehner, M. Interchannel error correlation associated with AIRS radiance observations: Inference and impact in data assimilation. *J. Appl. Meteorol. Climatol.* **2007**, *46*, 714–725.
16. Bormann, N.; Bauer, P. Estimates of spatial and interchannel observation-error characteristics for current sounder radiances for numerical weather prediction. I: Methods and application to ATOVS data. *Q. J. R. Meteorol. Soc.* **2010**, *136*, 1036–1050.
17. Monin, A.S.; Yaglom, A.M. *Statistical Fluid Mechanics, Volume II: Mechanics of Turbulence*; Dover Publications: Mineola, NY, USA, 2013; Volume 2.

18. Weaver, A.; Mirouze, I. On the diffusion equation and its application to isotropic and anisotropic correlation modelling in variational assimilation. *Q. J. R. Meteorol. Soc.* **2013**, *139*, 242–260.
19. Francis, X.; Shapiro, G.; Gonzalez-Ondina, J.M.; Wobus, F.; Maksymczuk, J.; Kenov, I.A.; While, J.; Asif, M. Comparison of skill between high and medium resolution ocean models in the North Indian Ocean. *Geophys. Res. Abstr.* **2019**, *21*. Available online: https://meetingorganizer.copernicus.org/EGU2019/EGU2019-5829.pdf (accessed on 8 December 2021).
20. Madec, G.; Delecluse, P.; Imbard, M.; Levy, C. *OPA 8 Ocean General Circulation Model—Reference Manual*; Technical Report, LODYC/IPSL Note 11; Laboratoire d'Océanographie DYnamique et de Climatologie: Paris, France, 1998.
21. Mogensen, K.; Balmaseda, M.; Weaver, A.; Martin, M.; Vidard, A. NEMOVAR: A variational data assimilation system for the NEMO ocean model. *ECMWF Newsl.* **2009**, *120*, 17–22.
22. Shapiro, G.; Luneva, M.; Pickering, J.; Storkey, D. The effect of various vertical discretization schemes and horizontal diffusion parameterization on the performance of a 3-D ocean model: The Black Sea case study. *Ocean. Sci.* **2013**, *9*, 377.
23. Weaver, A.; Courtier, P. Correlation modelling on the sphere using a generalized diffusion equation. *Q. J. R. Meteorol. Soc.* **2001**, *127*, 1815–1846.
24. Lam, S.K.; Pitrou, A.; Seibert, S. Numba: A llvm-based python jit compiler. In Proceedings of the Second Workshop on the LLVM Compiler Infrastructure in HPC, Austin, TX, USA, 15 November 2015; pp. 1–6.

Journal of
Marine Science and Engineering

MDPI

Article

A Spectral Method for Two-Dimensional Ocean Acoustic Propagation

Xian Ma , Yongxian Wang *, Xiaoqian Zhu *, Wei Liu , Qiang Lan and Wenbin Xiao

College of Meteorology and Oceanography, National University of Defense Technology, Changsha 410073, China; maxianxy@163.com (X.M.); liuwei@nudt.edu.cn (W.L.); lanqiang_nudt@163.com (Q.L.); xiaowenbin@nudt.edu.cn (W.X.)

* Correspondence: yxwang@nudt.edu.cn (Y.W.); zhu_xiaoqian@sina.com (X.Z.); Tel.: +86-1351-740-8284 (Y.W.); +86-1303-678-3184 (X.Z.)

Abstract: The accurate calculation of the sound field is one of the most concerning issues in hydroacoustics. The one-dimensional spectral method has been used to correctly solve simplified underwater acoustic propagation models, but it is difficult to solve actual ocean acoustic fields using this model due to its application conditions and approximation error. Therefore, it is necessary to develop a direct solution method for the two-dimensional Helmholtz equation of ocean acoustic propagation without using simplified models. Here, two commonly used spectral methods, Chebyshev–Galerkin and Chebyshev–collocation, are used to correctly solve the two-dimensional Helmholtz model equation. Since Chebyshev–collocation does not require harsh boundary conditions for the equation, it is then used to solve ocean acoustic propagation. The numerical calculation results are compared with analytical solutions to verify the correctness of the method. Compared with the mature Kraken program, the Chebyshev–collocation method exhibits higher numerical calculation accuracy. Therefore, the Chebyshev–collocation method can be used to directly solve the representative two-dimensional ocean acoustic propagation equation. Because there are no model constraints, the Chebyshev–collocation method has a wide range of applications and provides results with high accuracy, which is of great significance in the calculation of realistic ocean sound fields.

Keywords: ocean acoustic propagation; two-dimensional Helmholtz equation; Chebyshev–Galerkin spectral method; Chebyshev–collocation spectral method

Citation: Ma, X.; Wang, Y.; Zhu, X.; Liu, W.; Lan, Q.; Xiao, W. A Spectral Method for Two-Dimensional Ocean Acoustic Propagation. *J. Mar. Sci. Eng.* **2021**, *9*, 892. https://doi.org/10.3390/jmse9080892

Academic Editor: Mikhail Emelianov

Received: 21 July 2021
Accepted: 17 August 2021
Published: 19 August 2021

1. Introduction

In recent years, underwater acoustic technology has been widely used to detect underwater targets [1], describe underwater overviews [2], measure hydrological information [3], and perform other tasks [4]. Due to the influence of the water surface, water body, water bottom, and other features, sound wave propagation in the ocean presents a complicated distribution state [5]. According to the principle of ocean acoustic propagation, sound waves of a fixed frequency satisfy the elliptic partial differential Helmholtz equation. The sound pressure results throughout an entire space can be obtained after adding boundary conditions to constrain the Helmholtz equation [6]. Recently, Wang et al. [7] used the one-dimensional spectral method to correctly solve the parabolic equation model. Due to the high accuracy and fast convergence speed of the spectral method, it has greatly promoted the development of ocean acoustic calculation. However, there are still some problems in the process of calculation. For example, the beginning of this calculation method solves the simplified parabolic model of the Helmholtz equation. Although simplified models make numerical solutions easier to obtain, their scope of application is very narrow, and approximation errors are introduced. For example, the parabolic equation model replaces the elliptic equation with the parabolic equation. It is thus suitable only for situations in which the physical parameters change slowly with the horizontal distance and the far-field assumption is adopted [8–11]. The contribution of the reverse acoustic wave to the full-field wave is ignored, and only the one-way acoustic propagation problem can be solved [8]. The normal mode model is appropriate only for cases in which the physical parameters are constant, and this model cannot be used to solve problems in which these quantities change with distance, as in actual ocean environments [12]. The ray model can be applied to high-frequency situations, that is, when the wavelength of the sound wave is significantly smaller than any physical scale in the problems [6]. The mirror image method uses the superposition of the real sound source point and the mirrored sound source point to obtain the real sound field, but its requirements for sound speed are more stringent; thus, this method can only be used to determine ocean sound fields with a constant sound speed profile [13]. The wavenumber integration method is suitable only for horizontally layered media and uses numerical integration to determine the sound field [13]. As a result, the calculated sound does not reflect the real marine environment if the sea area does not meet the constraints of the applied simplified model. Most actual marine environments are relatively complex, and relatively few meet the constraints of various models. Therefore, it is necessary to develop a direct solution to the two-dimensional Helmholtz equation of acoustic propagation without using simplified models.

It is usually difficult to directly solve the partial differential Helmholtz equation's analytical solution. The finite difference method, finite element method and spectral method are commonly used discrete numerical methods for solving differential equations [14]. Among them, the finite difference and finite element methods are local approximation methods, and the spectral method is a global approximation method. The finite difference method is the most mature numerical method [15] and uses the difference quotient instead of the derivative. Its theoretical background and calculation are simple, and its format and structure are flexible. In 2007, a nine-point sixth-order accurate compact finite-difference method for solving the Helmholtz equation was developed and analyzed, but it can only solve the equation with the Neumann boundary condition [16]. In 2019, Zhu and Zhao proposed a high-order fast algorithm for solving the two-dimensional Helmholtz equation with Dirichlet and Neumann boundary conditions in polar coordinates [17]. Because the finite difference method has difficulty dealing with the Helmholtz equation with complex boundary conditions, it is commonly used to deal with the simplified partial differential equation of the Helmholtz equation for ocean acoustic propagation. At present, some mature ocean acoustic calculation programs have been developed using the finite difference method, such as RAM [18] based on the parabolic equation model and Kraken [19] based on the normal mode model. However, the finite difference method is not flexible enough to address complex boundary value problems in actual applications, and its approximation accuracy is limited by the format itself. Moreover, it is prone to instability. The finite element method decomposes a continuous area into a finite number of interconnected elements that can be polygons or other geometric shapes. This method easily addresses complex problems, but the accuracy of the approximation is also limited by the format itself [20]. Among the three commonly used numerical methods, the finite difference method and finite element method are both low-precision methods. With continued development of and improvement through scientific research, the study of small-scale and delicate problems has become increasingly important, and high-precision numerical methods have become a research focus of many scholars. The spectral method theoretically has infinite-order approximation accuracy and exponential-order convergence [21], and the spectral method can obtain high-precision results with a small degree of freedom, unlike the results of the finite difference method and the finite element method.

The classical numerical discrete methods are derived from the weighted residual method (MWR). The MWR linearly expands the solution under a set of trial functions, selects a set of test functions and approximate solutions as inner products, and finally forces the residual to zero. The spectral method is different from other numerical calculation methods, mainly in the selection of a class of globally smooth function clusters as trial functions and test functions. Commonly used function clusters include Chebyshev polynomials [22], Legendre polynomials [23], and Fourier functions [24]. The test function clusters are mainly divided into Galerkin, Tau and collocation types. Among them, the Galerkin and Tau types require that the trial function and test function are exactly the same, and the collocation type requires that the boundary conditions are met at the preselected grid points (i.e., collocation points). The name of each spectral method is described as an "A–B" type, where A is the type of trial function, such as Chebyshev, Legendre or Fourier, and B is the classification of the test function, such as Galerkin or collocation. Commonly used spectral methods include the Chebyshev–Galerkin spectral method and Chebyshev–collocation spectral method. In the 1980s, studies by Quartreroni, Canuto, Guo Benyu and others showed that the spectral method has infinite-order convergence, and the spectral method was applied to solve partial differential equations. Since then, the spectral method has been widely used to solve mathematical and physical problems [25–30]. The spectral method also has obvious advantages in solving atmospheric circulation problems [31]. In terms of the numerical eddy flow problem, the spectral method can be used to calculate the fluid flow in a straight channel, and the calculated laminar flow and turbulent flow results are in good agreement with the theoretical results [32]. The step-by-step pseudospectral method is used in the calculation of the water wave motion problem, and the pseudospectral method is established to numerically solve the two-dimensional tidal wave equation calculation model. As a result, the water wave motion caused by the perturbation of the water column of a uniform square pool is simulated and calculated [33]. In recent years, some scholars have begun to apply the spectral method to acoustic problems. Wise et al. [34] used the Fourier–collocation method to re-express the distribution of arbitrary sound sources and sensors, Wang et al. [35] proposed an improved sound propagation operator expansion scheme based on the spectral method, and Lan et al. [36] introduced an algorithm based on the spectral collocation method and were the first to explained the basic governing equations. In the calculation of ocean acoustics, Evans et al. [37] proposed a Legendre–Galerkin spectral method to solve the differential eigenvalue problem of complex and discontinuous coefficients in ocean acoustics.

In recent years, many scholars have studied direct solutions of two-dimensional partial differential equations using the spectral method. In 1995, Shen [38] used the Chebyshev–Galerkin spectral method to directly solve the two-dimensional Helmholtz equation with Dirichlet and Neumann boundary conditions, which achieved good results. In 2009, Wang et al. [39] used the Chebyshev-Tau and Chebyshev–Galerkin spectral methods to solve the two-dimensional Poisson equation. In 2019, Sahuck [40] successfully used the Chebyshev–Galerkin spectral method and quasi-inverse technique to directly solve the two-dimensional Poisson equation and Helmholtz equation with Dirichlet boundary conditions.

In this paper, the Chebyshev–Galerkin and Chebyshev–collocation spectral methods are used to directly solve the two-dimensional Helmholtz equation with Robin boundary conditions, and the results are compared with those of the corresponding analytical solutions. Because the Chebyshev–Galerkin spectral method requires trial functions to meet the boundary conditions, its boundary condition requirements are more stringent. In consideration of the complicated boundary conditions of the actual marine environment, the Chebyshev–collocation spectral method is used to solve the Helmholtz equation of two-dimensional ocean acoustic propagation. Similarly, the correctness of the method is verified by analytical solutions for an example of ocean acoustic propagation. Upon comparing the numerical calculation results of the collocation method with those of Kraken, the results show that the calculation accuracy of the collocation method is higher than that of Kraken. Therefore, it is feasible to use the spectral method to directly solve two-dimensional ocean acoustic propagation problems. The spectral method does not utilize simplified models, so there are no restrictions on the application conditions; thus, no approximation errors are introduced by the model. This method has a wide range of applications, high calculation accuracy, and high practical application value.

This paper is composed as follows. In Section 1, we introduce the current situation and main problems of ocean acoustic field solutions. Then, the Chebyshev–Galerkin and Chebyshev–collocation spectral methods are explained in Section 2 by applying them to solve the Helmholtz equation. In Section 3, we apply our method to the Helmholtz model equation and ocean acoustic program examples and analyze the results of the numerical calculations. Our conclusions are described in Section 4.

2. Solutions of the Two-Dimensional Helmholtz Equation

The basic idea of the spectral method is to approximately expand the solution of the problem on a cluster of smooth basis functions and convert the problem solved in the physical space into the problem of solving the expansion coefficient in the spectral space. The spectral method discretizes the differential equations in the entire solution domain into linear equations, and the core numerical algorithm is to solve the linear equations. In this section, we mainly study how to use the Chebyshev–Galerkin and Chebyshev–collocation spectral methods to directly solve the two-dimensional Helmholtz equation with Robin boundary conditions. Consider the two-dimensional boundary value problem:

$$\begin{gathered}\frac{\partial^2 u}{\partial x^2}+\frac{\partial^2 u}{\partial y^2}+\alpha\frac{\partial u}{\partial x}+\beta\frac{\partial u}{\partial y}+\gamma\, u=f,\ -1\le\ x,y\le 1\\ \left(u+k_1\frac{\partial u}{\partial x}\right)\bigg|_{|x|=1}=\left(u+k_2\frac{\partial u}{\partial y}\right)\bigg|_{|y|=1}=0\end{gathered}\tag{1}$$

where $u(x,y)$ is the quantity to be solved and the other variables are known. When solving the above partial differential equations, first, it is necessary to discretize the equations to form an algebraic equation group that can be solved by mature numerical calculation methods. Here, we use the global approximation method, that is, the spectral method, to discretize the original equations. The continuous and smooth functions $u(x,y)$ and $f(x,y)$ in Equation (1) can be approximated by the linear combination of the complete and linearly independent function cluster $\phi_{k,l}(x,y)$:

$$u(x,y)\approx\sum_{k=0}^{N}\sum_{l=0}^{N}\hat{u}_{k,l}\phi_{k,l}(x,y)\tag{2}$$

$$f(x,y)\approx\sum_{k=0}^{N}\sum_{l=0}^{N}\hat{f}_{k,l}\phi_{k,l}(x,y),\tag{3}$$

where $\hat{u}_{k,l}$ and $\hat{f}_{k,l}$ are coefficients of $u(x,y)$ and $f(x,y)$, respectively, and the function cluster $\phi_{k,l}(x,y)$ is a two-dimensional trial function. $\phi_{k,l}(x,y)$ can be expressed as the product of two one-dimensional trial functions:

$$\phi_{k,l}(x,y)=\phi_k(x)\phi_l(y).\tag{4}$$

In the Chebyshev–Galerkin and Chebyshev–collocation spectral methods, the two one-dimensional trial functions are both Chebyshev polynomials or their linear combinations.

2.1. The Definition and Properties of a One-Dimensional Chebyshev Polynomial

The Chebyshev polynomials $T_k(x)$ defined in the interval $[-1,+1]$ are:

$$T_k(x)=\cos\left(k\cos^{-1}(x)\right)\quad k=0,1,2\ldots\tag{5}$$

Definition 1 (inner product and orthogonality). *The Chebyshev polynomial cluster $\{T_k(x)\}$ satisfies the following (weighted) orthogonal relationship after introducing the following weighted inner product definition:*

$$\langle T_m,T_n\rangle_\omega:=\int_{-1}^{+1}T_m(x)T_n(x)w(x)dx=\begin{cases}0, & m\neq n\\ \pi, & m=n=0\\ \frac{\pi}{2}, & m=n\neq 0\end{cases}\tag{6}$$

where the weight function $w(x)$ is:

$$w(x) = \frac{1}{\sqrt{1-x^2}}. \tag{7}$$

Define the coefficient c_k:

$$c_k = \begin{cases} 2, & k = 0 \\ 1, & k > 0 \end{cases}. \tag{8}$$

Considering the convenience of writing, if the polynomials are, at most, of the N-th power, Equation (6) can be rewritten in matrix form as:

$$\mathbf{K} = diag\{k_{ii}\}, \quad k_{ii} := \frac{\pi}{2} c_k \tag{9}$$

where the matrix **K** *is a diagonal matrix with dimensions of $(N+1) \times (N+1)$ and the diagonal element value is determined by c_k in Equation (8).*

Definition 2 (Chebyshev transform). *The continuous smooth function $v(x)$ defined in the interval [−1,+1] can be approximately expanded with $T_k(x)$ as follows:*

$$v(x) \approx \sum_{k=0}^{N} \hat{v}_k T_k(x) \tag{10}$$

where N is the truncation order of the spectral method and $\hat{v}_k$ is the Chebyshev expansion coefficient of function v. The expansion coefficient is:

$$\begin{aligned} \hat{v}_k &= \frac{1}{d_k} \int_{-1}^{+1} v(x) T_k(x) w(x) dx, \\ d_k &= \int_{-1}^{+1} \frac{T_k^2(x)}{\sqrt{1-x^2}} dx = \begin{cases} \pi, & k = 0 \\ \frac{\pi}{2}, & k > 0 \end{cases} \end{aligned} \tag{11}$$

Equations (10) and (11) are called the inverse and forward Chebyshev transforms [41]. If the effect of the truncation order is not taken into account, a one-to-one correspondence can be established between the original function $v(x)$ and its Chebyshev expansion coefficient v_k using this pair of transformations, and they are located in the original physical space and spectral space.

Definition 3 (numerical integration method to calculate spectral expansion coefficient). *In the forward Chebyshev transform Equation (11), the Gauss–Lobatto integral is usually used to calculate the weighted integral formula:*

$$\hat{v}_k \approx \frac{1}{d_k} \sum_{i=0}^{N} v(x_i) T_k(x_i) w_i \tag{12}$$

where N is the number of integration nodes, which is taken as the truncation order of the spectral method. The integration nodes and weights are defined as:

$$x_0 = 1, x_N = -1, x_i = \cos\left(\frac{\pi i}{N}\right), w_0 = w_N = \frac{\pi}{2N}, w_i = \frac{\pi}{N} \tag{13}$$

Definition 4 (the spectral coefficient of the derivative function). *For any continuous smooth function $v(x)$ whose domain is [−1,+1], if the first derivative $v'(x)$ of $v(x)$ is still a smooth function, $v'(x)$ can also be expanded with $T_k(x)$ as:*

$$v'(x) \approx \sum_{k=0}^{N} \hat{v}'_k T_k(x) \tag{14}$$

It can be proven that the expansion coefficient $\hat{v}'_k$ of the first derivative of $v(x)$ and the expansion coefficient $\hat{v}_k$ of the original function satisfy the following relationship [29]:

$$\hat{v}'_k \approx \frac{2}{c_k} \sum_{\substack{p=k+1 \\ p+k=odd}}^{N} p\hat{v}_p, \quad k \geq 0 \tag{15}$$

where c_k is defined in Equation (8). Equation (15) can then be rewritten in matrix form:

$$\hat{\mathbf{v}}' = \mathbf{D}\hat{\mathbf{v}} \tag{16}$$

where $\hat{\mathbf{v}}'$ and $\hat{\mathbf{v}}$ are $1*(N+1)$ arrays defined as: $\hat{\mathbf{v}}' = [\hat{v}'_0, \hat{v}'_1 \ldots \hat{v}'_N]$ and $\hat{\mathbf{v}} = [\hat{v}_0, \hat{v}_1 \ldots \hat{v}_N]$. The matrix D is an upper triangular square matrix with dimensions of $(N+1) \times (N+1)$. Similarly, the expansion coefficient $\hat{\mathbf{v}}''$ of the second derivative $v''(x)$ can be written as:

$$\hat{\mathbf{v}}'' = \mathbf{D}^2\hat{\mathbf{v}} \tag{17}$$

2.2. The Construction and Properties of One-Dimensional Chebyshev–Galerkin Trial Functions

In the one-dimensional Chebyshev–Galerkin spectral method, the construction of the trial function needs to satisfy not only the orthogonality described in Section 2.1 but also the boundary conditions of the equation to be solved. Therefore, in practice, the trial function $\phi_k(x)$ is usually taken as the linear combination of the Chebyshev polynomial $T_k(x)$ and is called the Chebyshev–Galerkin trial function. Its general structure can be expressed as:

$$\phi_k(x) = T_k(x) + a_k T_{k+1}(x) + b_k T_{k+2}(x), \quad k = 0, 1, 2, \ldots, N-2 \tag{18}$$

where the undetermined combination coefficients a_k and b_k can be determined by specific boundary conditions. In matrix form, Equation (18) can be rewritten as:

$$\begin{bmatrix} \phi_0(x) \\ \phi_1(x) \\ \vdots \\ \phi_{N-2}(x) \end{bmatrix} = \mathbf{S}_x^T \begin{bmatrix} T_0(x) \\ T_1(x) \\ \vdots \\ T_N(x) \end{bmatrix} \tag{19}$$

$$\mathbf{S}_x^T = \begin{bmatrix} 1 & a_k & b_k & & & \\ & 1 & a_k & b_k & & \\ & & \ddots & \ddots & \ddots & \\ & & & 1 & a_k & b_k \end{bmatrix} \tag{20}$$

where the constant matrix $\mathbf{S}_x^T$ is a rectangular matrix with dimensions of $(N-1) \times (N+1)$.

Similar to the previous section, a series of properties of the Chebyshev–Galerkin trial functions can also be obtained. According to the orthogonality satisfied by the Chebyshev polynomials in Equations (6)–(9), the inner product of the Chebyshev–Galerkin trial functions can also be written in matrix form:

$$b_{k,j} = \langle \phi_j, \phi_k \rangle_\omega, \quad \mathbf{B}_x = \mathbf{S}_x^T \mathbf{K} \mathbf{S}_x \tag{21}$$

where $\mathbf{S}_x$ is the transpose of matrix $\mathbf{S}_x^T$. The inner product of the Chebyshev trial function and Galerkin trial function can be written as:

$$h_{k,j} = \langle T_j, \phi_k \rangle_\omega, \quad \mathbf{H}_x = \mathbf{K}\mathbf{S}_x \tag{22}$$

Similarly, the inner product of the Galerkin trial function and its derivative function can be written as:

$$c_{k,j} = \left\langle \phi'_j, \phi_k \right\rangle_\omega, \quad \mathbf{C}_x = \mathbf{S}_x^T \mathbf{K} \mathbf{D} \mathbf{S}_x \tag{23}$$

$$a_{k,j} = \left\langle \phi_j'', \phi_k \right\rangle_\omega, \quad \mathbf{A}_x = \mathbf{S}_x^T \mathbf{K} \mathbf{D}^2 \mathbf{S}_x \tag{24}$$

2.3. *Chebyshev–Galerkin Spectral Method for Solving the Two-Dimensional Helmholtz Equation*

We first determine the two-dimensional Chebyshev–Galerkin trial function using Equations (4) and (18) and then discuss how to discretize Equation (1). According to the principle of the MWR, it is necessary to approximate the value of the function u to be solved using the Chebyshev–Galerkin trial function for a finite N-term expansion and then make the residual of the equation orthogonal to the space of the test function; that is, the weighted inner products of the residual and each test function are zero. According to Equation (1), we can obtain:

$$\begin{aligned} &\left\langle \frac{\partial^2 u}{\partial x^2}, \varphi_{k,l} \right\rangle_\omega + \left\langle \frac{\partial^2 u}{\partial y^2}, \varphi_{k,l} \right\rangle_\omega + \alpha \left\langle \frac{\partial u}{\partial x}, \varphi_{k,l} \right\rangle_\omega + \\ &\beta \left\langle \frac{\partial u}{\partial y}, \varphi_{k,l} \right\rangle_\omega + \gamma \langle u, \varphi_{k,l} \rangle_\omega = \langle f, \varphi_{k,l} \rangle_\omega \end{aligned} \tag{25}$$

where $\varphi_{k,l}$ represents the two-dimensional test function. According to the different methods used in the test function, the spectral method discussed in this paper can be further divided into the Chebyshev–Galerkin spectral method and the Chebyshev–collocation spectral method.

In the Chebyshev–Galerkin spectral method, the test function is taken as the Chebyshev–Galerkin trial function itself. Therefore, each item in Equation (25) can be expressed in a concise matrix form. For instance, the inner product of the second-derivative term is:

$$\begin{aligned} &\left\langle \frac{\partial^2 u}{\partial x^2}, \varphi_{k,l} \right\rangle_\omega = \left\langle \sum_{m=0}^{N} \sum_{n=0}^{N} \hat{u}_{m,n} \, \phi_m''(x) \phi_n(y), \phi_k(x) \phi_l(y) \right\rangle_\omega \\ &= \sum_{m=0}^{N} \sum_{n=0}^{N} \hat{u}_{m,n} \iint_{-1}^{+1} \phi_m''(x) \phi_k(x) w(x) \phi_n(y) \phi_l(y) w(y) dx dy \\ &= \sum_{m=0}^{N} \sum_{n=0}^{N} \hat{u}_{m,n} \left\langle \phi_m''(x), \phi_k(x) \right\rangle_\omega \langle \phi_n(y), \phi_l(y) \rangle_\omega = \mathbf{A}_x \mathbf{U} \mathbf{B}_y^T. \end{aligned} \tag{26}$$

Similarly:

$$\left\langle \frac{\partial^2 u}{\partial y^2}, \varphi_{k,l} \right\rangle_\omega = \mathbf{B}_x \mathbf{U} \mathbf{A}_y^T \tag{27}$$

The inner product of the first-derivative term can be rewritten as:

$$\begin{aligned} &\left\langle \frac{\partial u}{\partial x}, \varphi_{k,l} \right\rangle_\omega = \left\langle \sum_{m=0}^{N} \sum_{n=0}^{N} \hat{u}_{m,n} \, \phi_m'(x) \phi_n(y), \phi_k(x) \phi_l(y) \right\rangle_\omega \\ &= \sum_{m=0}^{N} \sum_{n=0}^{N} \hat{u}_{m,n} \iint_{-1}^{+1} \phi_m'(x) \phi_k(x) w(x) \phi_n(y) \phi_l(y) w(y) dx dy \\ &= \sum_{m=0}^{N} \sum_{n=0}^{N} \hat{u}_{m,n} \left\langle \phi_m'(x), \phi_k(x) \right\rangle_\omega \langle \phi_n(y), \phi_l(y) \rangle_\omega = \mathbf{A}_x \mathbf{U} \mathbf{B}_y^T. \end{aligned} \tag{28}$$

Similarly:

$$\left\langle \frac{\partial u}{\partial y}, \varphi_{k,l} \right\rangle_\omega = \mathbf{B}_x \mathbf{U} \mathbf{C}_y^T \tag{29}$$

The inner product of the original-function term can be rewritten as:

$$\begin{aligned}\langle u,\varphi_{k,l}\rangle_\omega &= \left\langle \sum_{m=0}^{N}\sum_{n=0}^{N}\hat{u}_{m,n}\,\phi_m(x)\phi_n(y),\phi_k(x)\phi_l(y)\right\rangle_\omega \\ &= \sum_{m=0}^{N}\sum_{n=0}^{N}\hat{u}_{m,n}\iint_{-1}^{+1}\phi_m(x)\phi_k(x)w(x)\phi_n(y)\phi_l(y)w(y)dxdy \\ &= \sum_{m=0}^{N}\sum_{n=0}^{N}\hat{u}_{m,n}\,\langle\phi_m(x),\phi_k(x)\rangle_\omega\,\langle\phi_n(y),\phi_l(y)\rangle_\omega = \mathbf{B}_x\mathbf{U}\mathbf{B}_y^T.\end{aligned} \tag{30}$$

The inner product of the right-hand-side known-function term is:

$$\begin{aligned}\langle f,\varphi_{k,l}\rangle_\omega &= \sum_{m=0}^{N+2}\sum_{n=0}^{N+2}\hat{f}_{m,n}\,\langle T_m(x)T_n(x),\phi_k(x)\phi_l(y)\rangle_\omega \\ &= \sum_{m=0}^{N+2}\sum_{n=0}^{N+2}\hat{f}_{m,n}\,\langle T_m(x),\phi_k(x)\rangle_\omega\,\langle T_n(y),\phi_l(y)\rangle_\omega = \mathbf{H}_x^T\hat{\mathbf{f}}\mathbf{H}_y.\end{aligned} \tag{31}$$

where $\hat{\mathbf{f}}$ is the spectral expansion coefficient of the known function f under the Chebyshev trial function in Equation (1) and $\mathbf{U}$ is the spectral expansion coefficient of the unsolved quantity $u(x,y)$ under the Galerkin trial function. Substituting Equations (26)–(31) into Equation (25) gives:

$$\mathbf{A}_x\mathbf{U}\mathbf{B}_y^T + \mathbf{B}_x\mathbf{U}\mathbf{A}_y^T + \alpha\mathbf{C}_x\mathbf{U}\mathbf{B}_y^T + \beta\mathbf{B}_x\mathbf{U}\mathbf{C}_y^T + \gamma\mathbf{B}_x\mathbf{U}\mathbf{B}_y^T = \mathbf{H}_x^T\hat{\mathbf{f}}\mathbf{H}_y \tag{32}$$

To solve $\mathbf{U}$ conveniently, the two-dimensional matrices $\mathbf{U}$ and $\hat{\mathbf{f}}$ are compressed into one-dimensional vectors by columns, which are denoted as $\bar{\mathbf{u}}$ and $\bar{\mathbf{f}}$, respectively. Equation (32) can be rewritten as:

$$(\mathbf{B}_y\otimes\mathbf{A}_x + \mathbf{A}_y\otimes\mathbf{B}_x + \alpha\mathbf{B}_y\otimes\mathbf{C}_x + \beta\mathbf{C}_y\otimes\mathbf{B}_x + \gamma\mathbf{B}_y\otimes\mathbf{B}_x)\bar{\mathbf{u}} = \mathbf{H}_y^T\otimes\mathbf{H}_x^T\bar{\mathbf{f}} \tag{33}$$

where $\otimes$ is the Kronecker product. This is a linear equation system for the unknown vector $\bar{\mathbf{u}}$, which can be solved by using standard numerical methods to obtain the expansion coefficient $\bar{\mathbf{u}}$ of the function $u(x,y)$. Then, we can obtain the value of the function $u(x,y)$ according to Equation (2).

2.4. Chebyshev–Collocation Spectral Method for Solving the Two-Dimensional Helmholtz Equation

In Equation (25), if the test function cluster is selected as the following special function, it is called the Chebyshev–collocation spectral method. Taking the x direction as an example, specific $(N+1)$ points $\{x_j\}_{j=0}^{N}$ in the solution domain are selected as collocation points, and the δ function ($\delta(x)=0,(x\neq 0)$) is used as the weight function:

$$w_j(x) = \delta(x-x_j) \tag{34}$$

According to the principle of the MWR, the residual is required to be zero at the $(N+1)$ collocation points, which means that the differential equation is strictly established at the $(N+1)$ collocation points.

It is still possible to discretize a system of equations in spectral space, as discussed in Section 2.2, and then transform them back to physical space to obtain approximate solutions of the original equations. However, due to the feature of the test function, the Chebyshev–collocation spectral method can directly form a system of equations in the original physical space without having to resort to spectral space. Because of this feature, the collocation-type spectral method can easily address variable coefficients and nonlinear problems. We discuss this method and solution in detail as follows.

The calculation of the collocation-type spectral method is mainly carried out in grid-point space, and the calculation accuracy is related to the selection of the collocation point. For convenience, the Gauss–Lobatto node is selected as the grid collocation point, and the detailed definition is shown in Equation (13). First, we construct the Chebyshev derivative matrix. On the discrete point, x is used to represent the $(N+1)$ dimensional vector composed of discrete point coordinates, and $\mathbf{u}$ is used to represent the $(N+1)$ dimensional vector composed of the function values at these points. Then the $(N+1)\times(N+1)$ Chebyshev derivative matrix $\widetilde{\mathbf{D}}$ satisfies:

$$\mathbf{u}'(x) = \widetilde{\mathbf{D}}\mathbf{u} \tag{35}$$

Similarly, the second-order derivative matrix is $\widetilde{\mathbf{D}}^2$.

The function u in Equation (1) is discretized at the collocation points. Similar to the solution process of the Galerkin method, if the function u is compressed into a one-dimensional vector in columns, then Equation (1) on the left-hand side can be rewritten as:

$$\frac{\partial^2}{\partial x^2} \rightarrow \widetilde{\mathbf{D}}_N^2 \otimes \mathbf{I}_{M+1} \tag{36}$$

$$\frac{\partial^2}{\partial y^2} \rightarrow \mathbf{I}_{N+1} \otimes \widetilde{\mathbf{D}}_M^2 \tag{37}$$

$$\alpha\frac{\partial}{\partial x} \rightarrow \alpha\widetilde{\mathbf{D}}_N \otimes \mathbf{I}_{M+1} \tag{38}$$

$$\beta\frac{\partial}{\partial y} \rightarrow \beta\mathbf{I}_{N+1} \otimes \widetilde{\mathbf{D}}_M \tag{39}$$

$$\gamma \rightarrow \gamma\mathbf{I}_{(N+1)(M+1)} \tag{40}$$

where the number of discrete points in the x direction is $(N+1)$, that in the y direction is $(M+1)$, and $\mathbf{I}_M$ is an M-order identity matrix. The matrix forms of the boundary conditions $\mathbf{R}_x$ and $\mathbf{R}_y$ are as follows:

$$u + k_1\frac{\partial u}{\partial x} \rightarrow \mathbf{R}_x\mathbf{u}_{(N+1)(M+1)} = \mathbf{I}_{(N+1)(M+1)} + k_1\left(\widetilde{\mathbf{D}}_N \otimes \mathbf{I}_{M+1}\right) \tag{41}$$

$$u + k_2\frac{\partial u}{\partial y} \rightarrow \mathbf{R}_y\mathbf{u}_{(N+1)(M+1)} = \mathbf{I}_{(N+1)(M+1)} + k_2\left(\mathbf{I}_{N+1} \otimes \widetilde{\mathbf{D}}_M\right) \tag{42}$$

The rows of the matrices $\mathbf{R}_x$ and $\mathbf{R}_y$ corresponding to the boundaries $|x| = 1$ and $|y| = 1$ are placed into the corresponding positions in matrix $\mathbf{L}$ (Equation (43)).

$$\begin{gathered}\frac{\partial^2}{\partial x^2} + \frac{\partial^2}{\partial y^2} + \alpha\frac{\partial}{\partial x} + \beta\frac{\partial}{\partial y} + \gamma \rightarrow \mathbf{L} \\ \mathbf{L} = \widetilde{\mathbf{D}}_N^2 \otimes \mathbf{I}_{M+1} + \mathbf{I}_{N+1} \otimes \widetilde{\mathbf{D}}_M^2 + \alpha\widetilde{\mathbf{D}}_N \otimes \mathbf{I}_{M+1} + \beta\mathbf{I}_{N+1} \otimes \widetilde{\mathbf{D}}_M + \gamma\mathbf{I}_{(N+1)(M+1)}\end{gathered} \tag{43}$$

The symbol "-" above $\mathbf{L}$ indicates the modified matrix operator. Then, Equation (44) is solved to obtain $\mathbf{u}$.

$$\mathbf{u}_{(N+1)(M+1)\times 1} = \overline{\mathbf{L}}^{-1}\bar{\mathbf{f}} \tag{44}$$

where the vector $\mathbf{f}$ is a discretized description of $f(x,y)$. $\bar{\mathbf{f}}$ is modified as follows: the elements at the boundary $|x| = 1$ and $|y| = 1$ are taken as zero (according to the boundary conditions of Equation (1)).

3. Numerical Test and Results Analysis

In this section, we first apply the Chebyshev–Galerkin and Chebyshev–collocation spectral methods to solve the two-dimensional model Helmholtz equation and then compare the results and calculation times of the two numerical calculations. Since the

Chebyshev–collocation spectral method has a wide range of applications and does not require strict boundary conditions as Galerkin method, it is applied to solve actual ocean acoustic calculation examples, and analytical solutions are used to verify the correctness of the numerical results. Then, we compare the numerical calculation results of the collocation method with those of Kraken to improve the calculation accuracy. The truncation order N in the two directions is not required to be the same; however, to facilitate this test, the N values in the two directions are assumed to be the same in the numerical experiments. In the following, N refers to the truncation order in one direction unless otherwise specified. The numerical calculation program is written in GNU Octave code, and the test platform is a Dell XPS8930 desktop computer equipped with an Intel i7-8700K CPU and 32 GB of memory.

3.1. Two-Dimensional Model Helmholtz Equation

In this section, we take the solution of the following model equation as an example for numerical testing and analysis.

$$\begin{aligned} &\frac{\partial^2 u}{\partial x^2}+\frac{\partial^2 u}{\partial y^2}+\frac{\partial u}{\partial x}+\frac{\partial u}{\partial y}+u=f, \quad (x,y)\in(-1,1)\times(-1,1) \\ &u-\frac{1}{\pi}\frac{\partial u}{\partial x}\bigg|_{|x|=1}=u-\frac{1}{2\pi}\frac{\partial u}{\partial y}\bigg|_{|y|=1}=0 \\ &f(x,y)=\left(-5\pi^2+1\right)\sin\left(\pi x+\frac{\pi}{4}\right)\sin\left(2\pi y+\frac{\pi}{4}\right)+\pi\cos\left(\pi x+\frac{\pi}{4}\right)\sin\left(2\pi y+\frac{\pi}{4}\right) \\ &+2\pi\sin\left(\pi x+\frac{\pi}{4}\right)\cos\left(2\pi y+\frac{\pi}{4}\right) \end{aligned} \tag{45}$$

The exact solution of Equation (45) is $u=\sin\left(\pi x+\frac{\pi}{4}\right)\sin\left(2\pi y+\frac{\pi}{4}\right)$.

Equation (45) is solved by the Chebyshev–Galerkin and Chebyshev–collocation spectral methods, and the approximate solution images obtained are shown in Figure 1. The accuracy is determined by the relative L_2-norm of the error between the numerical solution and the exact solution and the definition of L_2-norm of vector $\mathbf{x}$ are:

$$\begin{aligned} \text{error} &= \frac{\|u-u_{\text{ana}}\|_2}{\|u_{\text{ana}}\|_2} \\ \|\mathbf{x}\|_2 &= \sqrt{|x_1|^2+|x_2|^2+|x_3|^2+\ldots+|x_n|^2} \end{aligned} \tag{46}$$

where $\|\cdot\|_2$ represents the L_2-norm of the error, u represents numerical results, and u_{ana} represents the exact solution.

Figure 1 shows the images of the exact solution and the numerical solutions of the two methods. This function has good smoothness and symmetry, but the image is more complex, and the gradient changes greatly. Table 1 shows that the two spectral methods show a similar order of accuracy until $N=24$. At $N=30$, all the numerical solutions nearly reach the machine accuracy ($\sim 10^{-16}$) under double precision. This shows that these two methods can obtain high-precision approximate solutions when the truncation order N is small, which verifies the high-precision approximation of the spectral method. The approximation accuracy of the two numerical methods increases rapidly with increasing truncation order. For almost every increase in order, the accuracy of the numerical solution increases by an order of magnitude, which verifies the rapid convergence of the spectral method. The Chebyshev–Galerkin method achieves the highest accuracy at $N=30$, when the error magnitude is 10^{-14}, and the Chebyshev–collocation method achieves the highest accuracy at $N=24$, when the error magnitude is 10^{-12}. By comparison, the two spectral methods are both high-precision calculation methods. However, the Chebyshev–collocation spectral method uses fewer orders to achieve the highest accuracy and can effectively reduce the amount of calculation required to solve complex and large-scale problems.

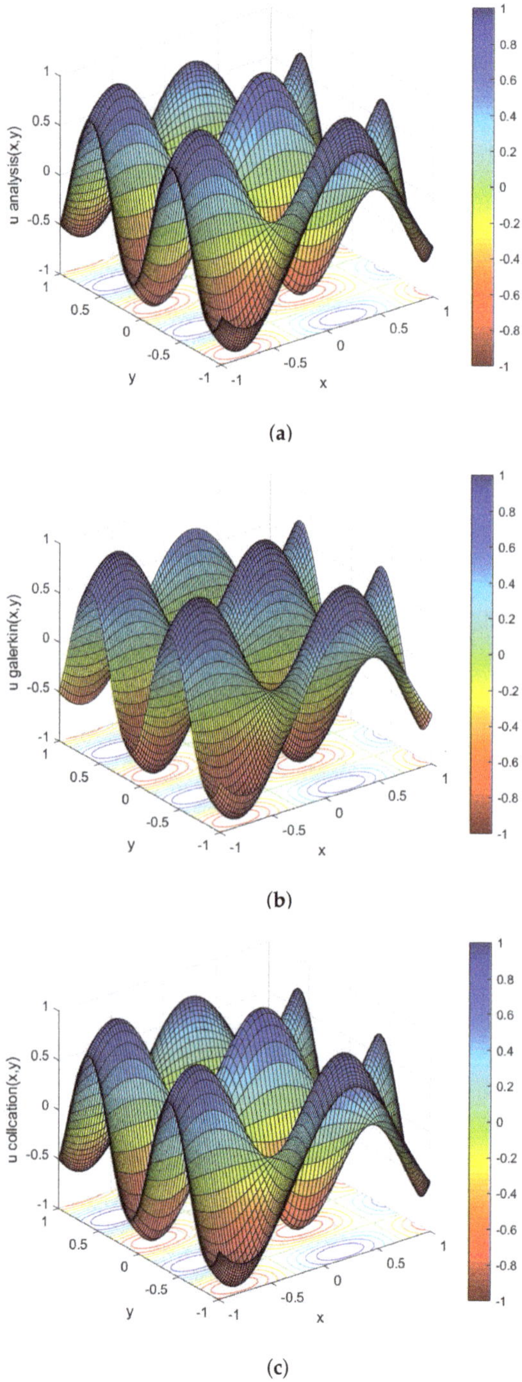

Figure 1. The results of Equation (45) obtained by analytical solution (**a**), Chebyshev–Galerkin spectral method (**b**) and Chebyshev–collocation spectral method (**c**).

Table 1. The error of the results calculated by the Chebyshev–Galerkin and Chebyshev–collocation spectral methods varies with the truncation order N.

N	Galerkin Error	Collocation Error
8	1.19×10^{-5}	5.60×10^{-2}
16	2.73×10^{-6}	2.68×10^{-6}
22	5.62×10^{-11}	6.24×10^{-11}
24	1.04×10^{-12}	2.84×10^{-12}
30	2.51×10^{-14}	8.15×10^{-12}
32	2.57×10^{-14}	5.51×10^{-12}

The calculation times of the Chebyshev–Galerkin and Chebyshev–collocation methods vary with N, as shown in Table 2. The results listed in the table are the average results of ten runs. The running time results of the two spectral methods show that the Chebyshev–collocation method runs faster than the Chebyshev–Galerkin spectral method. Comparing the calculation errors of these two methods, the Chebyshev–collocation method can achieve a higher calculation accuracy in a shorter time and with less truncation order. Therefore, compared to the Chebyshev–Galerkin method, the Chebyshev–collocation method can calculate more quickly and efficiently while ensuring accuracy when solving realistic problems of ocean acoustic propagation. In addition, the Galerkin-type spectral method requires trial functions to meet the boundary conditions, and the corresponding trial functions need to be designed for specific boundary conditions. However, when solving realistic ocean acoustic field problems, the construction of the Galerkin trial function is very difficult if the boundary conditions change with distance. Therefore, the next section introduces the application of the Chebyshev–collocation spectral method in solving realistic ocean acoustic fields.

Table 2. The calculation times of the Chebyshev–Galerkin and Chebyshev–collocation methods vary with N.

N	Galerkin Time (s)	Collocation Time (s)
8	0.007	0.003
16	0.014	0.005
22	0.020	0.008
24	0.023	0.010
30	0.032	0.022
32	0.042	0.025

3.2. Ocean Acoustic Propagation Example Test

The test in Section 3.1 verified that the Chebyshev–collocation spectral method directly solves the two-dimensional Helmholtz model equation with a high accuracy and fast convergence. Therefore, we apply it to solve the two-dimensional ocean acoustic propagation equation. Two ocean acoustic propagation examples, spherical waves and ideal fluid waveguides, with analytical solutions are selected for analysis and comparison to verify the effectiveness of the Chebyshev–collocation method in solving the ocean acoustic field.

When the ocean topography and ocean environment parameters are axisymmetric, the circumferential guide number in the cylindrical coordinate system is zero, that is, $\partial/\partial\theta = 0$. The Helmholtz equation in cylindrical coordinates can be simplified into a two-dimensional axisymmetric form [13]:

$$\frac{\partial}{\partial r}\left(\frac{1}{\rho}\frac{\partial P}{\partial r}\right) + \frac{1}{r\rho}\frac{\partial P}{\partial r} + \frac{\partial}{\partial z}\left(\frac{1}{\rho}\frac{\partial P}{\partial z}\right) + \frac{k^2 P}{\rho} = 0 \tag{47}$$

where P is the sound pressure, ρ is the density, k is the wavenumber, r is the horizontal axis, and z is the vertical axis. The boundary conditions are also required to obtain the specific sound pressure distribution. To simulate the sound field characteristics of a real large-scale sea area, it is often necessary to set the radiation boundary conditions, that is, the conditions that the solution of the wave equation needs to meet at infinity (the sound propagation direction at infinity can only be outward). The Sommerfeld radiation condition expression is:

$$\frac{\partial P}{\partial n} - ikP = 0 \tag{48}$$

This condition indicates that on the virtual boundary far from the sound source point, the normal derivative of sound pressure and the sound pressure itself satisfy Equation (48).

The premise of the radiation boundary conditions is that that they correspond to an infinite distance from the sound source point. However, due to the limitation of the calculation area, an "absorption layer" needs to be set at the boundary to avoid false reflections of the sound field by the radiation boundary conditions. The inner side of the absorption layer is connected to the sound field, and the radiation boundary condition is set on the outer side. The absorption coefficient in the absorption layer increases rapidly with increasing coordinate values, such that the sound pressure can be rapidly reduced within a short distance. The wavenumber expression in the absorption layer [13] is:

$$\begin{gathered} K^2 = k^2\left[1 + i\left(\gamma_r + \gamma_z + \frac{\beta}{27.287527}\right)\right] \\ \gamma_r = \begin{cases} 0, & r \le L \\ \mu\left(\frac{r-L}{\Delta L}\right)^2, & r > L \end{cases} \quad \gamma_z = \begin{cases} 0, & z \le H \\ \mu\left(\frac{z-H}{\Delta H}\right)^2, & z > H \end{cases} \end{gathered} \tag{49}$$

where the coefficient μ is usually zero, γ_r represents the change coefficient of the absorption layer in the horizontal direction, γ_z represents the change coefficient of the absorption layer in the vertical direction, L represents the horizontal coordinate of the inner side of the absorption layer in the horizontal direction, and H represents the vertical coordinate of the inner side of the absorption layer in the vertical direction. To simulate real ocean conditions, the absorption coefficient β is zero.

To show the calculation results of the sound field, the transmission loss of the sound pressure is defined as:

$$\mathrm{TL} = -20\log_{10}\left(\frac{|P|}{P_0}\right) \tag{50}$$

where P_0 refers to the sound pressure 1 m from the sound source. In displays of sound field results, TL is commonly used, and the unit of TL is decibels(dB).

3.2.1. Spherical Wave

The density of the ocean is uniform $\rho(z) = 1\ \mathrm{g/cm^3}$, the speed of underwater sound propagation is constant $c = 1500\ \mathrm{m/s}$, the sound source is set at the coordinate origin (0,0), and the frequency is 150 Hz. To avoid singularities in the calculation area, the horizontal solution area is (1, 100), and the vertical solution area is (−50, 50). The analytical solution of the spherical wave is:

$$P = \frac{e^{ikR}}{R}, \quad R = \sqrt{x^2 + y^2} \tag{51}$$

The upper boundary, lower boundary and right boundary of the solved rectangular area are all assigned radiation conditions, and the thicknesses of the three boundary absorption layers are all 20 m. The left boundary is described by the analytical solution formula.

The truncation order $N = 160$ is taken as an example to focus on the nonabsorbent layer area, that is, (1, 80) in the horizontal direction and (−30, 30) in the vertical direction. The corresponding calculation results are shown in Figure 2.

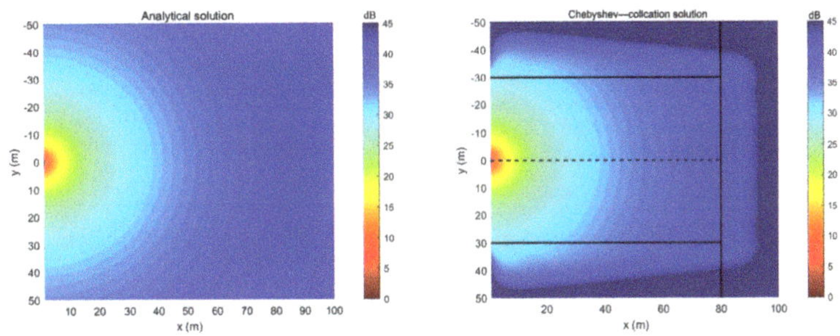

(**a**) (**b**)

Figure 2. TL field of a spherical wave: the result of the analytical solution (**a**); the result of the collocation method (**b**).

Figure 2 shows that the sound field value calculated by the Chebyshev–collocation spectral method in the nonabsorbent layer area is very similar to the results calculated by the analytical solution. However, due to the influence of the radiation conditions, strong reflections are generated near the boundary, resulting in a certain jitter in the sound field value near the inner boundary. Upon comparing the TLs at the sound source depth computed by the analytical solution (red solid line) with the results by the collocation method (black dashed line), as shown in Figure 3a, the TLs are basically the same. Figure 3b shows the error between the numerical solution and the analytical solution. Figure 3 shows that there is a slight difference in the calculation results of the collocation method in the area beyond 60 m due to the reflection of the radiation boundary.

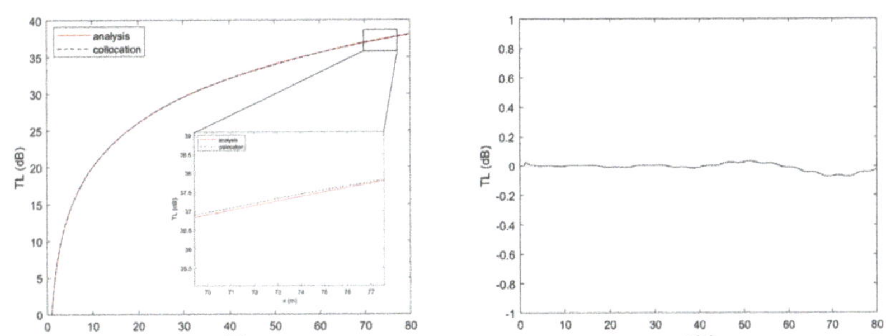

(**a**) (**b**)

Figure 3. The TL curve calculated by the analytical solution (red solid line) and Chebyshev–collocation spectral method (black dashed line) at the depth of the sound source (**a**); the error between the numerical solution and the analytical solution (**b**).

We use the analytical solution of spherical waves to analyze the accuracy of the numerical calculation results. The analysis is focused on the result in the nonabsorbent layer, so the absorption layer is ignored. The error between the numerically calculated sound field and the analytical solution is measured by the relative L_2-norm defined by Equation (46). Table 3 shows the error in the entire solution domain (except the absorption layer), and the depth of the sound source varies with the number of discrete points N. Table 3 also shows that with the increase in N, the calculation error of the Chebyshev–collocation spectral method gradually decreases, and the numerical calculation accuracy gradually improves.

Table 3. The numerical calculation error of the entire solution domain and the depth of the sound source vary with N.

N	Error in the Entire Domain	Error at the Depth of the Sound Source
60	$1.32700463 \times 10^{-1}$	$1.34443432 \times 10^{-1}$
80	$4.73780863 \times 10^{-3}$	$4.87139228 \times 10^{-3}$
100	$2.09898576 \times 10^{-3}$	$2.08030059 \times 10^{-3}$
120	$1.64301023 \times 10^{-3}$	$1.17537743 \times 10^{-3}$
140	$1.54173280 \times 10^{-3}$	$9.60149598 \times 10^{-4}$
160	$1.48522444 \times 10^{-3}$	$9.31331082 \times 10^{-4}$

We analyze the execution time of the spectral method as the scales of problem change. We guarantee that the calculation error in the whole solution domain is on the order of 10^{-3}. Then, the speed of sound and the size of the solution domain are fixed, the frequency of the sound source is changed, and the variation in the execution time with the wavenumber changing in the solution domain is analyzed. Under the premise of ensuring accuracy, Table 4 shows the change in the truncation order N and the execution time of the program with the frequency of the sound source. According to Table 4, as the frequency increases, the wavelength decreases, and the wavenumber contained in the solution domain increases. To ensure the calculation accuracy, the value of N must be increased, which also makes the scale of the discrete matrix $\mathbf{L}$ of Equation (43) further increase. Similarly, the scale of solving linear equations of Equation (44) increases, which leads to an increase in the execution time.

Table 4. The truncation order N and the execution time vary with the frequency of the sound source.

Frequency (Hz)	N	Execution Time (s)
150	100	14.1058
300	110	27.5516
400	130	77.5650
450	137	105.7993
500	145	145.8439

3.2.2. Ideal Fluid Waveguide

The density of the ocean is uniform at $\rho(z) = 1\ \mathrm{g/cm^3}$, the speed of underwater sound propagation is constant at $c = 1500\ \mathrm{m/s}$, the sound source is at a depth of 36 m, and the frequency is 20 Hz. The upper and lower boundaries of the solved rectangular area are the pressure release boundary conditions, the left boundary is described by analytical solutions, and the right boundary is described by the radiation boundary condition. The thickness of the absorption layer is 200 m. Therefore, we mainly analyze the calculation area within 1600 m in the horizontal direction. The truncation order $N = 120$ is considered, and the result is shown in Figure 4.

Figure 4 shows that the sound field solution of the nonabsorbent layer calculated by the collocation method is very similar to the analytical solution but that there is a small difference in the position of the sound shadow area. Figure 5 shows the TL curves of the two calculation methods at a sound source depth of 36 m. The calculation results of the collocation method and the analytical solution are very similar, but there is a certain error in the sound shadow area. The result in the sound shadow area is relatively complicated, and the calculation is difficult, which leads to errors in the calculation of the collocation method. The analytical solution of the ideal fluid wave is used to analyze the accuracy of the numerical calculation in the nonabsorbent layer region, and the results are shown in Table 5. The results in Table 5 indicate that when the value of N increases to 80, the accuracy of the numerical calculation can reach 10^{-4}. This example is also solved by the classic Kraken program based on the normal mode method. As shown in Table 5, even if the number of discrete points is increased, the magnitude of the error is still 10^{-1}, a very important reason is that Kraken program uses the normal mode approximation does not accurately reproduce the field near the sound source. Table 6 shows the errors of the two numerical calculation methods at the sound source depth of 36 m. The calculation accuracy of the collocation method is lower at $N = 70$, and the results of Kraken reach a higher accuracy of approximately 10^{-3}. The calculation accuracy of the collocation method is rapidly improved with increasing N. At $N = 120$, the accuracy of the collocation method exceeds that of Kraken by approximately 10^{-4}. Comparative analysis indicates that the accuracy of the Chebyshev–collocation method is higher than that of Kraken, which verifies the high-precision characteristics of the spectral method. For the above two calculation examples with complex boundary conditions, the Chebyshev–collocation spectral method can achieve a higher calculation accuracy when the truncation order is small. For ocean acoustic propagation examples with complex boundaries and propagation characteristics that are difficult to solve by other methods, the Chebyshev–collocation spectral method can be solved efficiently and accurately.

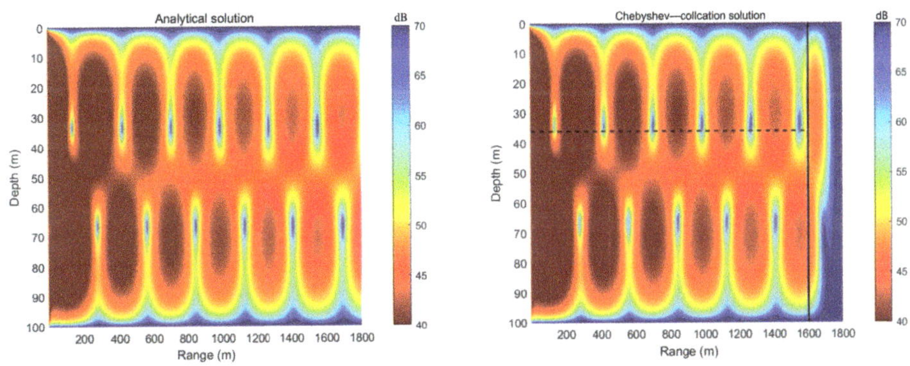

(a) (b)

Figure 4. TL field of an ideal fluid wave: the result of the analytical solution (**a**), the result of the collocation method (**b**).

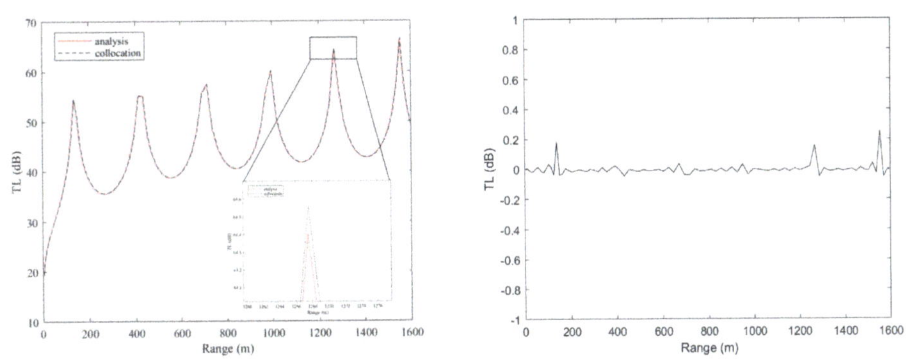

(a) (b)

Figure 5. The TL curve calculated by the analytical solution (red solid line) and the Chebyshev–collocation method (black dashed line) at the sound source depth (**a**); the error between the numerical solution and the analytical solution (**b**).

Table 5. The numerical calculation error in the entire solution domain varies with N.

N	Chebyshev–Collocation	Kraken
70	$2.06853038 \times 10^{-1}$	$2.41551098 \times 10^{-1}$
80	$5.67631508 \times 10^{-4}$	$2.41551093 \times 10^{-1}$
100	$7.09835092 \times 10^{-4}$	$2.41551112 \times 10^{-1}$
120	$4.75639831 \times 10^{-4}$	$2.41551129 \times 10^{-1}$

Table 6. The numerical calculation error at the sound source depth varies with N.

N	Chebyshev–Collocation	Kraken
70	$3.32036066 \times 10^{-1}$	$2.06786673 \times 10^{-3}$
80	$1.52415237 \times 10^{-3}$	$2.06475574 \times 10^{-3}$
100	$1.73634299 \times 10^{-3}$	$2.06635146 \times 10^{-3}$
120	$9.33660302 \times 10^{-4}$	$2.06476704 \times 10^{-3}$

4. Conclusions and Outlook

This paper uses the Chebyshev–Galerkin and Chebyshev–collocation spectral methods to directly solve two-dimensional Helmholtz model equations and uses an analytical solution to verify the correctness of the two methods. The Chebyshev–Galerkin spectral method first constructs a trial function that satisfies the boundary conditions, then interpolates at the Gauss–Lobatto node, forms a system of linear equations in the spectral space, and finally obtains the original function. However, the two-dimensional ocean acoustic propagation Helmholtz equation is relatively complicated due to the variable boundary conditions, and the trial function is difficult to construct. Therefore, realistic ocean acoustic examples can be solved by the more widely applicable Chebyshev–collocation spectral method. The Chebyshev–collocation spectral method approximates the function value at the Gauss–Lobatto node, interpolates physical quantities such as the boundary conditions and sound velocity to the Gauss–Lobatto node, and then directly solves the system of linear equations in the physical space to obtain the original function. The analytical solution is used to verify the correctness of the Chebyshev–collocation method, and compared with the classic Kraken program, the Chebyshev–collocation method has a higher calculation accuracy. The Chebyshev–collocation method can be used to directly solve the two-dimensional underwater acoustic propagation Helmholtz equation without being restricted and constrained by other simplified model application conditions and will not introduce model errors. This spectral method can be used to solve the situation where the sound velocity in the domain changes with the horizontal and vertical directions, and it is applicable to Dirichlet Neumann and Robin boundary conditions. The nonflat seabed and the multilayered medium with strong reflection of bottom sediments are still under further exploration. The Chebyshev–collocation method could solve the two-dimensional ocean acoustic Helmholtz equation and obtain high-precision results. It can effectively solve the ocean sound field when the marine environment does not satisfy the simplified model. Moreover, high-precision sound field calculation has a wide range of applications for human life, such as underwater topographic surveys, marine hydrological information measurements and animal protection.

This paper mainly introduces the use of the Chebyshev–collocation spectral method to solve relatively simple sound propagation problems in ocean media. The calculation of more complicated sound propagation examples such as non-flat seabed and the multilayered medium in the actual ocean and the use of parallel methods to increase the calculation speed are the next key research issues.

Author Contributions: Conceptualization, Y.W. and X.M.; methodology, X.Z. and X.M.; software, Y.W. and X.M.; validation, X.Z. and W.L.; formal analysis, W.X.; investigation, Q.L.; resources, W.L.; data curation, W.X.; writing—original draft preparation, Y.W. and X.M.; writing—review and editing, Y.W. and X.M.; visualization, W.L.; supervision, X.M.; project administration, Y.W. and X.M.; funding acquisition, Y.W. and X.Z. All authors have read and agreed to the published version of the manuscript.

Funding: This research was funded by the National Key Research and Development Program of China, grant number 2016YFC1401800, and the National Natural Science Foundation of China, grant numbers 61379056 and 61972406.

Institutional Review Board Statement: Not applicable.

Informed Consent Statement: Not applicable.

Data Availability Statement: Not applicable.

Acknowledgments: The authors are very grateful to Poter M for providing the Kraken program [19].

Conflicts of Interest: The authors declare no conflict of interest.

References

1. Erchang, S. Progress of geoacoustic inversion in underwater acoustics. *J. Appl. Acoust.* **2019**, *38*, 468–476.
2. Bhowal, A.; Kshetrimayum, R. Optical Improved Quadrature Spatial Modulation for Cooperative Underwater Wireless Communication under Weak Oceanic Turbulence Conditions. *IET Optoelectron.* **2020**, *14*, 434–439. [CrossRef]
3. Liu, X.; Ma, Y.; Li, Y.; Feng, T.; Yuan, X. Development of U.S. Underwater Unmanned Vehicles and Its Influence on U.S. Military Thinking. *Fly. Missile* **2020**, *6*, 12–19. (In Chinese)
4. Yang, K.; Lei, B.; Lu, Y. *Principle and Application of Typical Sound Field Model of Ocean Acoustics*; Northwestern Polytechnical University Press: Xi'an, China, 2018. (In Chinese)
5. Ya-fei, W.; Ying, Z.; Jing-hong, Y.; Hong-liang, Z. Research on Transmission of Acoustic Waves in Layered Media. *Piezoelectrics Acoustooptics* **2000**, *22*, 214–217.
6. Jensen, F.B.; Kuperman, W.A.; Porter, M.B.; Schmidt, H. *Computational Ocean Acoustics*; Springer: New York, NY, USA, 2011; [CrossRef]
7. Wang, Y.; Tu, H.; Liu, W.; Xiao, W.; Lan, Q. Application of a Chebyshev collocation method to solve a parabolic equation model of underwater acoustic propagation. *Acoust. Aust.* **2021**, *49*, 281–291 [CrossRef]
8. Tappert, F.T. *The Parabolic Equation Approximation Method in Wave Propagation and Underwater Acoustics*; Springer: New York, NY, USA, 1977; Volume 70.
9. Desanto, J.A. Relation between the solutions of the Helmholtz and parabolic equations for sound propagation. *J. Acoust. Soc. Am.* **1977**, *62*, 295–297. [CrossRef]
10. Estes, L.E.; Fain, G. Numerical technique for computing the wide-angle acoustic field in an ocean with rang-dependent velocity profiles. *J. Acoust. Soc. Am.* **1977**, *62*, 38–43. [CrossRef]
11. Mcdaniel, S.T.; Lee, D. A finite-difference treatment of interface conditions for the parabolic wave equation: The horizontal interface. *J. Acoust. Soc. Am.* **1982**, *71*, 855–858. [CrossRef]
12. Pekeris, C.L. Theory of propagation of explosive sound in shallow water. *Geol. Soc. Am. Mem.* **1948**, *27*, 1–117. [CrossRef]
13. Liu, W.; Wang, Y.; Zhang, L. *Numerical Ocean Acoustics*; Science Press: Beijing, China, 2019.
14. Tao, W. *Numerical Heat Transfer*, 2nd ed.; Xi'an Jiaotong University Press: Xi'an, China, 2001.
15. Thomas. *Numerical Partial Differential Equations-Finite-Difference Methods*; Springer: New York, NY, USA, 1977.
16. Nabavi, M.; Siddiqui, M.K.; Dargahi, J. A new 9-point sixth-order accurate compact finite-difference method for the Helmholtz equation. *J. Sound Vib.* **2007**, *307*, 972–982. [CrossRef]
17. Zhu, N.; Zhao, M. High-Order Finite Difference Method for Helmholtz Equation in polar Coordinates. *Am. J. Comput. Math.* **2019**, *3*, 174–186. [CrossRef]
18. Collins, M.D. *User's Guide for RAM Versions 1.0 and 1.0p*; Naval Research Lab.: Washington, DC, USA, 1999.
19. Porter, M.B. *The Kraken Normal Mode Program*; SACLANT Undersea Research Center: La Spezia, Italy, 2001.
20. Girault, V.; Raviart, P.A. *Finite Element Approximation of the Navier-Stokes Equations. Lecture Notes in Mathmetics*; Spring: New York, NY, USA, 1979.
21. Xiang, X. *Numerical Analysis of Spectral Methods*; Science Press: Beijing, China, 2000. (In Chinese)
22. Castillo, K.A.; de Jesus, M.N.B.; Petronilho, J.C.A. A note on orthogonal polynomials described by Chebyshev polynomials (Article). *J. Math. Anal. Appl.* **2021**, *497*, 124906. [CrossRef]
23. Aboites, V. Legendre Polynomials: A Simple Methodology. *J. Phys. Conf. Ser.* **2019**, *1221*, 012035. [CrossRef]
24. Shi, X.; Ametani, A.; Gole, A.M. A study on interpolation and weighting function for numerical Fourier transform. *Electr. Power Syst. Res.* **2021**, *195*, 107121. [CrossRef]
25. Huang, M.; Liu, H. *Galerkin Finite Element Method for Parabolic Problem*; Jilin University Press: Changchun, China, 1986. (In Chinese)
26. Quarteroni, A.; Valli, A. *Numerical Approximation of Partial Differential Equations*; Spring: New York, NY, USA, 1998.

27. Ma, H. Chebyshev-Legendre Viscosity Method for Nonlinear Conservation Laws. *SIAM J. Numer. Anal.* **1998**, *35*, 869–892. [CrossRef]
28. Bressan, N.; Quarteroni, A. Analysis of Chebyshev Collocation Methods for Parabolic Equations. *SIAM J. Numer. Anal.* **1986**, *23*, 1138–1154. [CrossRef]
29. Boyd, J.P. *Chebyshev and Fourier Spectral Methods*, 2nd ed.; Dover: New York, NY, USA, 2001.
30. Ben-Yu, G. *Spectral Methods and Their Applications*; World Scientific Publishing: Hackensack, NJ, USA; London, UK, 1988.
31. Qinglin, Z. Experiments on the 30-day long-range numerical weather predication in a seven-level spectral model. *J. Acad. Meteorol. Sci.* **1989**, *10*, 234–246.
32. Xu, J.; Ge, M. Numerical Simulation of Gas-solid Channel Flow with the Spectral Method. *J. Eng. Thermophilic* **1999**, *20*, 233–237.
33. Li, X.; He, K.; Li, H. Numerical Simulation of Tide-wave Nearby Daya Bay by Spectral Method. *Sci. Technol. Eng.* **2007**, *7*, 1847–1852.
34. Wise, E.S.; Cox, B.T.; Jaros, J.; Treeby, B.E. Representing arbitrary acoustic source and sensor distributions in Fourier collocation methods. *J. Acoust. Soc. Am.* **2019**, *146*, 278–288. [CrossRef] [PubMed]
35. Wang, J.B.; Pan, J. Acoustical wave propagator with modified Chebyshev expansion. *Comput. Phys. Commun.* **2006**, *174*, 187–190. [CrossRef]
36. Lan, B. Non-iterative, stable analysis of surface acoustic waves in anisotropic piezoelectric multilayers using spectral collocation method. *J. Sound Vib.* **2018**, *433*, 16–28. [CrossRef]
37. Evans, R.B. A Legendre-Galerkin Technique for Differential Eigenvalue Problems with Complex and Discontinuous Coefficients, Arising in Underwater Acoustics. 2016. Available online: http://oalib.hlsresearch.com/Modes/rimLG_oalib_2016/Legendre_Galerkin_Dec5_16.pdf (accessed on 17 August 2021).
38. Shen, J. Efficient spectral-Galerkin method II. Direct solvers of second-and fourth-order equations using Chebyshev polynomials. *SIAM J. Sci. Comput.* **1995**, *16*, 74–87. [CrossRef]
39. Wang, J. Research and Application of Spectral Method in Computational Fluid Dynamics. Master's Thesis, Northwestern Polytechnical University, Xi′ an, China, 2007. (In Chinese)
40. Oh, S. An Efficient Spectral Method to Solve Multi-Dimensional Linear Partial Different Equations Using Chebyshev Polynomials. *Mathematics* **2019**, *7*, 90. [CrossRef]
41. Shen, J.; Tang, T. *Spectral and High-Order Methods with Applications*; Science Press: Beijing, China, 2006.

Journal of
Marine Science and Engineering

MDPI

Article

A Vector Wavenumber Integration Model of Underwater Acoustic Propagation Based on the Matched Interface and Boundary Method

Wei Liu *, Lilun Zhang *, Yongxian Wang, Xinghua Cheng and Wenbin Xiao

College of Meteorology and Oceanography, National University of Defense Technology, Changsha 410073, China; yxwang@nudt.edu.cn (Y.W.); chengxinghua@nudt.edu.cn (X.C.); xiaowenbin@nudt.edu.cn (W.X.)
* Correspondence: liuwei@nudt.edu.cn (W.L.); zhanglilun@nudt.edu.cn (L.Z.)

Abstract: Acoustic particle velocities can provide additional energy flow information of the sound field; thus, the vector acoustic model is attracting increasing attention. In the current study, a vector wavenumber integration (VWI) model was established to provide benchmark solutions of ocean acoustic propagation. The depth-separated wave equation was solved using finite difference (FD) methods with second- and fourth-order accuracy, and the sound source singularity in this equation was treated using the matched interface and boundary method. Moreover, the particle velocity was calculated using the wavenumber integration method, consistent with the calculation of the sound pressure. Furthermore, the VWI model was verified using acoustic test cases of the free acoustic field, the ideal fluid waveguide, the Bucker waveguide, and the Munk waveguide by comparing the solutions of the VWI model, the analytical formula, and the image method. In the free acoustic field case, the errors of the second- and fourth-order FD schemes for solving the depth-separated equation were calculated, and the actual orders of accuracy of the FD schemes were tested. Moreover, the time-averaged sound intensity (TASI) was calculated using the pressure and particle velocity, and the TASI streamlines were traced to visualize the time-independent energy flow in the acoustic field and better understand the distribution of the acoustic transmission loss.

Keywords: acoustic particle velocity; vector acoustic model; sound intensity streamline; depth-separated equation; finite difference method

Citation: Liu, W.; Zhang, L.; Wang, Y.; Cheng, X.; Xiao, W. A Vector Wavenumber Integration Model of Underwater Acoustic Propagation Based on the Matched Interface and Boundary Method. *J. Mar. Sci. Eng.* **2021**, *9*, 1134. https://doi.org/10.3390/jmse9101134

Academic Editor: Mikhail Emelianov

Received: 28 August 2021
Accepted: 13 October 2021
Published: 15 October 2021

1. Introduction

Ocean acoustic waves propagate over long distances because their energy attenuates less in water than does that of electromagnetic waves; consequently, acoustic waves have been widely used in target detection, environmental monitoring, and underwater communication [1]. Underwater acoustic wave propagation excited by a time harmonic source can be described by the Helmholtz equation in the frequency domain, which can be solved using the wavenumber integration technique [2,3], the fast field program (FFP) [4–6], normal modes [7], ray theory [8], the parabolic equation [9], the finite difference method (FDM) [10], and the spectral method [11]. As these models for predicting the scalar sound level have reached a high level of development, our interest herein is in developing a vector model to accurately predict the particle velocity.

As a result of the growing interest in observing and investigating the underwater vector acoustic field [12], research is increasingly being conducted on the structure of the local vortices of the energy flux vector in acoustic waveguides [13]. Compared with experiments, vector acoustic simulations can more rapidly and economically provide more acoustic fields; accordingly, computer simulation and visualization techniques are highly important for studying the vector acoustic field [14–17]. Although the particle velocity can be obtained by calculating the pressure gradient using the FDM [18], the grid sizes in all coordinate directions must be sufficiently small to reduce numerical errors, resulting in a large number of grid points. Hence, other methods that can be used to analytically calculate some of the particle velocity components need to be developed, such as the analytical Lie-algebraic approach [19,20] and acoustic vector models that compute the particle velocity directly. Smith [21] introduced general numerical algorithms of the acoustic particle velocity field from the parabolic equation and normal mode models, and test cases of a Pekeris waveguide and range-dependent benchmark wedge were presented to verify the results of both models. Gulin [22] developed an analytical numerical acoustic model in terms of a scalar-vector description; the Helmholtz equation was replaced by two first-order partial differential equations that were used to govern the sound pressure and the velocity vector of fluid particles, and the normal mode method was utilized to simulate the acoustic scalar-vector field in layered stratified media. The FFP, which has become the most common wavenumber integration method in underwater acoustics, uses the far-field approximation and fast Fourier transform (FFT) to accelerate the computation. For instance, Zhu [23] developed a vector acoustic FFP model for an ocean-like environment with a sound speed profile of water and a multilayered elastic sediment bottom and analyzed the correlation between the sound pressure and the particle velocity.

The objective of this study was to obtain the most accurate vector acoustic model possible that can be used to provide benchmark solutions in range-independent environments for verifying other numerical acoustic models, such as FDM models [10] and spectral method models [11]. However, although the normal mode method, parabolic equation, and FFP mentioned above have the advantages of a small computational load and good numerical stability, they have difficulty effectively satisfying the required accuracy due to approximation errors. For instance, parabolic and far-field approximations should be used in the parabolic equation [24]; the branch cut integral is commonly neglected in normal modes [25]; and the far-field approximation needs to be applied to the FFP [26,27]. In contrast, the wavenumber integration method is attractive because it is basically an implementation of the integral transform technique for general range-independent media and directly evaluates the integrals of the solutions to the depth-separated wave equation (depth equation for short) by numerical quadrature; consequently, the field solution can be accurately calculated to any desired degree [2] (page 233). However, the wavenumber integration method incurs a relatively large computational load; nevertheless, for the model to generate benchmark solutions, guaranteeing the accuracy of the model is most often the greatest challenge, with reducing the computation time usually being a secondary priority [28].

To address the abovementioned challenges, a vector wavenumber integration (VWI) model was established in this study and verified using benchmark cases of underwater acoustic propagation. The primary contributions of this study are as follows:

(1) An FDM to solve the depth equation is proposed. Several methods exist for solving the depth equation, such as the propagator matrix approach [4] and the direct global matrix approach [29]; however, these approaches require that the density be approximated by the piecewise constant in each element along the depth direction, meaning that the range-independent environments are approximated as layered media. To further improve the solving accuracy of the depth equation solution, FD schemes with second-order (2nd-order) and fourth-order (4th-order) accuracy that can treat the density variation are proposed in this paper, and the recently developed matched interface and boundary (MIB) method is used to address the sound source singularity [30,31].

(2) A VWI model is established. In the present model, the particle velocity at any position in the sound field can be directly calculated using wavenumber integration (consistent with the calculation of the sound pressure sharing the zero-order Bessel function of the first kind) instead of calculating the spatial gradient of the pressure (two-dimensional grids are necessary, and the grid sizes in both directions need to be sufficiently small [18]).

(3) A sound field analysis method that displays both propagation loss contours and streamlines of the time-averaged sound intensity (TASI) is proposed. The TASI can be computed using the pressure and particle velocity and displayed using streamlines, thereby

bettering our understanding of the transmission loss (TL) distribution and the net transport of sound energy. For instance, TASI streamlines in the test cases show that the regions through which the energy flow is concentrated form zones of sound convergence.

2. The FDM for Solving the Depth Equation

Within a range-independent marine environment, because a simple point source is omnidirectional, the solutions of the steady-state acoustic field are symmetric about the z-axis and independent of the circumferential direction in a cylindrical coordinate system; thus, the Helmholtz equation in the frequency domain can be expressed as [2]:

$$\frac{1}{r}\frac{\partial}{\partial r}\left(r\frac{\partial P}{\partial r}\right)+\rho\frac{\partial}{\partial z}\left(\frac{1}{\rho}\frac{\partial P}{\partial z}\right)+k^2P=-\frac{2}{r}\delta(r)\delta(z-z_s) \tag{1}$$

where z is the vertical coordinate (in m), which is positive in the direction of increasing depth and passes through the source point; z_s is the source depth (z-coordinate of the source); r is the coordinate (in m) in the range direction from the source; $P(r,z)$ is the relatively complex sound pressure field (referenced to the pressure at a distance of 1 m from the sound source) and is assumed to have a time factor of $e^{-i\omega t}$; $\rho(z)$ is the medium density (in g/cm^3); δ is the Dirac delta function; and $k(z) = 2\pi f/c(z)$ is the wavenumber (f is the sound frequency in Hz and $c(z)$ is the speed of sound in the medium in m/s). We can apply a Hankel transform in the range direction to Equation (1) and define

$$\phi(k_r,z)=\int_0^{\infty}P(r,z)J_0(k_r r)rdr \tag{2}$$

where ϕ is the wavenumber kernel function (kernel for short), k_r is the horizontal wavenumber, and J_0 is the zero-order Bessel function of the first kind. Then, the sound pressure can be expressed in terms of the Sommerfeld integral:

$$P(r,z)=\int_0^{\infty}\phi(k_r,z)J_0(k_r r)k_r dk_r \tag{3}$$

Multiplying Equation (1) by $J_0(k_r r)rdr$ and integrating over r from 0 to ∞ ($\int_0^{\infty}(\cdot)J_0(k_r r)rdr$), we can obtain the depth equation:

$$\rho\frac{\partial}{\partial z}\left(\frac{1}{\rho}\frac{\partial\phi(k_r,z)}{\partial z}\right)+k_z^2\phi(k_r,z)=-2\delta(z-z_s) \tag{4}$$

where $k_z^2 = k^2 - k_r^2$. The wavenumber integration model consists of two successive steps: one is to obtain the kernel by solving Equation (4); the other is to attain the pressure by numerically integrating Equation (3).

2.1. The Boundary and Source Interface Conditions

As shown in Figure 1, the kernel (ϕ) is continuous at the upper boundary, and its derivative with respect to the depth (ϕ') satisfies

$$\rho_{0+}^{-1}\phi'_{0+}=\rho_{0-}^{-1}\phi'_{0-} \tag{5}$$

where the subscripts "0^-" and "0^+" represent the upper and lower surfaces, respectively, of the upper boundary (grid point 0). The derivative inside the water (ϕ'_{0+}) can be calculated using kernels on the interior grid points by the FDM. To calculate the derivative outside the water (ϕ'_{0-}), considering the homogeneous properties of the half-space above the water column, Equation (4) can be changed to the plane wave equation away from the source:

$$\partial^2\phi/\partial z^2+k_z^2\phi=0 \tag{6}$$

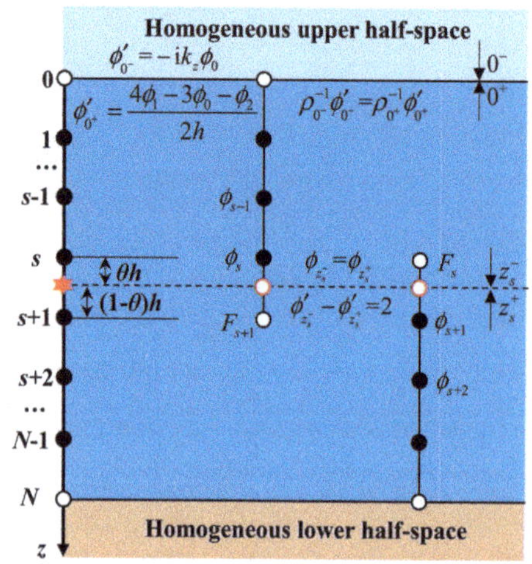

Figure 1. Schematic diagram of the boundary and source interface conditions.

Because only upgoing waves are present in the upper half-space, the solution of Equation (6) is:

$$\phi = Ce^{-ik_{z,0^-}z} \tag{7}$$

where C is a constant. Then, ϕ'_{0^-} can be expressed by the kernel on the upper boundary:

$$\phi'_{0^-} = -ik_{z,0^-}\phi_0 \tag{8}$$

The lower boundary conditions at grid point N can be treated as symmetric (only downgoing waves are present) to the upper boundary conditions.

The source interface conditions state that the kernel must be continuous ($\phi_{z_s^-} = \phi_{z_s^+}$), and the derivative of the kernel follows the jump condition. Let $z_s^- = z_s - \varepsilon$ and $z_s^+ = z_s + \varepsilon$, where $\varepsilon \to 0$. Then, Equation (4) is integrated over depth z from z_s^- to z_s^+ along the z-axis through the source. We can obtain the jump condition of the derivative of the kernel:

$$\phi'_{z_s^-} - \phi'_{z_s^+} = 2 \tag{9}$$

2.2. The FDM for the Depth Equaiton

As shown in Figure 1, equally-spaced grids in the depth direction are used in this paper, and h is the grid step size in meters. At the interior grid points far from the boundaries and source, the second derivatives in Equation (4) can be discretized using the standard 2nd-order accuracy FD scheme:

$$\left[\rho\frac{\partial}{\partial z}\left(\frac{1}{\rho}\frac{\partial\phi}{\partial z}\right)\right]_i = \frac{\rho_i}{h}\left[\left(\frac{1}{\rho}\frac{\partial\phi}{\partial z}\right)_{i+1/2} - \left(\frac{1}{\rho}\frac{\partial\phi}{\partial z}\right)_{i-1/2}\right] = \frac{\rho_i}{h}\left(\frac{\phi_{i+1}-\phi_i}{\rho_{i+1/2}h} - \frac{\phi_i-\phi_{i-1}}{\rho_{i-1/2}h}\right) \tag{10}$$

where $\rho_{i+1/2} = \rho(z_{i+1/2})$ and $\rho_{i-1/2} = \rho(z_{i-1/2})$. Then, the discretized depth equation at interior points can be expressed as:

$$\frac{1}{h^2}\frac{\rho_i}{\rho_{i-1/2}}\phi_{i-1} + \left[\left(k^2 - k_r^2\right) - \frac{1}{h^2}\left(\frac{\rho_i}{\rho_{i-1/2}} + \frac{\rho_i}{\rho_{i+1/2}}\right)\right]\phi_i + \frac{1}{h^2}\frac{\rho_i}{\rho_{i+1/2}}\phi_{i+1} = 0 \tag{11}$$

At the lower surface of the upper boundary, the derivative of the kernel (ϕ') can be approximated using the 2nd-order FD scheme:

$$\phi'_{0^+} = (-3\phi_0 + 4\phi_1 - \phi_2)/(2h) \tag{12}$$

After inserting Equations (5) and (8), we can establish the equation:

$$\rho_{0^+}^{-1}(-3\phi_0 + 4\phi_1 - \phi_2)/(2h) = -i\rho_{0^-}^{-1}k_{z,0^-}\phi_0 \tag{13}$$

and we can attain the following relationship between the kernel on the upper boundary ϕ_0 and the kernels at the interior points (ϕ_1 and ϕ_2):

$$\phi_0 = \rho_{0^-}(4\phi_1 - \phi_2)/(3\rho_{0^-} - 2ik_{z,0^-}h\rho_{0^+}) \tag{14}$$

Specifically, for the pressure-release upper boundary, we can set the density in the upper half-space to zero ($\rho_{0^-} = 0$, and thus, $\phi_0 = 0$). The lower boundary can be treated as symmetrical with respect to the upper boundary, and thus, we can derive the following relationship between the kernel on the lower boundary (ϕ_N) and the kernels at the interior points (ϕ_{N-1} and ϕ_{N-2}):

$$\phi_N = \rho_{N^+}(4\phi_{N-1} - \phi_{N-2})/(3\rho_{N^+} - 2ik_{z,N^+}h\rho_{N^-}) \tag{15}$$

At grid points near the source interface, the MIB method is used to form the FD schemes. The MIB is a numerical method that can treat discontinuous interfaces: the solution on each side of the source interface is smoothly extended beyond the interface by means of fictitious domains, and the fictitious values in the fictitious domains are determined simultaneously by enforcing the source interface conditions at the exact position of the interface [31,32]. Assuming that the source interface lies between the grid nodes s and $s+1$, the interface splits this grid step into two pieces with sizes of θh at the upper part and $(1-\theta)h$ at the lower part, where $0 \leq \theta < 1$. Because the derivative is discontinuous at the source interface, the MIB method smoothly extends the solution on the upper side of the source interface beyond the interface by means of fictitious domains and sets the fictitious kernel F_{s+1} at the location of grid point $s+1$; symmetrically, the fictitious kernel F_s is supplied at the location of grid point s to smoothly extend the solution from the lower side of the source interface to the upper side. Using Lagrange interpolation polynomials, the kernel at the source interface can be interpolated from the upper and lower sides:

$$\phi_{z_s^-} = \theta(\theta-1)\phi_{s-1}/2 + \left(1-\theta^2\right)\phi_s + \theta(\theta+1)F_{s+1}/2 \tag{16}$$

$$\phi_{z_s^+} = [1+\theta(\theta-3)/2]F_s - \theta(\theta-2)\phi_{s+1} + \theta(\theta-1)\phi_{s+2}/2 \tag{17}$$

Moreover, the derivatives of the kernel at the source interface can be approximated using the 2nd-order FD scheme from the upper and lower sides:

$$\phi'_{z_s^-} = (2h)^{-1}[(2\theta-1)\phi_{s-1} - 4\theta\phi_s + (1+2\theta)F_{s+1}] \tag{18}$$

$$\phi_{z_s^+} = (2h)^{-1}[(2\theta-3)F_s + 4(1-\theta)\phi_{s+1} + (2\theta-1)\phi_{s+2}] \tag{19}$$

Applying source interface conditions of $\phi_{z_s^-} = \phi_{z_s^+}$ and Equation (9), the fictitious kernels in the fictitious domains can be determined simultaneously:

$$\begin{cases} F_s = \frac{-\theta^2\phi_{s-1} + (\theta+1)^2\phi_s - 3\theta^2\phi_{s+1} + \theta^2\phi_{s+2} + 2\theta(\theta+1)h}{1+2\theta-2\theta^2} \\ F_{s+1} = \frac{(\theta-1)^2\phi_{s-1} - 3(\theta-1)^2\phi_s + (\theta-2)^2\phi_{s+1} - (\theta-1)^2\phi_{s+2} + 2(\theta-2)(\theta-1)h}{1+2\theta-2\theta^2} \end{cases} \tag{20}$$

Thus, by using the above FDM with 2nd-order accuracy, a linear system of the discretized depth equation at all grid points (excluding the boundary points 0 and N) can be obtained and solved using direct methods for solving linear systems:

$$A\phi = b,\ A \in \mathbb{C}^{(N-1)\times(N-1)}, \phi,\ b \in \mathbb{C}^{N-1} \tag{21}$$

where

$$b_i = \begin{cases} \frac{-1}{h}\frac{\rho_s}{\rho_{s+1/2}}\frac{2(\theta-2)(\theta-1)}{1+2\theta-2\theta^2}, & i = s \\ \frac{-1}{h}\frac{\rho_{s+1}}{\rho_{s+1/2}}\frac{2\theta(\theta+1)}{1+2\theta-2\theta^2}, & i = s+1 \\ 0, & else \end{cases} \tag{22}$$

and

$$A = \begin{pmatrix} a_{1,1} & a_{1,2} & & & & & & & \\ a_{2,1} & a_{2,2} & a_{2,3} & & & & & 0 & \\ & \ddots & \ddots & \ddots & & & & & \\ & & a_{s,s-1} & a_{s,s} & a_{s,s+1} & a_{s,s+2} & & & \\ & & a_{s+1,s-1} & a_{s+1,s} & a_{s+1,s+1} & a_{s+1,s+2} & & & \\ & & & & \ddots & \ddots & \ddots & & \\ & 0 & & & & a_{N-2,N-3} & a_{N-2,N-2} & a_{N-2,N-1} \\ & & & & & & a_{N-1,N-2} & a_{N-1,N-1} \end{pmatrix} \tag{23}$$

where

$$\begin{cases} a_{1,1} = k_z^2 + \frac{1}{h^2}\left(\frac{\rho_1}{\rho_{1/2}}\frac{\rho_{0-}+2ik_{z,0-}h\rho_{0+}}{3\rho_{0-}-2ik_{z,0-}h\rho_{0+}} - \frac{\rho_1}{\rho_{3/2}}\right) \\ a_{1,2} = \frac{1}{h^2}\left(\frac{\rho_1}{\rho_{3/2}} - \frac{\rho_1}{\rho_{1/2}}\frac{\rho_{0-}}{3\rho_{0-}-2ik_{z,0-}h\rho_{0+}}\right) \end{cases} \tag{24}$$

$$\begin{cases} a_{N-1,N-1} = k_z^2 + \frac{1}{h^2}\left(\frac{\rho_{N-1}}{\rho_{N-1/2}}\frac{\rho_{N+}+2ik_{z,N+}h\rho_{N-}}{3\rho_{N+}-2ik_{z,N+}h\rho_{N-}} - \frac{\rho_{N-1}}{\rho_{N-3/2}}\right) \\ a_{N-1,N-2} = \frac{1}{h^2}\left(\frac{\rho_{N-1}}{\rho_{N-3/2}} - \frac{\rho_{N-1}}{\rho_{N-1/2}}\frac{\rho_{N+}}{3\rho_{N+}-2ik_{z,N+}h\rho_{N-}}\right) \end{cases} \tag{25}$$

$$\begin{cases} a_{s,s-1} = \frac{1}{h^2}\left(\frac{\rho_s}{\rho_{s-1/2}} + \frac{\rho_s}{\rho_{s+1/2}}\frac{(\theta-1)^2}{1+2\theta-2\theta^2}\right) \\ a_{s,s} = k_z^2 - \frac{1}{h^2}\left(\frac{\rho_s}{\rho_{s-1/2}} + \frac{\rho_s}{\rho_{s+1/2}}\frac{(\theta-2)^2}{1+2\theta-2\theta^2}\right) \\ a_{s,s+1} = \frac{1}{h^2}\frac{\rho_s}{\rho_{s+1/2}}\frac{(\theta-2)^2}{1+2\theta-2\theta^2} \\ a_{s,s+2} = -\frac{1}{h^2}\frac{\rho_s}{\rho_{s+1/2}}\frac{(\theta-1)^2}{1+2\theta-2\theta^2} \end{cases} \tag{26}$$

$$\begin{cases} a_{s+1,s-1} = \frac{-1}{h^2}\frac{\rho_{s+1}}{\rho_{s+1/2}}\frac{\theta^2}{1+2\theta-2\theta^2} \\ a_{s+1,s} = \frac{1}{h^2}\frac{\rho_{s+1}}{\rho_{s+1/2}}\frac{(\theta+1)^2}{1+2\theta-2\theta^2} \\ a_{s+1,s+1} = k_z^2 - \frac{1}{h^2}\left[\frac{\rho_{s+1}}{\rho_{s+3/2}} + \frac{\rho_{s+1}}{\rho_{s+1/2}}\frac{(\theta+1)^2}{1+2\theta-2\theta^2}\right] \\ a_{s+1,s+2} = \frac{1}{h^2}\left(\frac{\rho_{s+1}}{\rho_{s+3/2}} + \frac{\rho_{s+1}}{\rho_{s+1/2}}\frac{\theta^2}{1+2\theta-2\theta^2}\right) \end{cases} \tag{27}$$

$$\begin{cases} a_{i,i-1} = \frac{1}{h^2}\frac{\rho_i}{\rho_{i-1/2}} \\ a_{i,i} = k_z^2 - \frac{1}{h^2}\left(\frac{\rho_i}{\rho_{i-1/2}} + \frac{\rho_i}{\rho_{i+1/2}}\right) \\ a_{i,i+1} = \frac{1}{h^2}\frac{\rho_i}{\rho_{i+1/2}} \end{cases}, \text{ where } 1 < i < s \text{ or } s+1 < i < N-1 \tag{28}$$

To improve the numerical accuracy of the kernels, a 4th-order FD scheme is also developed, which is described in Appendix A.

3. VWI Approaches

In addition to the sound pressure P, the VWI model also needs to calculate the acoustic particle velocity. According to Euler's equation, the acoustic particle velocity can be calculated from the spatial gradient of pressure [33]:

$$V(r,z) = [U(r,z),\ W(r,z)] = \frac{1}{i\rho\omega}\nabla P(r,z) \tag{29}$$

In the present VWI model, the particle velocity at any position in the acoustic field is directly calculated using the wavenumber integration method without a two-dimensional grid. Using Equations (3) and (29), the horizontal and vertical components of the particle velocity can be expressed as follows:

$$U = \frac{1}{i\rho\omega} P_r{}' = \frac{-1}{i\rho\omega} \int_0^\infty \phi(k_r, z) J_1(k_r r) k_r^2 dk_r \tag{30}$$

$$W = \frac{1}{i\rho\omega} P'_z = \int_0^\infty w(k_r, z) J_0(k_r r) k_r dk_r \tag{31}$$

where $P_r{}' = \partial P/\partial r$, $P_z{}' = \partial P/\partial z$, J_1 is the first-order Bessel function of the first kind ($\partial J_0(k_r r)/\partial r = -k_r J_1(k_r r)$), and $w = \phi'/(i\rho\omega)$ is the kernel of the vertical component of the particle velocity and can also be calculated using the FDM. For the 2nd-order FD scheme:

$$\phi'_i = \frac{1}{2h} \begin{cases} F_{s+1} - \phi_{s-1}, & i = s \\ \phi_{s+2} - F_s, & i = s+1 \\ \phi_{i+1} - \phi_{i-1}, & else \end{cases} \tag{32}$$

For the 4th-order FD scheme:

$$\phi'_i = \frac{1}{12h} \begin{cases} -3\phi_0 - 10\phi_1 + 18\phi_2 - 6\phi_3 + \phi_4, & i = 1 \\ 3F_{s+1} + 10\phi_s - 18\phi_{s-1} + 6\phi_{s-2} - \phi_{s-3}, & i = s \\ -3F_s - 10\phi_{s+1} + 18\phi_{s+2} - 6\phi_{s+3} + \phi_{s+4}, & i = s+1 \\ 3\phi_N + 10\phi_{N-1} - 18\phi_{N-2} + 6\phi_{N-3} - \phi_{N-4}, & i = N-1 \\ 8(\phi_{i+1} - \phi_{i-1}) - (\phi_{i+2} - \phi_{i-2}), & else \end{cases} \tag{33}$$

In the cycles of wavenumber integration, the zero-order Bessel function of the first kind $J_0(k_r r)$ can be shared by the integrals of the pressure P in Equation (3) and the vertical component of the particle velocity W in Equation (31).

3.1. Algorithm Parameters

To numerically evaluate the wavenumber integrals in Equations (3), (30) and (31), the kernel needs to be evaluated at discrete horizontal wavenumbers. In this paper, $k_r \in [0, k_{\max}]$ is discretized equidistantly to N_k cells, and the kernel at the midpoint of each cell is used as the average value of that cell for wavenumber integration. Thus, discretized horizontal wavenumbers used to solve the depth equation can be expressed as:

$$k_{r,n} = (n - 1/2)\Delta k_r,\ n = 1, 2, \ldots, N_k \tag{34}$$

where Δk_r is the horizontal wavenumber step size. To achieve more accurate integrals, Δk_r should be small enough to avoid aliasing and wrap-around effects [2], which means that a sufficient number of sampling points is needed at the maximum range $R_{\max}$. In this paper, the horizontal wavenumber step size is estimated using:

$$\Delta k_r = 2\pi/(n_R R_{\max}) \tag{35}$$

where n_R is the number of samples of the horizontal wavenumber in each 2π period (the Bessel functions exhibit periodic oscillations similar to the cosine function) at $r = R_{\max}$. Then, the total number of samples of the horizontal wavenumber can be calculated by $N_k = k_{\max}/\Delta k_r$, where $k_{\max}$ is the finite maximum value of the horizontal wavenumber used to truncate the range of the infinite Sommerfeld integrals.

In addition, to avoid the vertical wavenumber $k_z = 0$, the integration path is commonly moved into the complex plane [34], which means that:

$$k_n = k_{r,n} - i\varepsilon_k \tag{36}$$

where $\varepsilon_k = 3\Delta k_r/(2\pi\log_{10}\mathrm{e})$ and e is the natural constant. Then, the discrete wavenumber integrals of Equations (3) and (29) can be expressed as:

$$P(r,z) = \Delta k_r \sum_{n=1}^{N_k} \phi(k_n,z)J_0(k_n r)k_n \tag{37}$$

$$V(r,z) = \Delta k_r \left(\frac{\mathrm{i}}{\rho\omega} \sum_{n=1}^{N_k} \phi(k_n,z)J_1(k_n r)k_n^2,\ \sum_{n=1}^{N_k} w(k_n,z)J_0(k_n r)k_n \right) \tag{38}$$

3.2. Vector Sound Intensity

After the particle velocity field is calculated using the VWI method, the TASI streamlines can be traced to visualize the time-independent energy flow in the acoustic field. Although sound intensity streamlines have been applied in several studies [35–37], these streamlines are rarely shown together with the TL contours to better understand the TL distribution.

For a steady-state (time-harmonic) sound field, the TASI I can be obtained by temporally averaging the instantaneous intensity over one time period ($T = 2\pi/\omega$), which can be expressed as [35]:

$$I = \frac{1}{2}\mathrm{Re}\{PV^*\} = \frac{1}{2}\mathrm{Re}\{P^*V\} \tag{39}$$

Additionally, the two components of I can be expressed as:

$$I_r = \frac{1}{2}\mathrm{Re}\{PV_r^*\} \text{ and } I_z = \frac{1}{2}\mathrm{Re}\{PV_z^*\} \tag{40}$$

The TASI governs the net transport of energy, whose direction is proportional to the spatial gradient of the pressure phase and perpendicular to the resultant wave fronts (surfaces of the constant-pressure phase). The TASI streamline curves are tangent to the direction of the local TASI vector at every point [35,38], and these streamlines can describe the way in which sound energy is transferred, physically coming only from the sound source and propagating along the streamlines to infinity [39].

4. Benchmarking for Vector Acoustic Fields

Four benchmark cases were used to test the present VWI model: the free acoustic field, the ideal fluid waveguide, the Bucker waveguide, and the Munk waveguide. Moreover, the TASI vectors and streamlines are calculated and traced to visualize the time-averaged energy flow path, and the TL distribution is interpreted from the perspective of streamlines.

4.1. The Free Acoustic Field

If the acoustic environment is a homogeneous and unbounded medium (free space), the solution of the acoustic Helmholtz equation is analytical and can be expressed using the Sommerfeld–Weyl integral [2] (page 88):

$$P(r,z) = \frac{e^{\mathrm{i}kR}}{R} = \int_0^\infty \frac{\mathrm{i}}{k_z} \mathrm{e}^{\mathrm{i}k_z|z-z_s|} J_0(k_r r)k_r dk_r \tag{41}$$

where $R = \sqrt{r^2 + (z-z_s)^2}$ is the distance from the source point $(0, z_s)$ to the position (r, z). Moreover, the exact kernel is:

$$\phi^{\mathrm{Exact}} = \mathrm{i}\mathrm{e}^{\mathrm{i}k_z|z-z_s|}/k_z \tag{42}$$

which can be used to verify the numerical solution of the kernel solved by using the FDM.

Herein, the thickness of the uniform medium (water) is 5000 m, the sound speed and density of the medium are 1500 m/s and 1 g/cm^3, respectively, no attenuation occurs within the medium, and the properties of the medium in the upper and lower half-space are the same as those of the water. The depths of the sound source and the receiver are

2500 m and 20 m, respectively, the largest receiver range is 24 km, and the source frequency is 5 Hz. The interval step size in the range direction is 25 m, and n_R = 10 and $k_{\max}$ = 10 k_{ref}, where $k_{\text{ref}} = 2\pi f/c_{\text{ref}}$ is the reference wavenumber and c_{ref} = 1500 m/s is the reference sound speed.

First, the FDM is verified using Equation (42). Setting the grid step size in the depth direction to h = 20 m, comparisons of the kernel ($k_r = k_{r,1}$) obtained using the exact formula and the 2nd- and 4th-order FD schemes are shown in Figure 2a,b, demonstrating that the three results agree well and that the solutions of the 4th-order FD scheme are more accurate than those of the 2nd-order FD scheme.

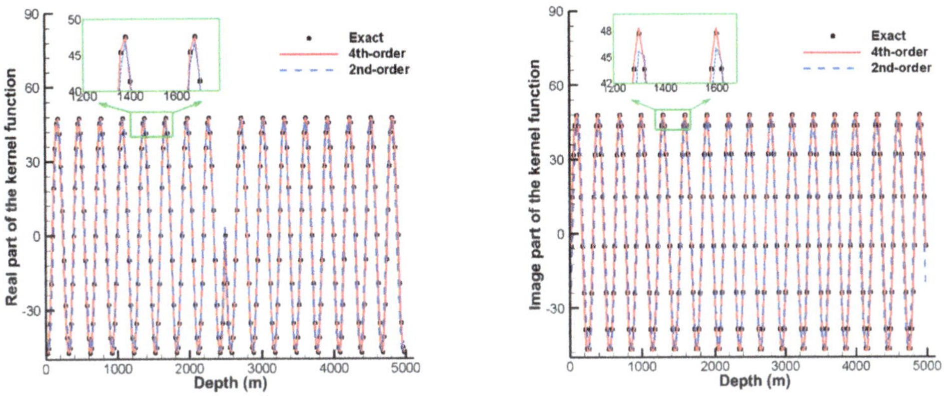

(**a**) the real part (**b**) the image part

Figure 2. Comparisons of the kernel ($k_r = k_{r,1}$, h = 20 m) obtained using the exact formula, and the 2nd- and 4th-order FD schemes.

Furthermore, the actual orders of the formal 2nd- and 4th-order FD schemes were tested. The relative error of the kernel at the ith grid point can be expressed as:

$$E_i(h) = \left|\phi_i^{\text{FDM}} - \phi_i^{\text{Exact}}\right| / \left|\phi_i^{\text{Exact}}\right| \tag{43}$$

where ϕ_i^{FDM} is the numerical solution of the kernel on grid point i and ϕ_i^{Exact} is the exact solution. In this paper, two grid size steps (in the depth direction) of h_1 = 20 m and h_2 = 10 m are used to solve the depth equation. Then, at the depth of $z = ih_1 = 2ih_2$, the actual orders of the FD schemes can be calculated using:

$$n(z) = \frac{\log_e[E_i(h_1)/E_{2i}(h_2)]}{\log_e(h_1/h_2)} \tag{44}$$

Figure 3a,b show the relative errors of the kernel and actual orders with respect to the formal 2nd- and 4th-order FD schemes ($k_r = k_{r,1}$), respectively, demonstrating that the relative errors can be decreased by reducing the grid step size; the relative errors near the source depth (2500 m) are smaller than those near the boundaries; the 4th-order FD scheme has smaller relative errors than the 2nd-order FD scheme; and the actual orders of the 2nd- and 4th-order FD schemes are close to their formal orders. It is worth pointing out that the MIB method exhibits good performance at the discontinuous source interface, meaning that the actual orders of the FD schemes do not decrease near the source interface.

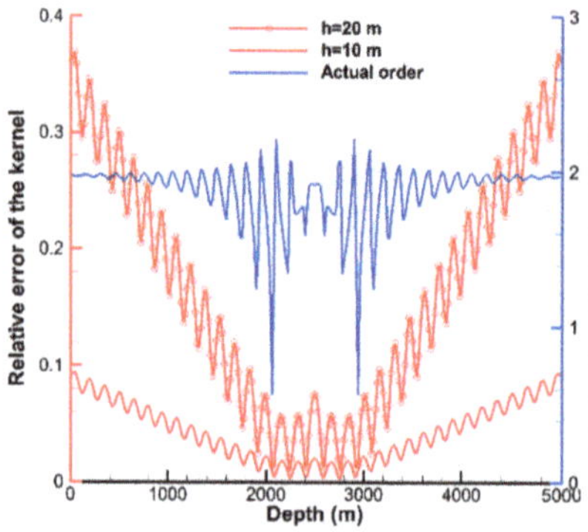

(**a**) the 2nd-order FD scheme

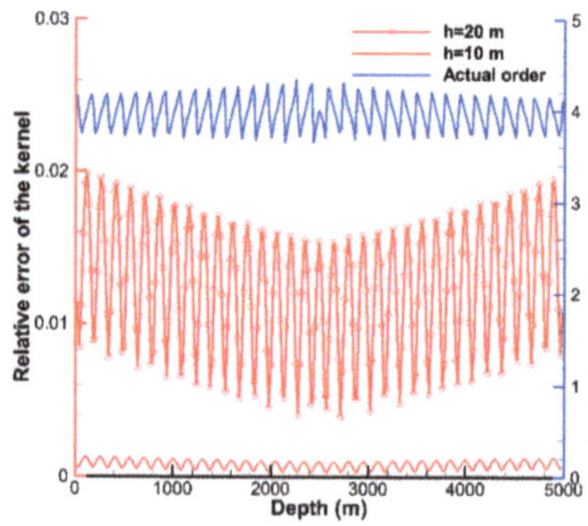

(**b**) the 4th-order FD scheme

Figure 3. Relative errors of the kernel and actual orders of the FD schemes ($k_r = k_{r,1}$).

Figure 4 shows the TL curves along the receiver line z = 20 m, where TL = −20 $\log_{10}|P|$ (in dB) and $|P|$ denotes the magnitude of the complex pressure P. The solutions of the 2nd- and 4th-order FD schemes (h = 20 m) match the exact analytical formula well in general, and the curve of the 4th-order FD scheme is closer to that of the exact formula than that of the 2nd-order FD scheme.

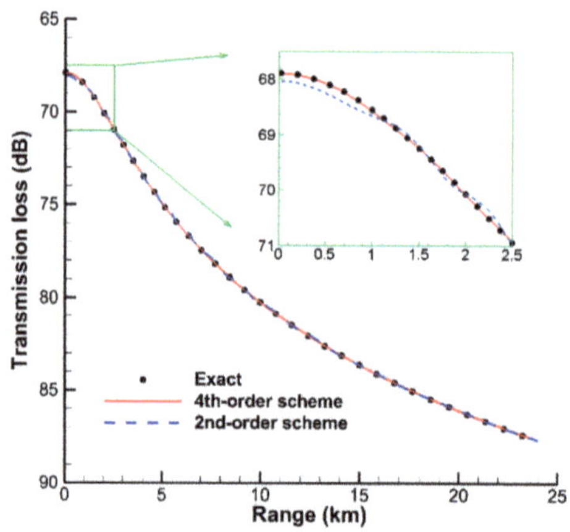

Figure 4. TL curves along z = 20 m for the free acoustic field (h = 20 m).

Figure 5 plots the TL contours and TASI streamlines obtained from the VWI model using the 2nd-order FD scheme (a) and the 4th-order FD scheme (b). In both figures, the TL contour lines are approximately concentric circles (the 4th-order scheme is more accurate than the 2nd-order scheme), whereas the TASI streamlines are straight lines from the source point to infinity, which is consistent with the characteristics of spherical waves in free space.

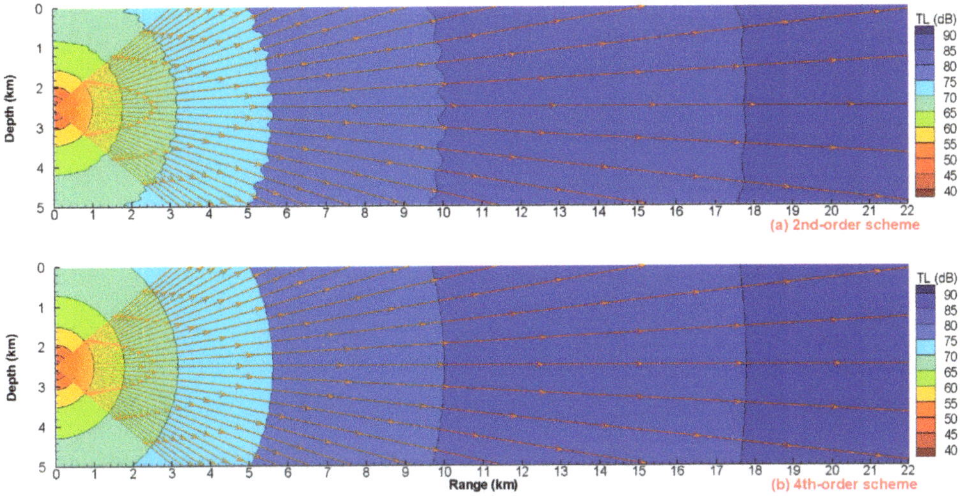

Figure 5. TL contours and TASI streamlines obtained from the VWI model (h = 20 m) for the free acoustic field using the 2nd-order FD scheme and the 4th-order FD scheme.

4.2. The Ideal Fluid Waveguide

The ideal fluid waveguide is represented by a range-independent homogeneous water column with pressure-release upper and lower boundaries [2] (page 102). The density and sound speed of the water are $\rho_w = 1\ \mathrm{g/cm^3}$ and $c_w = 1500\ \mathrm{m/s}$, respectively, and the densities of the medium in the upper and lower half-spaces are both $0\ \mathrm{g/cm^3}$. The parameters of the sound field are as follows: the depth of the seabed is 100 m, the depths of the sound source and receiver line are 35 m and 25 m, respectively, the largest receiver range is 3 km, and the source frequency is 20 Hz. In the VWI model, n_R = 10 and $k_{\max}$ = 10 k_{ref}. The interval step sizes in the range and depth directions are both 5 m.

The image method can be used to calculate the theoretical solutions of the ideal fluid waveguide, and the acoustic pressure can be expressed using the summation of the physical source and an infinite number of image sources as follows (in this paper, the summations of the image method are truncated at m = 500):

$$P(r,z) = \sum_{m=1}^{\infty}\left(\frac{e^{ikR_{m1}}}{R_{m1}} - \frac{e^{ikR_{m2}}}{R_{m2}} - \frac{e^{ikR_{m3}}}{R_{m3}} + \frac{e^{ikR_{m4}}}{R_{m4}}\right) \tag{45}$$

with

$$R_{mn} = \sqrt{r^2 + (z - z_{mn})^2},\ n = 1,2,3,4$$

where

$$z_{m1} = 2(m-1)H + z_s,\ z_{m2} = 2mH - z_s,\ z_{m3} = -z_{m1},\ \text{and} z_{m4} = -z_{m2}$$

and H is the depth of the seabed in meters.

Figure 6 shows the TL curves along the receiver line calculated by the image method and the present VWI model using the 2nd- and 4th-order FD schemes. The three curves generally match well. Moreover, at the peaks and troughs of the curves, the solutions of the 4th-order FDM are closer to those of the image method than those of the 2nd-order FDM.

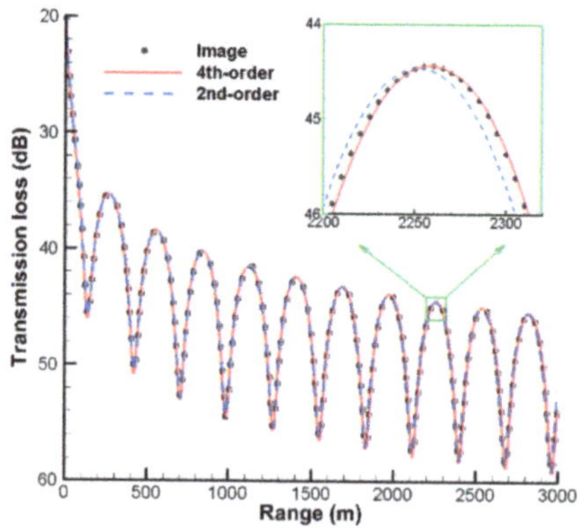

Figure 6. TL curves along the receiver line (z = 25 m) for the ideal fluid waveguide.

The TL contours and TASI streamlines obtained from the present VWI model using the 2nd- and 4th-order FD schemes are shown in Figure 7a,b, respectively, and they match well throughout the whole field. The main characteristic of the sound field is that the TASI streamlines are periodic in the range direction.

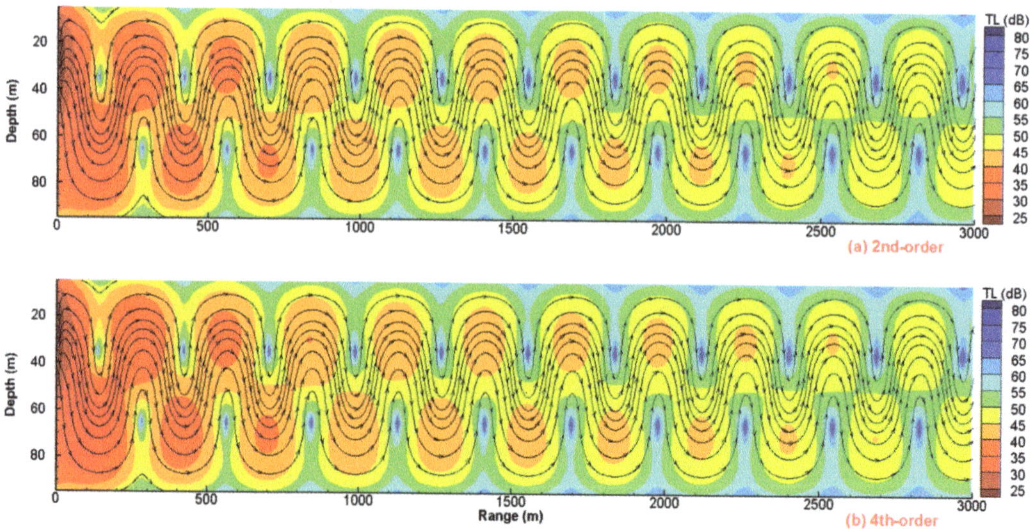

Figure 7. TL contours and TASI streamlines obtained from the VWI model for the ideal fluid waveguide using the 2nd-order FDM and the 4th-order FDM.

4.3. The Bucker Waveguide

The Bucker waveguide consists of a shallow water waveguide containing a bilinear sound speed (in m/s) profile in the water column [40]:

$$c(z) = 1498 + |120 - z|/60 \quad (46)$$

The depth of the water is 240 m, and the density of the water is ρ_w = 1.0 g/cm^3. The density of the medium in the upper half-space is 0 g/cm^3, whereas the homogeneous lower half-space of the penetrable sediment layer has a sound speed of 1505 m/s and a density of 2.1 g/cm^3, and no attenuation occurs in this problem. The depths of the sound source and the receiver line are 30 m and 90 m, respectively, the largest receiver range is 2 km, and the source frequency is 100 Hz. In the VWI model, $n_R = 10$ and $k_{\max} = 10\, k_{\text{ref}}$. The interval step sizes in the range and depth directions are both 1 m. Because the density contrast in this waveguide yields a significant number of virtual modes close to the real wavenumber axis, normal mode models ignoring the continuous spectrum cannot provide accurate solutions. Alternatively, wavenumber integration models have no restrictions on the density contrast and are therefore capable of providing exact solutions for this environment [2] (page 300).

Figure 8 shows the TL curves along the receiver line calculated by the present VWI model using the 2nd- and 4th-order FD schemes, demonstrating that the two curves match well in all ranges. Figure 9 shows the TL contours obtained from the present VWI model using the 4th-order FDM, revealing that the acoustic regions with denser TASI streamlines have a smaller TL, which means that the energy distribution in the sound field is not uniform.

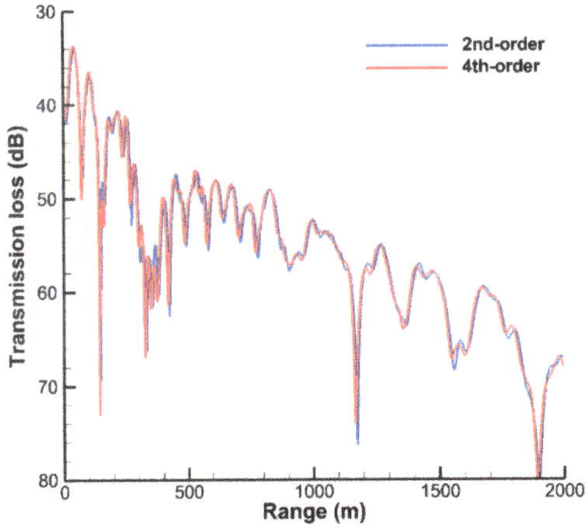

Figure 8. TL curves along $z = 90$ m for the Bucker waveguide.

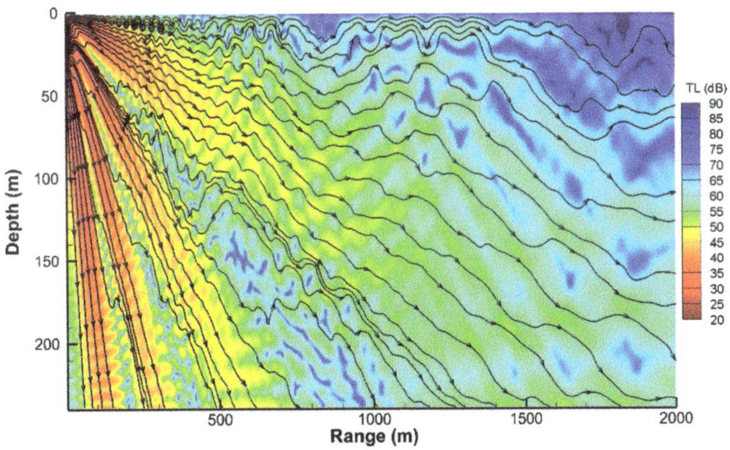

Figure 9. TL contours and TASI streamlines for the Bucker waveguide (4th-order FDM).

4.4. The Munk Waveguide

The Munk waveguide has the following analytical sound speed (in m/s) profile in water [2] (page 356):

$$c(z) = 1500\left[1 + 0.00737\left(\tilde{z} - 1 + e^{-\tilde{z}}\right)\right] \tag{47}$$

where the scaled depth $\tilde{z} = 2(z - 1300)/1300$. The depth of the water is 5000 m, and the density of the water is ρ_w = 1.0 g/cm^3. The density of the medium in the upper half-space is 0 g/cm^3, whereas the homogeneous lower half-space of the penetrable sediment layer has a sound speed of 1600 m/s, a density of 1.8 g/cm^3, and an attenuation rate of 0.8 dB/λ. The depths of the sound source and the receiver line are both 100 m, and the source frequency is 50 Hz. In the VWI model, n_R = 8 and k_{max} = k_{ref}. The interval step sizes in the range and depth directions are 300 m and 3.75 m, respectively.

Figure 10 shows the TL contours obtained from the present VWI model using the 2nd-order FDM and the acoustic rays calculated using the BELLHOP model in the acoustic toolbox [41]. Figure 11 shows the TL contours and TASI streamlines obtained from the VWI model using the 4th-order FDM, indicating that the acoustic regions with denser TASI streamlines have a smaller TL, and the regions through which the energy flow is concentrated form zones of sound convergence.

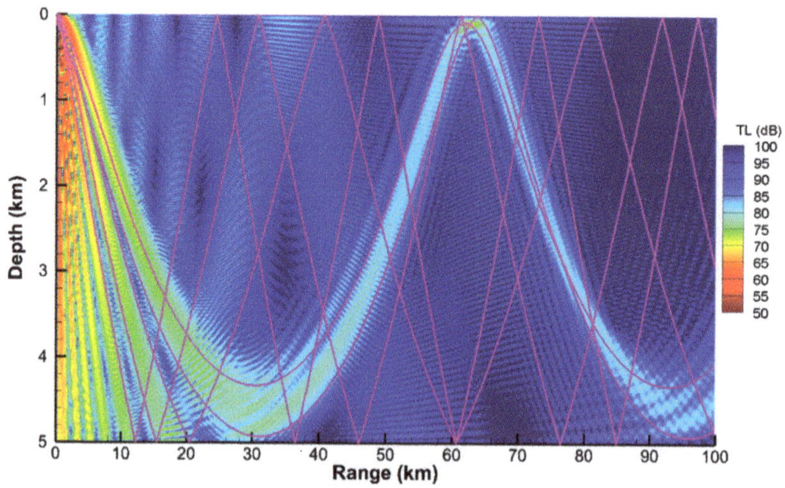

Figure 10. TL contours and rays obtained from the VWI model using the 2nd-order FDM and the BELLHOP model, respectively.

Moreover, the TL contours in Figures 10 and 11 match well, but the acoustic rays are significantly different from the TASI streamlines due to the following reasons. First, the TASI streamlines are calculated using the sound field solutions of the vector acoustic model, but the geometric rays are traced to calculate the solutions of the acoustic field. Second, because only one TASI vector direction may exist at any point in the field, the TASI streamlines cannot intersect; in contrast, the geometric rays can intersect among the multipath arrivals of propagating wavefronts in an ocean-like waveguide. Third, the TASI streamlines can accurately visualize the time-averaged energy flux, whereas the geometric rays intuitively visualize the approximate propagation of an acoustic field from the source to a receiver and do not include the diffracted components of the field.

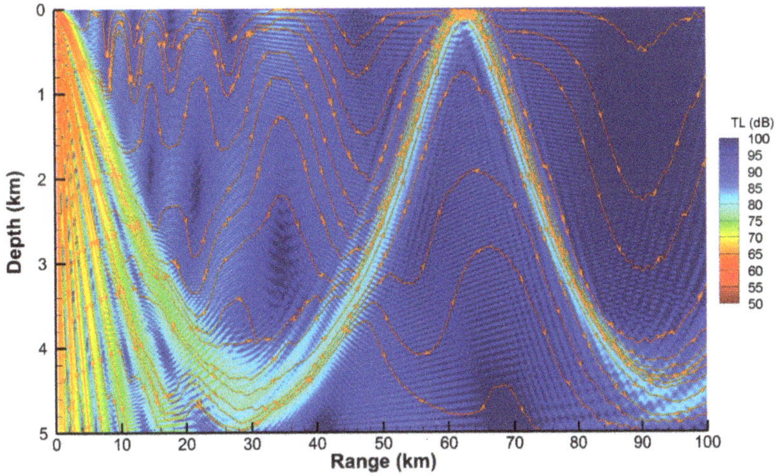

Figure 11. TL contours and TASI streamlines obtained from the VWI model using the 4th-order FDM.

Figure 12 shows the TL curves along the receiver line calculated by the present VWI model using the 2nd- and 4th-order FD schemes, and the two curves match well in all ranges. To test the effect of the density variation with depth on the acoustic field, a density profile in water is used for the Munk waveguide (other properties remain unchanged):

$$\rho_{\mathrm{w}}(z) = 1.0 + 1.2z/H \left(\text{in g/cm}^3\right) \tag{48}$$

where H = 5000 m is the seabed depth. Figure 13 shows the TL curves along the receiver line for the Munk waveguide with different density profiles in water calculated by the present VWI model using the 4th-order FDM, demonstrating that the density variation of water has an important influence on the peaks and troughs of the TL curves.

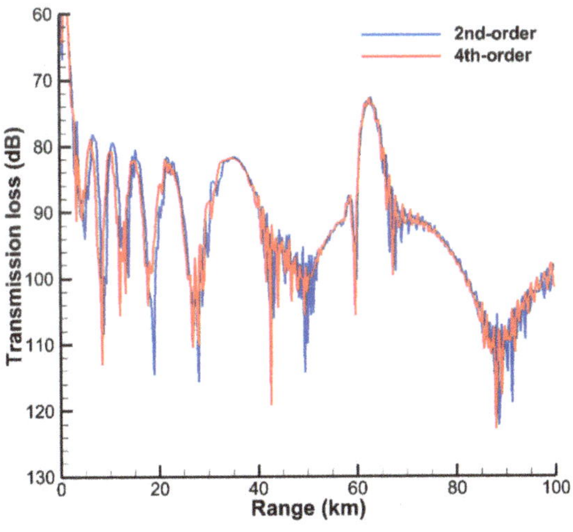

Figure 12. TL curves along z = 100 m for the Munk waveguide (ρ_{w} = 1.0 g/cm^3).

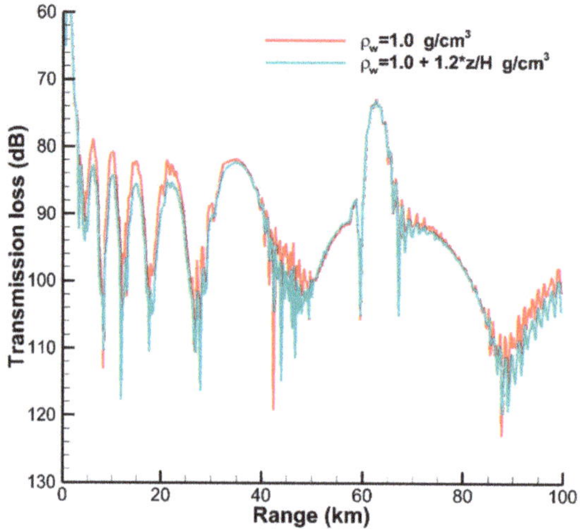

Figure 13. TL curves along z = 100 m for the Munk waveguide (4th-order FDM).

5. Conclusions and Discussion

The present range-independent VWI model can serve as a benchmark vector model for other acoustic models, such as FDM models and spectral method models. Moreover, by applying the particle velocity, the TASI streamlines can be computed and displayed, thereby bettering our understanding of the TL distribution, and these TASI streamlines can also be applied to other vector acoustic models. Furthermore, the particle velocity can be used to investigate the vortex-like structure and Stokes parameters [42] of the vector acoustic field.

It should be mentioned that the present VWI model is computationally intensive because it includes three Sommerfeld integrals (the pressure and two velocity components), and the linear system of the discretized depth equation using the FDM is solved directly because it is difficult to use the efficient forward elimination and backward substitution algorithm since matrix A is not a diagonal banded form as a result of the MIB method. Thus, parallel computing techniques [43] should be applied to the VWI model in future research to accelerate the computation.

In addition to the FDM combined with the MIB method in this paper, finite element methods and spectral methods [7] with high orders of accuracy can also be used to solve the depth equation, and a performance comparison among these methods may be studied in the future. Furthermore, the MIB method shows good performance in addressing the discontinuous source interface of the one-dimensional depth equation, and thus, it is very attractive to apply the MIB method to treat seabed discontinuities in FDM models [44].

Author Contributions: Conceptualization, W.L. and L.Z.; methodology, W.L. and Y.W.; software, W.L. and X.C.; validation, W.L. and W.X.; writing—original draft preparation, W.L.; writing—review and editing, W.L., X.C. and W.X.; visualization, W.L. and W.X.; supervision, L.Z.; project administration, Y.W.; funding acquisition, W.L., Y.W. and L.Z. All authors have read and agreed to the published version of the manuscript.

Funding: This work was supported in part by the National Natural Science Foundation of China (Grant Nos. 61972406, 61972411), in part by the National Key Research and Development Program of China (Grant No. 2016YFC1401800), and in part by the Aerospace Support Technology Fund of China (2014).

Institutional Review Board Statement: Not applicable.

Informed Consent Statement: Not applicable.

Data Availability Statement: Not applicable.

Acknowledgments: The authors are grateful to M. Porter for providing the BELLHOP program [41].

Conflicts of Interest: The authors declare no conflict of interest.

Appendix A. The FD Scheme with 4th-Order Accuracy

The derivation of the FD scheme with 4th-order accuracy is similar to that of the FD scheme with 2nd-order accuracy, but the computing stencils are wider. The computational stencil coefficients of the derivatives can be calculated using Taylor series and the method of undetermined coefficients [45] (or Lagrange interpolation polynomials).

Appendix A.1. Interior Grid Points

At the interior grid points away from the boundaries and source interface, the second derivatives in Equation (4) can be discretized using the cell-centered FD scheme with 4th-order accuracy [44,46]:

$$\left[\rho\frac{\partial}{\partial z}\left(\frac{1}{\rho}\frac{\partial\phi}{\partial z}\right)\right]_i=\frac{\rho_i}{24h}\left[27\left(\frac{\phi'_{i+1/2}}{\rho_{i+1/2}}-\frac{\phi'_{i-1/2}}{\rho_{i-1/2}}\right)-\frac{\phi'_{i+3/2}}{\rho_{i+3/2}}+\frac{\phi'_{i-3/2}}{\rho_{i-3/2}}\right] \quad \text{(A1)}$$

and the first derivatives at the half-integer points can be discretized using the same FD scheme:

$$\phi'_{i+1/2}=\frac{27(\phi_{i+1}-\phi_i)-(\phi_{i+2}-\phi_{i-1})}{24h} \quad \text{(A2)}$$

Then, the discretized depth equation at the interior points can be expressed as:

$$\begin{aligned}&\frac{\rho_i}{576h^2}\left[\frac{1}{\rho_{i+3/2}}\phi_{i+3}-\left(\frac{27}{\rho_{i+3/2}}+\frac{27}{\rho_{i+1/2}}\right)\phi_{i+2}+\left(\frac{27^2}{\rho_{i+1/2}}+\frac{27}{\rho_{i-1/2}}+\frac{27}{\rho_{i+3/2}}\right)\phi_{i+1}\right.\\&\left.+\left(\frac{27^2}{\rho_{i-1/2}}+\frac{27}{\rho_{i+1/2}}+\frac{27}{\rho_{i-3/2}}\right)\phi_{i-1}-\left(\frac{27}{\rho_{i-1/2}}+\frac{27}{\rho_{i-3/2}}\right)\phi_{i-2}+\frac{1}{\rho_{i-3/2}}\phi_{i-3}\right]\\&+\left[k_z^2-\frac{\rho_i}{576h^2}\left(\frac{27^2}{\rho_{i+1/2}}+\frac{27^2}{\rho_{i-1/2}}+\frac{1}{\rho_{i+3/2}}+\frac{1}{\rho_{i-3/2}}\right)\phi_i\right]=0\end{aligned} \quad \text{(A3)}$$

Appendix A.2. Grid Points near the Boundaries

At the lower surface of the upper boundary, the derivative of ϕ can be approximated using the 4th-order FD scheme:

$$\phi'_{0^+}=(-25\phi_0+48\phi_1-36\phi_2+16\phi_3-3\phi_4)/(12h) \quad \text{(A4)}$$

After inserting Equations (5) and (8), we can obtain the following relationship:

$$\phi_0=\rho_{0^-}(48\phi_1-36\phi_2+16\phi_3-3\phi_4)/(25\rho_{0^-}-12ik_{z,0^-}h\rho_{0^+}) \quad \text{(A5)}$$

The 4th-order FD scheme of the second derivative at grid point 1 is:

$$\left[\rho\frac{\partial}{\partial z}\left(\frac{\phi'}{\rho}\right)\right]_{i=1}=\frac{\rho_1}{840h}\left(128\frac{\phi'_{0^+}}{\rho_{0^+}}-1085\frac{\phi'_{1/2}}{\rho_{1/2}}+1015\frac{\phi'_{3/2}}{\rho_{3/2}}-63\frac{\phi'_{5/2}}{\rho_{5/2}}+5\frac{\phi'_{7/2}}{\rho_{7/2}}\right) \quad \text{(A6)}$$

where [47]

$$\phi'_{1/2}=(-22\phi_0+17\phi_1+9\phi_2-5\phi_3+\phi_4)/(24h) \quad \text{(A7)}$$

The stencil coefficients of the FD scheme at the lower boundary point are anti-symmetric to those at the upper boundary point.

Appendix A.3. Grid Points near the Source Interface

The kernels at the source interface can be interpolated from the upper and lower sides:

$$\phi_{z_s^-} = \sum_{i=s-3}^{s} w_{0,i}^- \phi_i + w_{0,s+1}^- F_{s+1} \tag{A8}$$

$$\phi_{z_s^+} = w_{0,s}^+ F_s + \sum_{i=s+1}^{s+4} w_{0,i}^+ \phi_i \tag{A9}$$

where

$$\begin{cases} w_{0,s-3}^- = \theta(-2-\theta+2\theta^2+\theta^3)/24 \\ w_{0,s-2}^- = \theta(3+\theta-3\theta^2-\theta^3)/6 \\ w_{0,s-1}^- = \theta(-6+\theta+4\theta^2+\theta^3)/4 \\ w_{0,s}^- = 1+\theta(5-5\theta-5\theta^2-\theta^3)/6 \\ w_{0,s+1}^- = \theta(6+11\theta+6\theta^2+\theta^3)/24 \end{cases} \tag{A10}$$

$$\begin{cases} w_{0,s}^+ = 1+\theta(-50+35\theta-10\theta^2+\theta^3)/24 \\ w_{0,s+1}^+ = \theta(24-26\theta+9\theta^2-\theta^3)/6 \\ w_{0,s+2}^+ = \theta(-12+19\theta-8\theta^2+\theta^3)/4 \\ w_{0,s+3}^+ = \theta(8-14\theta+7\theta^2-\theta^3)/6 \\ w_{0,s+4}^+ = \theta(-6+11\theta-6\theta^2+\theta^3)/24 \end{cases} \tag{A11}$$

The derivatives of the kernel at the source interface can be approximated using the 4th-order FD scheme from the upper and lower sides:

$$\phi'_{z_s^-} = \sum_{i=s-3}^{s} w_{1,i}^- \phi_i + w_{1,s+1}^- F_{s+1} \tag{A12}$$

$$\phi'_{z_s^+} = w_{1,s}^+ F_s + \sum_{i=s+1}^{s+4} w_{1,i}^+ \phi_i \tag{A13}$$

where

$$\begin{cases} w_{1,s-3}^- = \frac{1}{12h}(-1-\theta+3\theta^2+2\theta^3) \\ w_{1,s-2}^- = \frac{1}{6h}(3+2\theta-9\theta^2-4\theta^3) \\ w_{1,s-1}^- = \frac{1}{2h}(-3+\theta+6\theta^2+2\theta^3) \\ w_{1,s}^- = \frac{1}{6h}(5-10\theta-15\theta^2-4\theta^3) \\ w_{1,s+1}^- = \frac{1}{12h}(3+11\theta+9\theta^2+2\theta^3) \end{cases} \tag{A14}$$

$$\begin{cases} w_{1,s}^+ = \frac{1}{12h}(-25+35\theta-15\theta^2+2\theta^3) \\ w_{1,s+1}^+ = \frac{1}{6h}(24-52\theta+27\theta^2-4\theta^3) \\ w_{1,s+2}^+ = \frac{1}{2h}(-6+19\theta-12\theta^2+2\theta^3) \\ w_{1,s+3}^+ = \frac{1}{6h}(8-28\theta+21\theta^2-4\theta^3) \\ w_{1,s+4}^+ = \frac{1}{12h}(-3+11\theta-9\theta^2+2\theta^3) \end{cases} \tag{A15}$$

Applying the source interface conditions of Equation (9) and $\phi_{z_s^-} = \phi_{z_s^+}$, we can obtain:

$$\begin{cases} F_s = \frac{1}{\gamma}\left[\sum_{i=s-3}^{s}\left(w_{0,s+1}^- w_{1,i}^- - w_{1,s+1}^- w_{0,i}^-\right)\phi_i - \sum_{i=s+1}^{s+4}\left(w_{0,s+1}^- w_{1,i}^+ - w_{1,s+1}^- w_{0,i}^+\right)\phi_i - 2w_{0,s+1}^-\right] \\ F_{s+1} = \frac{1}{\gamma}\left[\sum_{i=s-3}^{s}\left(w_{0,s}^+ w_{1,i}^- - w_{1,s}^+ w_{0,i}^-\right)\phi_i - \sum_{i=s+1}^{s+4}\left(w_{0,s}^+ w_{1,i}^+ - w_{1,s}^+ w_{0,i}^+\right)\phi_i - 2w_{0,s}^+\right] \end{cases} \tag{A16}$$

where $\gamma = w_{1,s}^+ w_{0,s+1}^- - w_{1,s+1}^- w_{0,s}^+$.

The 4th-order FD schemes of the second derivative at grid points s and $s+1$ are:

$$\left[\rho\frac{\partial}{\partial z}\left(\frac{1}{\rho}\frac{\partial\phi}{\partial z}\right)\right]_s = \frac{\rho_s}{24h}\left(-\frac{\phi'_{s-7/2}}{\rho_{s-7/2}} + 5\frac{\phi'_{s-5/2}}{\rho_{s-5/2}} - 9\frac{\phi'_{s-3/2}}{\rho_{s-3/2}} - 17\frac{\phi'_{s-1/2}}{\rho_{s-1/2}} + 22\frac{\phi'_{(s+1/2)^-}}{\rho_{s+1/2}}\right) \tag{A17}$$

$$\left[\rho\frac{\partial}{\partial z}\left(\frac{1}{\rho}\frac{\partial\phi}{\partial z}\right)\right]_{s+1}=\frac{\rho_{s+1}}{24h}\left(\frac{\phi'_{s+9/2}}{\rho_{s+9/2}}-5\frac{\phi'_{s+7/2}}{\rho_{s+7/2}}+9\frac{\phi'_{s+5/2}}{\rho_{s+5/2}}+17\frac{\phi'_{s+3/2}}{\rho_{s+3/2}}-22\frac{\phi'_{(s+1/2)^+}}{\rho_{s+1/2}}\right) \tag{A18}$$

where

$$\phi'_{(s+1/2)^-}=(22F_{s+1}-17\phi_s-9\phi_{s-1}+5\phi_{s-2}-\phi_{s-3})/(24h) \tag{A19}$$

$$\phi'_{(s+1/2)^+}=(-22F_s+17\phi_{s+1}+9\phi_{s+2}-5\phi_{s+3}+\phi_{s+4})/(24h) \tag{A20}$$

The other first derivatives can be approximated using Equation (A2).

Appendix A.4. Linear System of the 4th-Order FD Scheme

Thus, a linear system of the discretized depth equation at all grid points can be obtained using the 4th-order FDM:

$$A_{4\text{th}}\phi=b_{4\text{th}},\ A_{4\text{th}}\in\mathrm{C}^{(N-1)\times(N-1)},\phi,\ b_{4\text{th}}\in\mathrm{C}^{N-1} \tag{A21}$$

where

$$b_{4\text{th},i}=\begin{cases}\frac{1}{576h^2}\frac{\rho_{s-2}}{\rho_{s-1/2}}\frac{2w_{0,s}^+}{\gamma}, & i=s-2\\ \frac{-1}{576h^2}\left(\frac{27\rho_{s-1}}{\rho_{s-1/2}}+\frac{22\rho_{s-1}}{\rho_{s+1/2}}\right)\frac{2w_{0,s}^+}{\gamma}, & i=s-1\\ \frac{1}{576h^2}\left(\frac{17\rho_s}{\rho_{s-1/2}}+\frac{484\rho_s}{\rho_{s+1/2}}\right)\frac{2w_{0,s}^+}{\gamma}, & i=s\\ \frac{1}{576h^2}\left(\frac{17\rho_s}{\rho_{s-1/2}}+\frac{484\rho_s}{\rho_{s+1/2}}\right)\frac{2w_{0,s}^+}{\gamma}, & i=s\\ \frac{1}{576h^2}\left(\frac{17\rho_{s+1}}{\rho_{s-1/2}}+\frac{484\rho_{s+1}}{\rho_{s+1/2}}\right)\frac{2w_{0,s+1}^-}{\gamma}, & i=s+1\\ \frac{-1}{576h^2}\left(\frac{27\rho_{s+2}}{\rho_{s+3/2}}+\frac{22\rho_{s+2}}{\rho_{s+1/2}}\right)\frac{2w_{0,s+1}^-}{\gamma}, & i=s+2\\ \frac{1}{576h^2}\frac{\rho_{s+3}}{\rho_{s+3/2}}\frac{2w_{0,s+1}^-}{\gamma}, & i=s+3\\ 0, & else\end{cases},\ 1\le i\le N-1 \tag{A22}$$

and

$$A_{4\text{th}}=(m_{i,j}),\quad 1\le i,j\le N-1 \tag{A23}$$

The nonzero elements in the first row of $A_{4\text{th}}$ include

$$\begin{cases}m_{1,1}=k_z^2+\frac{1}{20160h^2}\left[48\alpha_1-\left(18445\frac{\rho_1}{\rho_{1/2}}+27405\frac{\rho_1}{\rho_{3/2}}+63\frac{\rho_1}{\rho_{5/2}}\right)\right]\\ m_{1,2}=\frac{1}{20160h^2}\left(-36\alpha_1-9765\frac{\rho_1}{\rho_{1/2}}+27405\frac{\rho_1}{\rho_{3/2}}+1701\frac{\rho_1}{\rho_{5/2}}+5\frac{\rho_1}{\rho_{7/2}}\right)\\ m_{1,3}=\frac{1}{20160h^2}\left(16\alpha_1+5425\frac{\rho_1}{\rho_{1/2}}-1015\frac{\rho_1}{\rho_{3/2}}-1701\frac{\rho_1}{\rho_{5/2}}-135\frac{\rho_1}{\rho_{7/2}}\right)\\ m_{1,4}=\frac{1}{20160h^2}\left(-3\alpha_1-1085\frac{\rho_1}{\rho_{1/2}}+63\frac{\rho_1}{\rho_{5/2}}+135\frac{\rho_1}{\rho_{7/2}}\right)\\ m_{1,5}=\frac{1}{20160h^2}\left(-5\frac{\rho_1}{\rho_{7/2}}\right)\end{cases} \tag{A24}$$

where $\alpha_1=\frac{(23870\rho_1/\rho_{1/2}+1015\rho_1/\rho_{3/2})\rho_{0-}-3072\mathrm{i}k_{z,0-}h\rho_1}{25\rho_{0-}-12\mathrm{i}k_{z,0-}h\rho_{0+}}$.

The nonzero elements in the second row of $A_{4\text{th}}$ include

$$\begin{cases}m_{2,1}=\frac{1}{576h^2}\left(48\alpha_2+17\frac{\rho_2}{\rho_{1/2}}+729\frac{\rho_2}{\rho_{3/2}}+27\frac{\rho_2}{\rho_{5/2}}\right)\\ m_{2,2}=k_z^2+\frac{1}{576h^2}\left(-36\alpha_2+9\frac{\rho_2}{\rho_{1/2}}-729\frac{\rho_2}{\rho_{3/2}}-729\frac{\rho_2}{\rho_{5/2}}-\frac{\rho_2}{\rho_{7/2}}\right)\\ m_{2,3}=\frac{1}{576h^2}\left(16\alpha_2-5\frac{\rho_2}{\rho_{1/2}}+27\frac{\rho_2}{\rho_{3/2}}+729\frac{\rho_2}{\rho_{5/2}}+27\frac{\rho_2}{\rho_{7/2}}\right)\\ m_{2,4}=\frac{1}{576h^2}\left(-3\alpha_2+\frac{\rho_2}{\rho_{1/2}}-27\frac{\rho_2}{\rho_{5/2}}-27\frac{\rho_2}{\rho_{7/2}}\right)\\ m_{2,5}=\frac{1}{576h^2}\frac{\rho_2}{\rho_{7/2}}\end{cases} \tag{A25}$$

where $\alpha_2 = -\left(22\frac{\rho_2}{\rho_{1/2}} + 27\frac{\rho_2}{\rho_{3/2}}\right)\frac{\rho_{0-}}{25\rho_{0-} - 12ik_{z,0-}h\rho_{0+}}$.

The nonzero elements in the third row of A_{4th} include

$$\begin{cases} m_{3,1} = \frac{1}{576h^2}\left(48\alpha_3 - 27\frac{\rho_3}{\rho_{5/2}} - 27\frac{\rho_3}{\rho_{3/2}}\right) \\ m_{3,2} = \frac{1}{576h^2}\left(-36\alpha_3 + 27^2\frac{\rho_3}{\rho_{5/2}} + 27\frac{\rho_3}{\rho_{7/2}} + 27\frac{\rho_3}{\rho_{3/2}}\right) \\ m_{3,3} = k_z^2 + \frac{1}{576h^2}\left(16\alpha_3 - 27^2\frac{\rho_3}{\rho_{7/2}} - 27^2\frac{\rho_3}{\rho_{5/2}} - \frac{\rho_3}{\rho_{9/2}} - \frac{\rho_3}{\rho_{3/2}}\right) \\ m_{3,4} = \frac{1}{576h^2}\left(-3\alpha_3 + 27^2\frac{\rho_3}{\rho_{7/2}} + 27\frac{\rho_3}{\rho_{5/2}} + 27\frac{\rho_3}{\rho_{9/2}}\right) \\ m_{3,5} = \frac{1}{576h^2}\left(-27\frac{\rho_3}{\rho_{9/2}} - 27\frac{\rho_3}{\rho_{7/2}}\right) \\ m_{3,6} = \frac{1}{576h^2}\frac{\rho_3}{\rho_{9/2}} \end{cases} \tag{A26}$$

where $\alpha_3 = \frac{\rho_3}{\rho_{3/2}}\frac{\rho_{0-}}{25\rho_{0-} - 12ik_{z,0-}h\rho_{0+}}$.

The nonzero elements in the $(s-2)$th row of A_{4th} include

$$\begin{cases} m_{s-2,s-5} = \frac{1}{576h^2}\frac{\rho_{s-2}}{\rho_{s-7/2}} \\ m_{s-2,s-4} = \frac{1}{576h^2}\left(-27\frac{\rho_{s-2}}{\rho_{s-5/2}} - 27\frac{\rho_{s-2}}{\rho_{s-7/2}}\right) \\ m_{s-2,s-3} = \frac{1}{576h^2}\left[27^2\frac{\rho_{s-2}}{\rho_{s-5/2}} + 27\frac{\rho_{s-2}}{\rho_{s-3/2}} + 27\frac{\rho_{s-2}}{\rho_{s-7/2}} + \alpha_{s-2}\left(w_{0,s}^+ w_{1,s-3}^- - w_{1,s}^+ w_{0,s-3}^-\right)\right] \\ m_{s-2,s-2} = k_z^2 + \frac{1}{576h^2}\left[\alpha_{s-2}\left(w_{0,s}^+ w_{1,s-2}^- - w_{1,s}^+ w_{0,s-2}^- - \gamma\right) - \frac{27^2\rho_{s-2}}{\rho_{s-3/2}} - \frac{27^2\rho_{s-2}}{\rho_{s-5/2}} - \frac{\rho_{s-2}}{\rho_{s-7/2}} - \frac{\rho_{s-2}}{\rho_{s-1/2}}\right] \\ m_{s-2,s-1} = \frac{1}{576h^2}\left[27^2\frac{\rho_{s-2}}{\rho_{s-3/2}} + 27\frac{\rho_{s-2}}{\rho_{s-5/2}} + 27\frac{\rho_{s-2}}{\rho_{s-1/2}} + \alpha_{s-2}\left(w_{0,s}^+ w_{1,s-1}^- - w_{1,s}^+ w_{0,s-1}^-\right)\right] \\ m_{s-2,s} = \frac{1}{576h^2}\left[-27\frac{\rho_{s-2}}{\rho_{s-1/2}} - 27\frac{\rho_{s-2}}{\rho_{s-3/2}} + \alpha_{s-2}\left(w_{0,s}^+ w_{1,s}^- - w_{1,s}^+ w_{0,s}^-\right)\right] \\ m_{s-2,j} = \frac{1}{576h^2}\alpha_{s-2}\left(w_{1,s}^+ w_{0,j}^+ - w_{0,s}^+ w_{1,j}^+\right), \quad s+1 \le j \le s+4 \end{cases} \tag{A27}$$

where $\alpha_{s-2} = \frac{\rho_{s-2}}{\rho_{s-1/2}}\frac{1}{\gamma}$.

The nonzero elements in the $(s-1)$th row of A_{4th} include

$$\begin{cases} m_{s-1,s-4} = \frac{1}{576h^2}\frac{\rho_{s-1}}{\rho_{s-5/2}} \\ m_{s-1,s-3} = \frac{1}{576h^2}\left[-\frac{27\rho_{s-1}}{\rho_{s-5/2}} - \frac{27\rho_{s-1}}{\rho_{s-3/2}} + \frac{\rho_{s-1}}{\rho_{s+1/2}} + \alpha_{s-1}\left(w_{0,s}^+ w_{1,s-3}^- - w_{1,s}^+ w_{0,s-3}^-\right)\right] \\ m_{s-1,s-2} = \frac{1}{576h^2}\left[\frac{27\rho_{s-1}}{\rho_{s-5/2}} + \frac{27^2\rho_{s-1}}{\rho_{s-3/2}} + \frac{27\rho_{s-1}}{\rho_{s-1/2}} - \frac{5\rho_{s-1}}{\rho_{s+1/2}} + \alpha_{s-1}\left(w_{0,s}^+ w_{1,s-2}^- - w_{1,s}^+ w_{0,s-2}^-\right)\right] \\ m_{s-1,s-1} = k_z^2 + \frac{1}{576h^2}\left[\alpha_{s-1}\left(w_{0,s}^+ w_{1,s-1}^- - w_{1,s}^+ w_{0,s-1}^-\right) - \frac{\rho_{s-1}}{\rho_{s-5/2}} - \frac{27^2\rho_{s-1}}{\rho_{s-3/2}} - \frac{27^2\rho_{s-1}}{\rho_{s-1/2}} + \frac{9\rho_{s-1}}{\rho_{s+1/2}}\right] \\ m_{s-1,s} = \frac{1}{576h^2}\left[\frac{27\rho_{s-1}}{\rho_{s-3/2}} + \frac{27^2\rho_{s-1}}{\rho_{s-1/2}} + \frac{17\rho_{s-1}}{\rho_{s+1/2}} + \alpha_{s-1}\left(w_{0,s}^+ w_{1,s}^- - w_{1,s}^+ w_{0,s}^-\right)\right] \\ m_{s-1,j} = \frac{1}{576h^2}\alpha_{s-1}\left(w_{1,s}^+ w_{0,j}^+ - w_{0,s}^+ w_{1,j}^+\right), \quad s+1 \le j \le s+4 \end{cases} \tag{A28}$$

where $\alpha_{s-1} = -\left(\frac{27\rho_{s-1}}{\rho_{s-1/2}} + \frac{22\rho_{s-1}}{\rho_{s+1/2}}\right)\frac{1}{\gamma}$.

The nonzero elements in the sth row of A_{4th} include

$$\begin{cases} m_{s,s-5} = \frac{1}{576h^2}\frac{-\rho_s}{\rho_{s-7/2}} \\ m_{s,s-4} = \frac{1}{576h^2}\left(\frac{27\rho_s}{\rho_{s-7/2}} + \frac{5\rho_s}{\rho_{s-5/2}}\right) \\ m_{s,s-3} = \frac{1}{576h^2}\left[\alpha_s\left(w_{0,s}^+ w_{1,s-3}^- - w_{1,s}^+ w_{0,s-3}^-\right) - \frac{27\rho_s}{\rho_{s-7/2}} - \frac{135\rho_s}{\rho_{s-5/2}} - \frac{9\rho_s}{\rho_{s-3/2}} - \frac{22\rho_s}{\rho_{s+1/2}}\right] \\ m_{s,s-2} = \frac{1}{576h^2}\left[\alpha_s\left(w_{0,s}^+ w_{1,s-2}^- - w_{1,s}^+ w_{0,s-2}^-\right) + \frac{\rho_s}{\rho_{s-7/2}} + \frac{135\rho_s}{\rho_{s-5/2}} + \frac{243\rho_s}{\rho_{s-3/2}} - \frac{17\rho_s}{\rho_{s-1/2}} + \frac{110\rho_s}{\rho_{s+1/2}}\right] \\ m_{s,s-1} = \frac{1}{576h^2}\left[\alpha_s\left(w_{0,s}^+ w_{1,s-1}^- - w_{1,s}^+ w_{0,s-1}^-\right) - \frac{5\rho_s}{\rho_{s-5/2}} - \frac{243\rho_s}{\rho_{s-3/2}} + \frac{459\rho_s}{\rho_{s-1/2}} - \frac{198\rho_s}{\rho_{s+1/2}}\right] \\ m_{s,s} = k_z^2 + \frac{1}{576h^2}\left[\alpha_s\left(w_{0,s}^+ w_{1,s}^- - w_{1,s}^+ w_{0,s}^-\right) + \frac{9\rho_s}{\rho_{s-3/2}} - \frac{459\rho_s}{\rho_{s-1/2}} - \frac{374\rho_s}{\rho_{s+1/2}}\right] \\ m_{s,j} = \frac{1}{576h^2}\alpha_s\left(w_{1,s}^+ w_{0,j}^+ - w_{0,s}^+ w_{1,j}^+\right), \quad s+1 \le j \le s+4 \end{cases} \tag{A29}$$

where $\alpha_s = \left(\frac{17\rho_s}{\rho_{s-1/2}} + \frac{484\rho_s}{\rho_{s+1/2}}\right)\frac{1}{\gamma}$.

The nonzero elements in the $(s+1)$th row of A_{4th} include

$$\begin{cases} m_{s+1,s+6} = \frac{\rho_{s+1}}{576h^2}\frac{-1}{\rho_{s+9/2}} \\ m_{s+1,s+5} = \frac{\rho_{s+1}}{576h^2}\left(\frac{27}{\rho_{s+9/2}} + \frac{5}{\rho_{s+7/2}}\right) \\ m_{s+1,s+4} = \frac{\rho_{s+1}}{576h^2}\left[-\frac{27}{\rho_{s+9/2}} - \frac{135}{\rho_{s+7/2}} - \frac{9}{\rho_{s+5/2}} - \frac{22}{\rho_{s+1/2}} - \alpha_{s+1}\left(w^-_{0,s+1}w^+_{1,s+4} - w^-_{1,s+1}w^+_{0,s+4}\right)\right] \\ m_{s+1,s+3} = \frac{\rho_{s+1}}{576h^2}\left[\frac{1}{\rho_{s+9/2}} + \frac{135}{\rho_{s+7/2}} + \frac{243}{\rho_{s+5/2}} - \frac{17}{\rho_{s+3/2}} + \frac{110}{\rho_{s+1/2}} - \alpha_{s+1}\left(w^-_{0,s+1}w^+_{1,s+3} - w^-_{1,s+1}w^+_{0,s+3}\right)\right] \\ m_{s+1,s+2} = \frac{\rho_{s+1}}{576h^2}\left[-\frac{5}{\rho_{s+7/2}} - \frac{243}{\rho_{s+5/2}} + \frac{459}{\rho_{s+3/2}} - \frac{198}{\rho_{s+1/2}} - \alpha_{s+1}\left(w^-_{0,s+1}w^+_{1,s+2} - w^-_{1,s+1}w^+_{0,s+2}\right)\right] \\ m_{s+1,s+1} = k_z^2 + \frac{\rho_{s+1}}{576h^2}\left[\frac{9}{\rho_{s+5/2}} - \frac{459}{\rho_{s+3/2}} - \frac{374}{\rho_{s+1/2}} - \alpha_{s+1}\left(w^-_{0,s+1}w^+_{1,s+1} - w^-_{1,s+1}w^+_{0,s+1}\right)\right] \\ m_{s+1,j} = \frac{\rho_{s+1}}{576h^2}\alpha_{s+1}\left(w^-_{0,s+1}w^-_{1,i} - w^-_{1,s+1}w^-_{0,i}\right), \quad s-3 \le j \le s \end{cases} \tag{A30}$$

where $\alpha_{s+1} = \left(\frac{17}{\rho_{s+3/2}} + \frac{484}{\rho_{s+1/2}}\right)\frac{1}{\gamma}$.

The nonzero elements in the $(s+2)$th row of $A_{4\text{th}}$ include

$$\begin{cases} m_{s+2,s+5} = \frac{\rho_{s+2}}{576h^2}\frac{1}{\rho_{s+7/2}} \\ m_{s+2,s+4} = \frac{\rho_{s+2}}{576h^2}\left[-\frac{27}{\rho_{s+7/2}} - \frac{27}{\rho_{s+5/2}} + \frac{1}{\rho_{s+1/2}} - \alpha_{s+2}\left(w^-_{0,s+1}w^+_{1,s+4} - w^-_{1,s+1}w^+_{0,s+4}\right)\right] \\ m_{s+2,s+3} = \frac{\rho_{s+2}}{576h^2}\left[\frac{27}{\rho_{s+7/2}} + \frac{27^2}{\rho_{s+5/2}} + \frac{27}{\rho_{s+3/2}} - \frac{5}{\rho_{s+1/2}} - \alpha_{s+2}\left(w^-_{0,s+1}w^+_{1,s+3} - w^-_{1,s+1}w^+_{0,s+3}\right)\right] \\ m_{s+2,s+2} = k_z^2 + \frac{\rho_{s+2}}{576h^2}\left[\frac{9}{\rho_{s+1/2}} - \frac{1}{\rho_{s+7/2}} - \frac{27^2}{\rho_{s+5/2}} - \frac{27^2}{\rho_{s+3/2}} - \alpha_{s+2}\left(w^-_{0,s+1}w^+_{1,s+2} - w^-_{1,s+1}w^+_{0,s+2}\right)\right] \\ m_{s+2,s+1} = \frac{\rho_{s+2}}{576h^2}\left[\left(\frac{27}{\rho_{s+5/2}} + \frac{27^2}{\rho_{s+3/2}} + \frac{17}{\rho_{s+1/2}}\right) - \alpha_{s+2}\left(w^-_{0,s+1}w^+_{1,s+1} - w^-_{1,s+1}w^+_{0,s+1}\right)\right] \\ m_{s+2,j} = \frac{\rho_{s+2}}{576h^2}\alpha_{s+2}\left(w^-_{0,s+1}w^-_{1,i} - w^-_{1,s+1}w^-_{0,i}\right), \quad s-3 \le j \le s \end{cases} \tag{A31}$$

where $\alpha_{s+2} = -\left(\frac{27}{\rho_{s+3/2}} + \frac{22}{\rho_{s+1/2}}\right)\frac{1}{\gamma}$.

The nonzero elements in the $(s+3)$th row of $A_{4\text{th}}$ include

$$\begin{cases} m_{s+3,s+6} = \frac{\rho_{s+3}}{576h^2}\frac{1}{\rho_{s+9/2}} \\ m_{s+3,s+5} = \frac{\rho_{s+3}}{576h^2}\left(-\frac{27}{\rho_{s+9/2}} - \frac{27}{\rho_{s+7/2}}\right) \\ m_{s+3,s+4} = \frac{\rho_{s+3}}{576h^2}\left[\frac{27^2}{\rho_{s+7/2}} + \frac{27}{\rho_{s+5/2}} + \frac{27}{\rho_{s+9/2}} - \alpha_{s+3}\left(w^-_{0,s+1}w^+_{1,s+4} - w^-_{1,s+1}w^+_{0,s+4}\right)\right] \\ m_{s+3,s+3} = k_z^2 + \frac{\rho_{s+3}}{576h^2}\left[-\frac{27^2}{\rho_{s+7/2}} - \frac{27^2}{\rho_{s+5/2}} - \frac{1}{\rho_{s+9/2}} - \frac{1}{\rho_{s+3/2}} - \alpha_{s+3}\left(w^-_{0,s+1}w^+_{1,s+3} - w^-_{1,s+1}w^+_{0,s+3}\right)\right] \\ m_{s+3,s+2} = \frac{\rho_{s+3}}{576h^2}\left[\frac{27^2}{\rho_{s+5/2}} + \frac{27}{\rho_{s+7/2}} + \frac{27}{\rho_{s+3/2}} - \alpha_{s+3}\left(w^-_{0,s+1}w^+_{1,s+2} - w^-_{1,s+1}w^+_{0,s+2}\right)\right] \\ m_{s+3,s+1} = \frac{\rho_{s+3}}{576h^2}\left[-\frac{27}{\rho_{s+5/2}} - \frac{27}{\rho_{s+3/2}} - \alpha_{s+3}\left(w^-_{0,s+1}w^+_{1,s+1} - w^-_{1,s+1}w^+_{0,s+1}\right)\right] \\ m_{s+3,j} = \frac{\rho_{s+3}}{576h^2}\alpha_{s+3}\left(w^-_{0,s+1}w^-_{1,j} - w^-_{1,s+1}w^-_{0,j}\right), \quad s-3 \le j \le s \end{cases} \tag{A32}$$

where $\alpha_{s+3} = \frac{1}{\rho_{s+3/2}}\frac{1}{\gamma}$.

The nonzero elements in the $(N-3)$th row of $A_{4\text{th}}$ include

$$\begin{cases} m_{N-3,N-1} = \frac{1}{576h^2}\left(48\alpha_{N-3} - 27\frac{\rho_{N-3}}{\rho_{N-5/2}} - 27\frac{\rho_{N-3}}{\rho_{N-3/2}}\right) \\ m_{N-3,N-2} = \frac{1}{576h^2}\left(-36\alpha_{N-3} + 27^2\frac{\rho_{N-3}}{\rho_{N-5/2}} + 27\frac{\rho_{N-3}}{\rho_{N-7/2}} + 27\frac{\rho_{N-3}}{\rho_{N-3/2}}\right) \\ m_{N-3,N-3} = k_z^2 + \frac{1}{576h^2}\left(16\alpha_{N-3} - 27^2\frac{\rho_{N-3}}{\rho_{N-7/2}} - 27^2\frac{\rho_{N-3}}{\rho_{N-5/2}} - \frac{\rho_{N-3}}{\rho_{N-9/2}} - \frac{\rho_{N-3}}{\rho_{N-3/2}}\right) \\ m_{N-3,N-4} = \frac{1}{576h^2}\left(-3\alpha_{N-3} + 27^2\frac{\rho_{N-3}}{\rho_{N-7/2}} + 27\frac{\rho_{N-3}}{\rho_{N-5/2}} + 27\frac{\rho_{N-3}}{\rho_{N-9/2}}\right) \\ m_{N-3,N-5} = \frac{1}{576h^2}\left(-27\frac{\rho_{N-3}}{\rho_{N-9/2}} - 27\frac{\rho_{N-3}}{\rho_{N-7/2}}\right) \\ m_{N-3,N-6} = \frac{1}{576h^2}\frac{\rho_{N-3}}{\rho_{N-9/2}} \end{cases} \tag{A33}$$

where $\alpha_{N-3} = \frac{\rho_{N-3}}{\rho_{N-3/2}}\frac{\rho_{N+}}{25\rho_{N+} - 12ik_{z,N+}h\rho_{N-}}$.

The nonzero elements in the $(N-2)$th row of $A_{4\text{th}}$ include

$$
\begin{cases}
m_{N-2,N-1} = \frac{1}{576h^2}\left(48\alpha_{N-2} + 17\frac{\rho_{N-2}}{\rho_{N-1/2}} + 729\frac{\rho_{N-2}}{\rho_{N-3/2}} + 27\frac{\rho_{N-2}}{\rho_{N-5/2}}\right) \\
m_{N-2,N-2} = k_z^2 + \frac{1}{576h^2}\left(-36\alpha_{N-2} + 9\frac{\rho_{N-2}}{\rho_{N-1/2}} - 729\frac{\rho_{N-2}}{\rho_{N-3/2}} - 729\frac{\rho_{N-2}}{\rho_{N-5/2}} - \frac{\rho_{N-2}}{\rho_{N-7/2}}\right) \\
m_{N-2,N-3} = \frac{1}{576h^2}\left(16\alpha_{N-2} - 5\frac{\rho_{N-2}}{\rho_{N-1/2}} + 27\frac{\rho_{N-2}}{\rho_{N-3/2}} + 729\frac{\rho_{N-2}}{\rho_{N-5/2}} + 27\frac{\rho_{N-2}}{\rho_{N-7/2}}\right) \\
m_{N-2,N-4} = \frac{1}{576h^2}\left(-3\alpha_{N-2} + \frac{\rho_{N-2}}{\rho_{N-1/2}} - 27\frac{\rho_{N-2}}{\rho_{N-5/2}} - 27\frac{\rho_{N-2}}{\rho_{N-7/2}}\right) \\
m_{N-2,N-5} = \frac{1}{576h^2}\frac{\rho_{N-2}}{\rho_{N-7/2}}
\end{cases}
\tag{A34}
$$

where $\alpha_{N-2} = -\left(\frac{22}{\rho_{N-1/2}} + \frac{27}{\rho_{N-3/2}}\right)\frac{\rho_{N+}}{25\rho_{N+} - 12\mathrm{i}k_{z,N+}h\rho_{N-}}$.

The nonzero elements in the $(N-1)$th row of $\boldsymbol{A}_{4\text{th}}$ include

$$
\begin{cases}
m_{N-1,N-1} = k_z^2 + \frac{1}{20160h^2}\left[48\alpha_{N-1} - \left(18445\frac{\rho_{N-1}}{\rho_{N-1/2}} + 27405\frac{\rho_{N-1}}{\rho_{N-3/2}} + 63\frac{\rho_{N-1}}{\rho_{N-5/2}}\right)\right] \\
m_{N-1,N-2} = \frac{1}{20160h^2}\left(-36\alpha_{N-1} - 9765\frac{\rho_{N-1}}{\rho_{N-1/2}} + 27405\frac{\rho_{N-1}}{\rho_{N-3/2}} + 1701\frac{\rho_{N-1}}{\rho_{N-5/2}} + 5\frac{\rho_{N-1}}{\rho_{N-7/2}}\right) \\
m_{N-1,N-3} = \frac{1}{20160h^2}\left(16\alpha_{N-1} + 5425\frac{\rho_{N-1}}{\rho_{N-1/2}} - 1015\frac{\rho_{N-1}}{\rho_{N-3/2}} - 1701\frac{\rho_{N-1}}{\rho_{N-5/2}} - 135\frac{\rho_{N-1}}{\rho_{N-7/2}}\right) \\
m_{N-1,N-4} = \frac{1}{20160h^2}\left(-3\alpha_{N-1} - 1085\frac{\rho_{N-1}}{\rho_{N-1/2}} + 63\frac{\rho_{N-1}}{\rho_{N-5/2}} + 135\frac{\rho_{N-1}}{\rho_{N-7/2}}\right) \\
m_{N-1,N-5} = \frac{1}{20160h^2}\left(-5\frac{\rho_{N-1}}{\rho_{N-7/2}}\right)
\end{cases}
\tag{A35}
$$

where $\alpha_{N-1} = \frac{(23870\rho_{N-1}/\rho_{N-1/2} + 1015\rho_{N-1}/\rho_{N-3/2})\rho_{N+} - 3072\mathrm{i}k_{z,N+}h\rho_{N-1}}{25\rho_{N+} - 12\mathrm{i}k_{z,N+}h\rho_{N-}}$.

The nonzero elements in the other rows ($3 < i < s-2$ or $s+3 < i < N-3$) of $\boldsymbol{A}_{4\text{th}}$ include

$$
\begin{cases}
m_{i,i\pm3} = \frac{\rho_i}{576h^2}\frac{1}{\rho_{i\pm3/2}} \\
m_{i,i\pm2} = \frac{\rho_i}{576h^2}\left(-\frac{27}{\rho_{i\pm3/2}} - \frac{27}{\rho_{i\pm1/2}}\right) \\
m_{i,i\pm1} = \frac{\rho_i}{576h^2}\left(\frac{27^2}{\rho_{i\pm1/2}} + \frac{27}{\rho_{i\mp1/2}} + \frac{27}{\rho_{i\pm3/2}}\right) \\
m_{i,i} = k_z^2 + \frac{\rho_i}{576h^2}\left(-\frac{27^2}{\rho_{i+1/2}} - \frac{27^2}{\rho_{i-1/2}} - \frac{1}{\rho_{i+3/2}} - \frac{1}{\rho_{i-3/2}}\right)
\end{cases}
\tag{A36}
$$

References

1. Etter, P.C. *Underwater Acoustic Modeling and Simulation*, 5th ed.; CRC Press: Boca Raton, FL, USA, 2018.
2. Jensen, F.B.; Kuperman, W.A.; Porter, M.B.; Schmidt, H. *Computational Ocean Acoustics*, 2nd ed.; Springer: New York, NY, USA, 2011.
3. Lippert, T.; Lippert, S. Modelling of pile driving noise by means of wavenumber integration. *Acoust. Aust.* **2012**, *40*, 178–182.
4. West, M.; Sack, R.A.; Walkden, F. The fast field program (FFP). A second tutorial: Application to long range sound propagation in the atmosphere. *Appl. Acoust.* **1991**, *33*, 199–228.
5. Li, Y.L.; White, M.J. New fast field programs for a multiple—Source system. *J. Acoust. Soc. Am.* **1994**, *96*, 3277. [CrossRef]
6. Franke, S.J.; Swenson, G.W. A brief tutorial on the fast field program (FFP) as applied to sound propagation in the air. *Appl. Acoust.* **1989**, *27*, 203–215. [CrossRef]
7. Tu, H.; Wang, Y.; Lan, Q.; Liu, W.; Xiao, W.; Ma, S. A Chebyshev-Tau spectral method for normal modes of underwater sound propagation with a layered marine environment. *J. Sound Vib.* **2021**, *492*, 115784. [CrossRef]
8. Porter, M.B. Beam tracing for two- and three-dimensional problems in ocean acoustics. *J. Acoust. Soc. Am.* **2019**, *146*, 2016–2029. [CrossRef]
9. Wang, Y.; Tu, H.; Liu, W.; Xiao, W.; Lan, Q. Application of a Chebyshev collocation method to solve a parabolic equation model of underwater acoustic propagation. *Acoust. Aust.* **2021**, *49*, 281–291. [CrossRef]
10. Liu, W.; Cheng, X.; Xiao, W.; Wang, Y.; Zhang, L. A direct high-order finite difference solution to the Helmholtz equation in ocean acoustics. In Proceedings of the OCEANS 2018, Charleston, SC, USA, 22–25 October 2018.
11. Ma, X.; Wang, Y.; Zhu, X.; Liu, W.; Lan, Q.; Xiao, W. A spectral method for two-dimensional ocean acoustic propagation. *J. Mar. Sci. Eng.* **2021**, *9*, 892. [CrossRef]
12. Eliseevnin, V.A.; Tuzhilkin, Y.I. Acoustic power flux in a waveguide. *Acoust. Phys.* **2001**, *47*, 688–694. [CrossRef]
13. Shchurov, V.A. Peculiarities of real shallow sea wave-guide vortex structure. *J. Acoust. Soc. Am.* **2019**, *145*, 525–530. [CrossRef]
14. Deal, T.J.; Smith, K.B. Reciprocity relationships in vector acoustics and their application to vector field calculations. *J. Acoust. Soc. Am.* **2017**, *142*, 523–529. [CrossRef]
15. Yu, Y.; Ling, Q.; Xu, J. Pressure and velocity cross-spectrum of normal modes in low-frequency acoustic vector field of shallow water and its application. *J. Syst. Eng. Electron.* **2015**, *26*, 241–249. [CrossRef]

16. Dall'Osto, D.R.; Dahl, P.H.; Woong Choi, J. Properties of the acoustic intensity vector field in a shallow water waveguide. *J. Acoust. Soc. Am.* **2012**, *131*, 2023–2035. [CrossRef]
17. Wang, W.; Yang, D.; Shi, J. A prediction method for acoustic intensity vector field of elastic structure in shallow water waveguide. *Shock. Vib.* **2020**, *2020*, 5389719. [CrossRef]
18. Sheng, X.L.; Mei, J.D.; Guo, Y.; Yin, J.W.; Guo, L.X. The application of vertical sound intensity in shallow and low frequency sound field. In Proceedings of the OCEANS 2011, Waikoloa, HI, USA, 19–22 September 2011.
19. Petrov, P.S.; Prants, S.V.; Petrova, T.N. Analytical Lie-algebraic solution of a 3D sound propagation problem in the ocean. *Phys. Lett. A* **2017**, *381*, 1921–1925. [CrossRef]
20. Petrov, P.N.; Petrov, P.S. Asymptotic solution for the problem of sound propagation in a shallow sea with the bathymetry described by a parametric quadratic function. *J. Acoust. Soc. Am.* **2019**, *146*, 1946–1955. [CrossRef]
21. Smith, K.B. Validating range-dependent, full-field models of the acoustic vector field in shallow water environments. *J. Comput. Acoust.* **2008**, *16*, 471–486.
22. Gulin, O.E.; Yang, D. On the certain semi-analytical models of low-frequency acoustic fields in terms of scalar-vector description. *Chin. J. Acoust.* **2004**, *23*, 58–70.
23. Han-hao, Z.; Sheng-chun, P.; Wei, L.; Hai-gang, Z. Research on acoustic vector properties in shallow water for geoacoustic parameters inversion. In Proceedings of the IEEE/OES Acoustics in Underwater Geosciences Symposium, Rio de Janeiro, Brazil, 24–26 July 2013; pp. 1–6.
24. Tu, H.; Wang, Y.; Liu, W.; Ma, X.; Xiao, W.; Lan, Q. A Chebyshev spectral method for normal mode and parabolic equation models in underwater acoustics. *Math. Probl. Eng.* **2020**, *2020*, 7461314. [CrossRef]
25. Westwood, E.K.; Koch, R.A. Elimination of branch cuts from the normal-mode solution using gradient half spaces. *J. Acoust. Soc. Am.* **1999**, *106*, 2513–2523. [CrossRef]
26. Li, Y.L.; White, M.J. A note on using the fast field program. *J. Acoust. Soc. Am.* **1994**, *95*, 3100–3102. [CrossRef]
27. Raspet, R.; Lee, S.W.; Gilbert, R.; Bong, N.; Kuester, E.; Chang, D.C.; Richards, W.F. A fast-field program for sound propagation in a layered atmosphere above an impedance ground. *Acoust. Soc. Am. J.* **1985**, *77*, 345–352. [CrossRef]
28. Robertson, J.S. A classification scheme for computational ocean acoustic benchmark problems. *Appl. Acoust.* **1989**, *27*, 65–68. [CrossRef]
29. Schmidt, H.; Tango, G. Efficient global matrix approach to the computation of synthetic seismograms. *Geophys. J. R. Astron. Soc.* **1986**, *84*, 331–359. [CrossRef]
30. Zhou, Y.C.; Wei, G.W. On the fictitious-domain and interpolation formulations of the matched interface and boundary (MIB) method. *J. Comput. Phys.* **2006**, *219*, 228–246. [CrossRef]
31. Pan, K.; Tan, Y.; Hu, H. An interpolation matched interface and boundary method for elliptic interface problems. *J. Comput. Appl. Math.* **2010**, *234*, 73–94. [CrossRef]
32. Kreiss, H.O.; Petersson, N.A. An embedded boundary method for the wave equation with discontinuous coefficients. *SIAM J. Sci. Comput.* **2006**, *28*, 2054–2074. [CrossRef]
33. Deal, T.; Smith, K. Modeling Acoustic Vector Fields for Inverse Problems. In Proceedings of the 2016 IEEE/OES China Ocean Acoustics (COA), Harbin, China, 9–11 January 2016; pp. 1–5.
34. Luo, W.; Yu, X.; Yang, X.; Zhang, R. Analytical solution based on the wavenumber integration method for the acoustic field in a Pekeris waveguide. *Chin. Phys. B* **2016**, *25*, 044302.
35. Chapman, D.M.F. Using streamlines to visualize acoustic energy flow across boundaries. *J. Acoust. Soc. Am.* **2008**, *124*, 48–56. [CrossRef]
36. Mann, J.A.; Tichy, J.; Romano, A.J. Instantaneous and time-averaged energy transfer in acoustic fields. *J. Acoust. Soc. Am.* **1987**, *82*, 17–30. [CrossRef]
37. Chien, C.F.; Waterhouse, R.V. Singular points of intensity streamlines in two-dimensional sound fields. *J. Acoust. Soc. Am.* **1997**, *101*, 705–712. [CrossRef]
38. Waterhouse, R.V.; Yates, T.W.; Feit, D.; Liu, Y.N. Energy streamlines of a sound source. *J. Acoust. Soc. Am.* **1985**, *78*, 758–762. [CrossRef]
39. Zhe, J. Intensity streamlines and vorticity streamlines in three-dimensional sound fields. *J. Acoust. Soc. Am.* **2000**, *107*, 725–730. [CrossRef]
40. Brooke, G.H.; Thomson, D.J. Non-local boundary conditions for high-order parabolic equation algorithms. *Wave Motion* **2000**, *31*, 117–129. [CrossRef]
41. Porter, M.B. Acoustics Toolbox. 2020. Available online: https://oalib-acoustics.org/AcousticsToolbox/index_at.html (accessed on 15 October 2021).
42. Bonnel, J.; Flamant, J.; Dall'Osto, D.R.; Le Bihan, N.; Dahl, P.H. Polarization of ocean acoustic normal modes. *J. Acoust. Soc. Am.* **2021**, *150*, 1897–1911. [CrossRef]
43. Hope, G.; Schmidt, H. A parallelization of the wavenumber integration acoustic modelling package OASES. *Comput. Geosci.* **2019**, *23*, 777–792. [CrossRef]
44. Liu, W.; Zhang, L.; Wang, W.; Wang, Y.; Ma, S.; Cheng, X.; Xiao, W. A three-dimensional finite difference model for ocean acoustic propagation and benchmarking for topographic effects. *J. Acoust. Soc. Am.* **2021**, *150*, 1140–1156. [CrossRef]

45. LeVeque, R.J. *Finite Difference Methods for Ordinary and Partial Differential Equations;* Society for Industrial and Applied Mathematics (SIAM): Philadelphia, PA, USA, 2007; p. 342.
46. Lele, S.K. Compact finite difference schemes with spectral-like resolution. *J. Comput. Phys.* **1992**, *103*, 16–42. [CrossRef]
47. Pan, G.; Abubakar, A.; Habashy, T.M. An effective perfectly matched layer design for acoustic fourth-order frequency-domain finite-difference scheme. *Geophys. J. Int.* **2012**, *188*, 211–222. [CrossRef]

Journal of
Marine Science and Engineering

MDPI

Article

A New Compression and Storage Method for High-Resolution SSP Data Based-on Dictionary Learning

Kaizhuang Yan, Yongxian Wang * and Wenbin Xiao

College of Meteorology and Oceanography, National University of Defense Technology, Changsha 410073, China
* Correspondence: yxwang@nudt.edu.cn

Abstract: The sound speed profile data of seawater provide an important basis for carrying out underwater acoustic modeling and analysis, sonar performance evaluation, and underwater acoustic assistant decision-making. The data volume of the high-resolution sound speed profile is vast, and the demand for data storage space is high, which severely limits the analysis and application of the high-resolution sound speed profile data in the field of marine acoustics. This paper uses the dictionary learning method to achieve sparse coding of the high-resolution sound speed profile and uses a compressed sparse row method to compress and store the sparse characteristics of the data matrix. The influence of related parameters on the compression rate and recovery data error is analyzed and discussed, as are different scenarios and the difference in compression processing methods. Through comparative experiments, the average error of the sound speed profile data compressed is less than 0.5 m/s, the maximum error is less than 3 m/s, and the data volume is about 10% to 15% of the original data volume. This method significantly reduces the storage capacity of high-resolution sound speed profile data and ensures the accuracy of the data, providing technical support for efficient and convenient access to high-resolution sound speed profiles.

Keywords: dictionary learning; sparse representation; sound speed profile; compressed storage

Citation: Yan, K.; Wang, Y.; Xiao, W. A New Compression and Storage Method for High-Resolution SSP Data Based-on Dictionary Learning. *J. Mar. Sci. Eng.* **2022**, *10*, 1095. https://doi.org/10.3390/jmse10081095

Academic Editors: Mikhail Emelianov and Georgy I. Shapiro

Received: 30 June 2022
Accepted: 4 August 2022
Published: 10 August 2022

1. Introduction

The ocean has been an important strategic resource for a long time, and countries worldwide are paying increasing attention to research on the marine environment. In the field of marine acoustic application technology, the sound speed in seawater is an environmental parameter that has an important impact on sound propagation in the ocean. Research on the mechanism of the underwater sound field, the detection of underwater acoustic targets, and the calculation and evaluation of underwater sound performance all need to be based on sound speed profile (SSP) data. The marine environment is treacherous and variable, and the sound speed in adjacent seas varies greatly. Low-resolution SSP data cannot meet the accuracy requirements of marine acoustic research. High-resolution SSP data can not only improve the accuracy of marine acoustic applications but also provide strong guarantees for the characterization of the marine environment.

The volume of high-resolution SSP data is large, and the data size at a single data sampling time is on the order of GB. The global sea area is vast and deep, so storing the SSP data of the whole sea places high requirements on the data storage space, which constrains large-scale ocean acoustic research based on the SSP information of the global sea area and is not conducive to using high-resolution SSP data. Therefore, while compressing the storage space required for data and ensuring the accuracy of data recovery, it is necessary to find a reliable and efficient high-resolution SSP data storage method so that the mass data can better meet the needs on standard laptop or desktop computers.

In recent years, dictionary learning [1] has been widely used in the field of sound velocity profile data processing as a feature extraction method. In 2018, Michael Bianco [2] used dictionary learning to extract features and reconstruct ocean SSP data to model ocean

SSPs. Then, Kuo et al. [3] proposed range-dependent dictionary learning to retrieve SSPs in order to achieve classification and reconstruction; Sijia et al. [4] compared the modeling capabilities of EOF and dictionary learning to verify that dictionary learning has higher-precision reconstruction capabilities; Qianqian [5] used dictionary learning to invert the sound field of the SSP, reducing the error of the sound field inversion. Existing studies mainly focus on using dictionary learning to invert and reconstruct SSP data, but there is a lack of research and discussion on exploiting the sparse processing and compression capabilities of dictionary learning.

This paper proposes a method for compressing and storing high-resolution SSP data using dictionary learning. The sparse processing of the data is optimized through a dictionary composed of a non-orthogonal basis, and the high-resolution SSP is greatly compressed under the premise of ensuring the recovery accuracy. The method not only saves storage space, but also provides the possibility for the wide application of high-resolution SSP data.

2. Theory and Method

2.1. Mathematical Foundation of Dictionary Learning

A mathematical description of dictionary learning is shown as follows [6]: A data matrix $\mathbf{Y}_{m\times n} = (\mathbf{y}_1, \mathbf{y}_2, \cdots, \mathbf{y}_n) \in \mathbb{R}^{m\times n}$ can be decomposed into matrix $\mathbf{D}_{m\times k} = (\mathbf{d}_1, \mathbf{d}_2, \cdots, \mathbf{d}_k) \in \mathbb{R}^{m\times k}$ and matrix $\mathbf{X}_{k\times n} = (\mathbf{x}_1, \mathbf{x}_2, \cdots, \mathbf{x}_n) \in \mathbb{R}^{k\times n}$.$\mathbf{D}$ is a dictionary matrix, consisting of $k(k < n)$ normalized column vectors (also called atoms); $\mathbf{X}$ is a sparse matrix with a sparsity T (the number of non-zero elements in a column vector of the sparse matrix). In order to make the matrix $\mathbf{X}$ more sparse, $\mathbf{Y}$ can be approximately represented by $\mathbf{Y} \approx \mathbf{DX}$, satisfying $\|\mathbf{Y} - \mathbf{DX}\|_F^2 < \epsilon$ [7]. Thus, the goal of dictionary learning is to solve the dictionary $\mathbf{D}$ under the limited sparsity T, so that ϵ reaches the minimum value. The dictionary learning processor is written as

$$\min_{\mathbf{D},\mathbf{X}} \|\mathbf{Y} - \mathbf{DX}\|_F^2, \ \ s.t. \ \forall i, \|x_i\|_0 \leq T, \tag{1}$$

where $\|\mathbf{x_i}\|_0$ is the number of non-zero elements of a vector $\mathbf{x}_i$. T is a preset sparsity. The ultimate goal is to minimise the residual, $\|\mathbf{Y} - \mathbf{DX}\|_F^2$, where the error is described by the F-norm of the residual matrix, where the F-norm of any matrix $\mathbf{A}_{m\times n}$ is defined as

$$\|\mathbf{A}\|_F = \sqrt{\sum_{\mathrm{i}=1}^{\mathrm{m}} \sum_{j=1}^{\mathrm{n}} |a_{ij}|^2}, \tag{2}$$

The constraint condition to be satisfied for optimisation is $\|\mathbf{x}_i\|_0 \leq T$. Since there are two unknowns in the optimisation problem, namely, the dictionary matrix $\mathbf{D}$ and the sparse matrix $\mathbf{X}$, it is necessary to fix one variable, solve the other variable, and then alternately optimise. Thus, dictionary learning is generally divided into two steps: sparse coding and dictionary update.

2.1.1. Sparse Coding

In Equation (1), when the dictionary matrix $\mathbf{D}$ is determined, the sparse matrix $\mathbf{X}$ is solved. This process is defined as sparse coding [8]. Sparseness is reflected in the fact that each vector $\mathbf{x}_i$ has only a few non-zero elements. Sparse coding can be solved by algorithms such as matching pursuit(MP) and basis pursuit(BP) [9]. In this paper, sparse coding is realised by the orthogonal matching pursuit (OMP) algorithm.

The OMP [10] algorithm uses the atomic selection method of the MP algorithm, which has a faster convergence rate. In this algorithm, the residual vector always remains orthogonal to the selected dictionary atom, so it is guaranteed that the same atom will not be reused, and the result can converge in a finite time. To achieve the effect of orthogonality, the Schmidt orthogonal method needs to be introduced to calculate the residuals [11].

The residual $\mathbf{r}$ of the original vector $\mathbf{y}_i$ can be calculated as

$$\mathbf{r} = \mathbf{y}_i - \mathbf{A}(\mathbf{A}^T\mathbf{A})^{-1}\mathbf{A}^T\mathbf{y}_i, \tag{3}$$

where matrix $\mathbf{A}$ is made of k different atoms that have been selected from the dictionary matrix $\mathbf{D}$. In summary, the algorithm for OMP is as follows:

1. Set iteration time $j = 1$, residual $\mathbf{r}_{j-1} = \mathbf{y}_i$, preset sparsity T, dictionary matrix $\mathbf{D}_{m\times k} = (\mathbf{d}_1, \mathbf{d}_2, \cdots, \mathbf{d}_k)$, index sequence vector $\mathbf{Idx} = 0$, matrix $\mathbf{A} = \varnothing$ and original vector $\mathbf{y}_i$.
2. Compute $\mathbf{p} = \mathbf{D}^T\mathbf{r}_{j-1}$, select the element with the largest absolute value in $\mathbf{p}$, and denote its row number as *pos* and add it to $\mathbf{Idx}$.
3. Add $\mathbf{d}_{pos}$ to the new last column of matrix $\mathbf{A}$, and compute the sparse coefficient vector $\mathbf{s} = (\mathbf{A}^T\mathbf{A})^{-1}\mathbf{A}^T\mathbf{y}_i$.
4. Compute the residual $\mathbf{r}_j = \mathbf{y}_i - \mathbf{A}\mathbf{s}$. If $\mathbf{r}_j$ meets the accuracy requirements or $k = T$, stop and generate $\mathbf{x}_i$ by using the elements in $\mathbf{Idx}$ as the row index to fill in the values in $\mathbf{s}$ into the corresponding rows of the vector $\mathbf{x_i}$ in turn, and filling all the remaining positions with 0. Otherwise, $j = j + 1$, and go to step 2.

2.1.2. Dictionary Update

Dictionary update refers to the process of obtaining a better dictionary under the premise of a fixed sparse representation. K-single value decomposition (K-SVD) is a commonly used dictionary update algorithm [7,12]. It mainly uses singular value decomposition (SVD) to sequentially optimize K-th dictionary atoms and coefficients.

Assuming the sparse matrix $\mathbf{X}$ is known, update the dictionary column by column. When updating the k-th column of the dictionary, denote $\mathbf{x}_T^k$ as the k-th row vector of the sparse matrix $\mathbf{X}$. $\|\mathbf{Y} - \mathbf{D}\mathbf{X}\|_F^2$ can be written as

$$\begin{aligned}\|\mathbf{Y} - \mathbf{D}\mathbf{X}\|_F^2 &= \left\|\mathbf{Y} - \sum_{j=1}^{K}\mathbf{d}_j\mathbf{x}_T^j\right\|_F^2 \\ &= \left\|\left(\mathbf{Y} - \sum_{j\neq k}\mathbf{d}_j\mathbf{x}_T^j\right) - \mathbf{d}_k\mathbf{x}_T^k\right\|_F^2 \\ &= \left\|\mathbf{E}_k - \mathbf{d}_k\mathbf{x}_T^k\right\|_F^2,\end{aligned} \tag{4}$$

where

$$\mathbf{E}_k = \mathbf{Y} - \sum_{j\neq k}\mathbf{d}_j\mathbf{x}_T^j. \tag{5}$$

Thus, the optimization problem can be described as

$$\min_{\mathbf{d}_k,\mathbf{x}_T^k}\left\|\mathbf{E}_k - \mathbf{d}_k\mathbf{x}_T^k\right\|_F^2. \tag{6}$$

Equation (6) is a least squares problem. To obtain appropriate $\mathbf{x}_T^k$ and $\mathbf{d}_k$, the SVD method [13] is introduced here to solve the problem, as shown in

$$\mathbf{E}_k = \mathbf{U}(\mathbf{\Sigma}\mathbf{V}^\mathbf{T}). \tag{7}$$

$\mathbf{U}$ is composed of the eigenvectors of matrix $\mathbf{A}\mathbf{A}^\mathbf{T}$. $\mathbf{\Sigma}$ is a singular value matrix, where the diagonals are singular values σ and the rest are 0. $\mathbf{\Sigma}^2$ is eigenvalues of the matrix $\mathbf{A}^\mathbf{T}\mathbf{A}$ and σ takes the positive square root. $\mathbf{V}$ is composed of the eigenvectors of matrix $\mathbf{A}^\mathbf{T}\mathbf{A}$. The obtained $\mathbf{U}$ is updated into the dictionary matrix $\mathbf{D}$ as a new $\mathbf{d}_k$, and $\mathbf{\Sigma}\mathbf{V}^\mathbf{T}$ is updated into the sparse matrix $\mathbf{X}$ as a new $\mathbf{x}_T^k$. Update each $\mathbf{x}_T^k$ by using SVD and keep $\mathbf{E}_k$ shrinking until it meets the error threshold or the iteration number t. Error threshold can be set according to the required accuracy, whose calculation can refer to Equation (10).

2.2. Method and Process of 3D Ocean Sound Speed Data Compression

The ocean SSP is a vertical section formed by the sound speed in the sea changing with depth. Spatially, ocean SSPs are different at different latitude and longitude positions. The SSPs under the same latitude and longitude also vary according to the time interval. At the same time, the denser the latitude and longitude grid points are, the higher the resolution of the SSP data, and the more accurate the data. Therefore, global SSP data generally are large scale and high resolution and involve long-term series. According to these typical characteristics, dictionary learning is naturally introduced to compress high-resolution global SSP data to save storage space [14].

First, we select a part from the SSP data as the training sample and then randomly select a certain amount of data from the sample to form the initial **D**. The dictionary learning method is used to train and update the **D**, and obtain a new $\mathbf{D}_{new}$ that meets the error threshold or reaches the iteration time t which can be freely selected according to the actual compression requirements, and, generally, should be lower than 10. Finally, use $\mathbf{D}_{new}$ to sparse the original data to obtain **X** by OMP to realize the compression of the effective data of the original SSP and store the sparse matrix through a compressed sparse row (CSR), which is one of the mainstream compressed storage methods for the sparse matrix. The sparse coding and compression storage process of the general ocean SSP are shown in Figure 1.

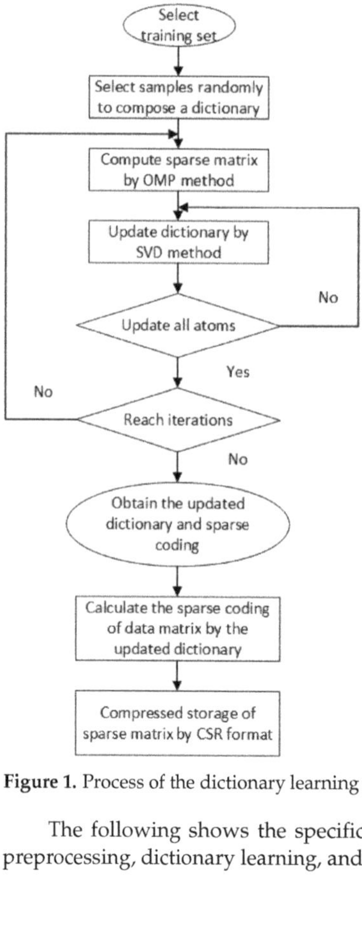

Figure 1. Process of the dictionary learning compression storage.

The following shows the specific implementation steps from three aspects: data preprocessing, dictionary learning, and CSR storage.

2.2.1. Data Preprocessing

The experimental data come from the "Global High-Resolution Ocean Reanalysis System Development and Product Development" project. The sound speed in the ocean is a function of temperature, salinity, and pressure. Use the empirical formula of the SSP to calculate global high-resolution ocean SSP data, according to the longitude, latitude, ocean temperature, salinity, and water depth data of the global seas [15] provided by the global high-resolution ocean reanalysis products.

First, the annual ocean SSP data of each longitude and latitude point are processed separately. For the annual ocean SSP data of a certain latitude and longitude, we combine the sound speed data corresponding to different depths at a single sampling time point into a column vector and then sort by column using the sampling time as a clue to obtain the annual high resolution corresponding to a single latitude and longitude point SSP matrix. Then, all the obtained latitude and longitude point matrix blocks are sorted by column to obtain the high-resolution SSP matrix of the global ocean area. Denote y_{ij} as the sound speed at the jth time point at the ith layer depth. Data matrix construction is shown in Figure 2.

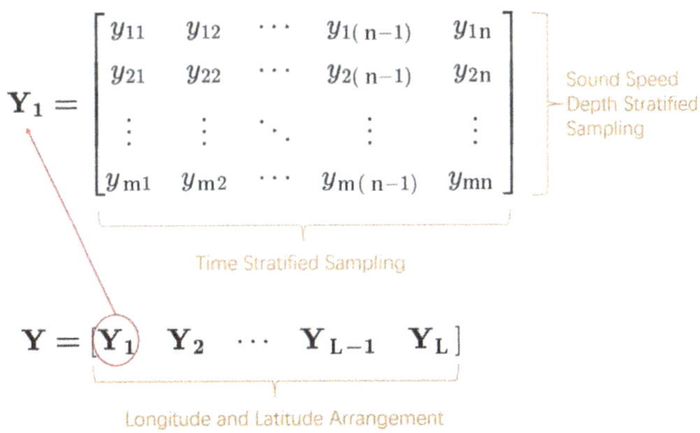

Figure 2. Data matrix construction.

2.2.2. Dictionary Learning

To better train the dictionary, the global sea area high-resolution SSP matrix $\mathbf{Y}_{m\times n}$ is used as the training sample. We randomly select k samples from it to form the initial dictionary $\mathbf{D}_{m\times k}$, where k is generally between $0.02n$ and $0.2n$. Based on the initial dictionary and training samples, we set an appropriate sparsity T which can be selected according to the required compression ratio, generally between 2 and $0.1m$. Use the OMP algorithm as described in Section 2.1.1 to sparsely encode the training samples [16].

The K-SVD algorithm as described in Section 2.1.2 is used to update the initial overcomplete dictionary column by column, and the residual is calculated. If the residual exceeds the set threshold, OMP algorithm and K-SVD algorithm are repeatedly used to iterate until the error threshold is met or iteration time t is reached. After dictionary learning, the trained dictionary matrix $\mathbf{D}_{m\times k}$ and the corresponding global high-resolution SSP data sparse matrix $\mathbf{X}_{k\times n}$ are obtained.

2.2.3. CSR Storage

The CSR storage method is a method for efficiently storing sparse matrices, which is called row format storage. When storing a sparse matrix of order k in the CSR format, assuming that the sparse matrix $\mathbf{X}_{k\times n}$ has a total of Tn non-zero elements, the non-zero elements can be stored in the row vector H_3 in the order of row first. Then, the column numbers corresponding to the non-zero elements of the sparse matrix in the row vector H_2 are stored in sequence, and the position of the first non-zero element in each row in the row vector H_3 in the progressive row vector H_1 is stored. The last bit stores the total number of non-zero elements. Take $\mathbf{X}_{k\times n}$ as an example in Figure 3, where $k = 3$, $n = 4$, $T = 2$.

$$\mathbf{X_{3\times 4}} = \begin{bmatrix} x_{11} & 0 & x_{13} & 0 \\ 0 & x_{22} & x_{23} & x_{24} \\ 0 & x_{32} & 0 & x_{34} \end{bmatrix}$$
$$\mathbf{H_1} = \begin{bmatrix} 1 & 3 & 6 & 7 \end{bmatrix}$$
$$\mathbf{H_2} = \begin{bmatrix} 1 & 3 & 2 & 3 & 4 & 2 & 4 \end{bmatrix}$$
$$\mathbf{H_3} = \begin{bmatrix} x_{11} & x_{13} & x_{22} & x_{23} & x_{24} & x_{32} & x_{34} \end{bmatrix}$$

Figure 3. Data matrix construction.

The ratio δ_{CSR} of the amount of CSR stored data to the amount of stored data of the original sparse matrix is shown in

$$\delta_{CSR} = \frac{num(\mathbf{H}_1) + num(\mathbf{H}_2) + num(\mathbf{H}_3)}{num(\mathbf{X})} \leq \frac{2Tn + k + 1}{kn}. \tag{8}$$

$num(\cdot)$ is the number of elements in the matrix. The maximum number of $num(\mathbf{H}_2)$ and $num(\mathbf{H}_3)$ is Tn. The maximum number of $num(\mathbf{H}_1)$ is $k+1$. The number of $num(\mathbf{X})$ is kn, which is the number of data in the original sparse matrix. In Figure 3, δ_{CSR} is $\frac{2\times2\times4+3+1}{3\times4} = \frac{5}{3}$. Because T and k are similar, $\mathbf{X}_{3\times4}$ is not a sparse matrix and cannot take advantage of the storage advantages of CSR. When $1 \leq T \ll k$ and $1 < n$, $\delta_{CSR} \approx \frac{k+1}{kn} < 1$, a better compression storage effect can be achieved for the sparse matrix.

2.3. Performance Evaluation Method

For the compressed storage of high-resolution SSP data, not only the compression capacity but also the error size of the data recovery after compression must be considered to evaluate the reliability of the compression. Therefore, the compression performance is evaluated by the performance evaluation index of the compression ratio δ_{comp} and recovery data error.

2.3.1. Compression Ratio

Through dictionary learning, the dense data matrix $\mathbf{Y}_{m\times n}$ is converted into dictionary matrix $\mathbf{D}_{m\times k}$ and sparse matrix $\mathbf{X}_{k\times n}$, where $\mathbf{X}_{k\times n}$ is the sparse matrix trained when the sparsity is T, so the maximum number of non-zero elements is Tn. The compression ratio δ_{comp} can be expressed as the ratio of the sum of the data volume of the three row vectors stored in the compressed dictionary matrix and CSR form to the data volume of the original dense data matrix, as shown in

$$\delta_{comp} = \frac{num(\mathbf{H}_1) + num(\mathbf{H}_2) + num(\mathbf{H}_3) + num(\mathbf{D})}{num(\mathbf{Y})} \leq \frac{2Tn + k + 1 + mk}{mn}. \tag{9}$$

In the case of a large-scale data matrix, n is much larger than m, T, and k, and Equation (9) can be simplified to

$$\delta_{comp} \approx \frac{2T}{m}. \tag{10}$$

According to Equation (10), δ_{comp} mainly depends on the proportional relationship between T and m. The smaller $\frac{T}{m}$ is, the better the compression effect.

2.3.2. Recovered Data Error

The sparsity T is set as the threshold when generating the sparse matrix. Therefore, there is a certain error in the data after compression and storage of the high-resolution SSP data based on dictionary learning. The control error in a reasonable range is also an important criterion for compression performance considerations.

To better compare the size of the error, we choose the average absolute error and the maximum absolute error as the error measurement standard. The average absolute error is abbreviated as MAD, that is, the absolute value of the error of each point is summed and averaged, and the expression is shown in

$$MAD = \frac{1}{mn}\sum_{i=1}^{m}\sum_{j=1}^{n}\left|\mathbf{y}_{ij} - \mathbf{y}'_{ij}\right|. \tag{11}$$

where $\mathbf{y}'_{ij}$ are the elements in the restored data matrix $\mathbf{Y}'_{m\times n}$ obtained by multiplying the dictionary matrix $\mathbf{D}_{m\times k}$ and the sparse matrix $\mathbf{X}_{k\times n}$. Here, mn is the total number of elements contained in the matrix $\mathbf{Y}_{m\times n}$.

The maximum value of absolute error is the absolute value of the maximum error in all data. Since the average value of ocean sound velocity is approximately 1500 m/s, when the absolute error is less than 3 m/s, that is, the relative error is less than 0.2%, the accuracy requirements of most ocean SSPs are met. Therefore, in the process of dictionary learning and compression storage, it is necessary to monitor the average absolute error and the maximum absolute error to ensure that the result meets the set error range.

3. Numerical Results and Analysis

To demonstrate the compression performance of dictionary learning on the high-resolution SSP matrix and to explore the influence of different parameter selections on the δ_{comp} and MAD of this method, two locations of 10.95° N, 112.95° E and 0.95° N, 4.95° E were selected to form two different $\mathbf{Y}_1$. For the high-resolution SSP data of these coordinate points in the whole year of 2014, the sampling time interval is 3 h, so the matrix has a total of $n = 8 \times 365 = 2920$ sampling times and $m = 50$. The schematic diagram of the SSP of the two coordinate points is as follows: Figure 4.

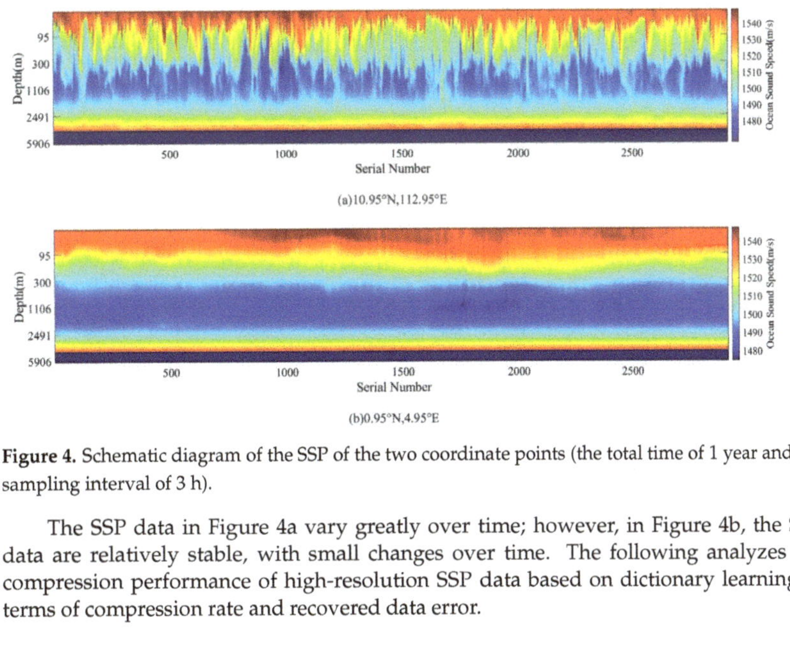

Figure 4. Schematic diagram of the SSP of the two coordinate points (the total time of 1 year and the sampling interval of 3 h).

The SSP data in Figure 4a vary greatly over time; however, in Figure 4b, the SSP data are relatively stable, with small changes over time. The following analyzes the compression performance of high-resolution SSP data based on dictionary learning in terms of compression rate and recovered data error.

3.1. *Influence of Each Parameter in the Algorithm*

3.1.1. Actual Impact of Parameters on the Compression Ratio

The two locations of SSP data are compressed by dictionary learning, and δ_{comp} under different T and different k are calculated using Equation (9). The results are shown in Figure 5.

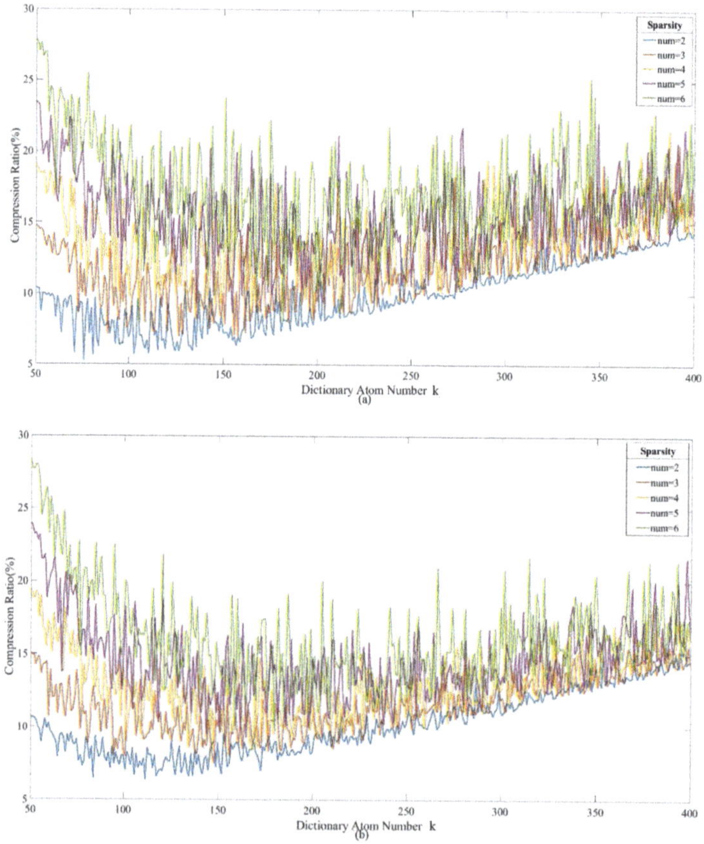

Figure 5. The effect of dictionary atom number k and sparsity T on compression ratio δ_{comp} ((**a**) is 10.95° N, 112.95° E; (**b**) is 0.95° N, 4.95° E).

Figure 5 is the compression result of dictionary learning with two selected samples. The figure shows the influence of dictionary atom number k and the sparsity T on the compression ratio δ_{comp} of high-resolution ocean SSP data. In Figure 5, T significantly affects δ_{comp}, when k is the same; the smaller T is, the smaller δ_{comp}, and the better the compression performance. When T is the same, the minimum of δ_{comp} is mainly concentrated in $k \in [100, 150]$. Since the SSP Figure 5a fluctuates greatly, δ_{comp} varies greatly, while, in Figure 5b, the SSP changes gently, so δ_{comp} is relatively stable.

3.1.2. Actual Influence of Parameters on the Error of Recovered Data

Recovery data error tests are performed on the two selected samples, Equation (11) is used to calculate MAD under different T and k, and Figure 6 is obtained.

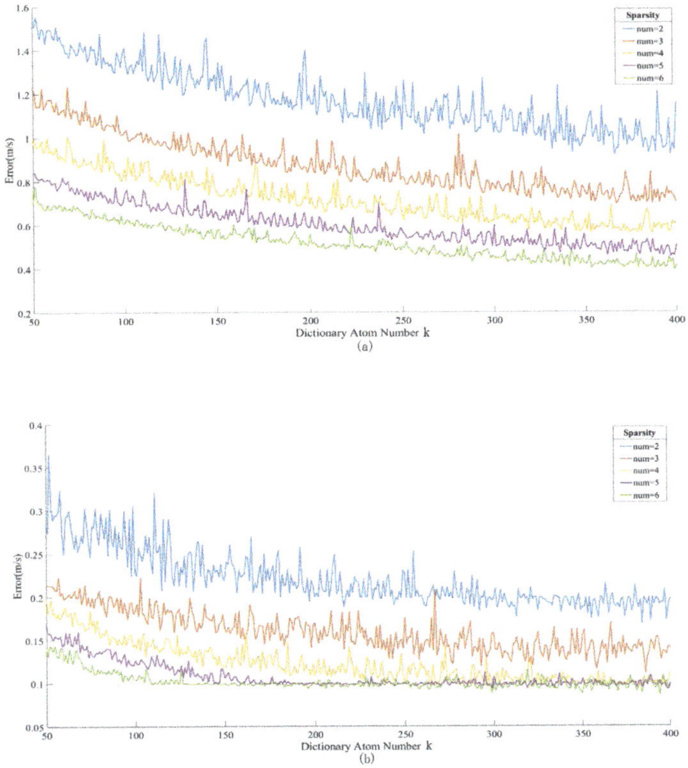

Figure 6. The effect of dictionary atom number k and sparsity T on Error MAD ((**a**) is 10.95° N, 112.95° E; (**b**) is 0.95° N, 4.95° E).

Figure 5 shows the influence of dictionary atom number k and the sparsity T on the error MAD. The MAD shows a downward trend as k increases when T is the same. When k is the same, the smaller T is, the larger MAD is. Since the SSP of Figure 6a varies greatly, MAD of Figure 6a is larger than that of Figure 6b under the same conditions. When $k > 100$, MAD is basically stable.

3.1.3. Selection of Compression Performance Parameters

Through comparative experiments of more samples, the following conclusions can be drawn:

When only focusing on δ_{comp}, a small T should be selected, and the selection range of k is in the range of 0.03–$0.05n$. When MAD is concerned, a larger T should be chosen, and the selection range of k should be larger than $0.03n$. Thus, the choice of T depends on the balance between δ_{comp} and MAD.

Therefore, before using dictionary learning to compress high-resolution ocean SSP data, it is necessary to analyze the original data scale n and select an appropriate k and T according to application requirements and scenarios to ensure that both compression ratio and recovery accuracy are taken into account.

When the ocean SSP data n are large, due to the limitation of computer calculation performance, k is selected according to the corresponding calculation matrix ability. The larger k is, the more accurate the final recovered data to better ensure that the choice of δ_{comp} and T is related to m. When the t is approximately 0.02–$0.1n$, it is easier to train a good dictionary matrix and ensure the low MAD and high δ_{comp} of the data.

3.2. Impact of Data Matrix Organization on Performance

In the actual application process, the depths of various sea areas around the world are different, resulting in different numbers of sample elements of the formed high-resolution SSP matrix and different n of the data matrix, which makes it difficult to directly compress and store dictionary learning. Therefore, this paper proposes two data matrix construction schemes and implementation methods to solve the problem of matrix scale mismatch in the whole sea area. The selected experimental data samples are ocean data of a certain area of the Pacific Ocean (400 km $\times$ 400 km) and converted into SSP data using the empirical sound speed formula. The experimental data have a total of 1600 latitude and longitude points, each of which has 2920 time sampling points throughout the year, and the maximum depth is 5906 m.

3.2.1. Spatio-Temporal Structure

The spatio-temporal combination structure is a data matrix construction method that arranges spatio-temporal information in an orderly manner to form a super-large-scale matrix. That is, the size of the generated matrix is 50 $\times$ 4,672,000. Due to the large number of samples, the method of selecting a training set to train a complete dictionary is used to obtain a dictionary matrix that matches the data matrix. This method is suitable for use on machines with high computing performance and large memory capacity.

A typical SSP is divided into three layers: a mixed layer, a maincline layer, and a deep-water isothermal layer. In the deep-water isothermal layer, the sound speed increases with a positive gradient. The analysis shows that the data of this sea area have reached the respective deep-water isothermal layer. Therefore, they can be interpolated and extended according to their different changing trend to make them unified to 5906 m, which satisfies the largest 50-layer data structure of the matrix, and thus a high-resolution SSP matrix with a scale of 50 $\times$ 4,672,000 can be obtained.

We randomly sample this matrix to ensure robustness of the process, select n = 10,000 samples to form the training set, and select $k = 2000$ to form the initial dictionary for training. According to that described in Section 3.1, we set $T = 3$ to better obtain compression gains and error data analysis. After the shrinking is complete, we restore the data as described in Section 2.2 for comparison, as shown in Figure 7.

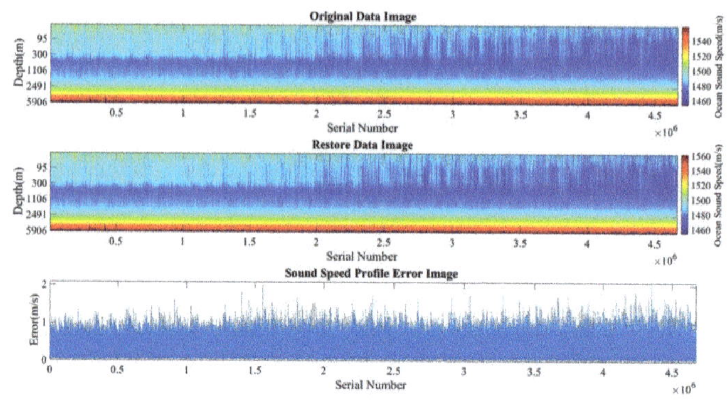

Figure 7. Performance of dictionary learning compression under spatial discrete structure ($k = 100$, $T = 3$).

Table 1 shows seven sets of experimental data of dictionary learning compression performance parameters under the spatio-temporal combination structure.

Under $n = 10,000$, $k = 2000$, and $T = 3$, δ_{comp} of using dictionary learning to compress and store the high-resolution ocean SSP matrix is stable at approximately 12%, MAD is less than 0.5 m/s, the maximum error does not exceed 3 m/s, and the relative error is approximately 0.2%, which meets the accuracy requirements of the SSP data.

Table 1. Dictionary learning compression performance parameter under the spatio-temporal combination structure.

Average Error (m/s)	Maximum Error (m/s)	Compression Rate (%)
0.4578	2.7455	13.45
0.4871	2.3451	13.22
0.4534	2.2457	12.47
0.4965	2.9874	11.87
0.4658	2.9122	13.45
0.4178	2.4257	12.49
0.4844	2.5567	11.51

3.2.2. Spatial Discrete Structure

When computing performance and memory is low, the spatial discrete structure can be used. The spatial discrete structure uses dictionary learning to compress all latitude and longitude points, which solves the problem of the inability to establish a unified matrix due to the different depths of each point in SSP data. At the same time, since the change of SSP at the same location is much smaller than change of different locations, separate dictionary learning for each latitude and longitude point can obtain a matrix dictionary that is more in line with the characteristics of the sample and improve the recovery accuracy of large-scale high-resolution ocean SSP data. Due to the small amount of data, it is not necessary to distinguish the training set and the test set to update the dictionary for all samples.

We perform dictionary learning with $T = 3$ for 1600 latitude and longitude points and select $K = 100$. The effect is shown in Figure 8.

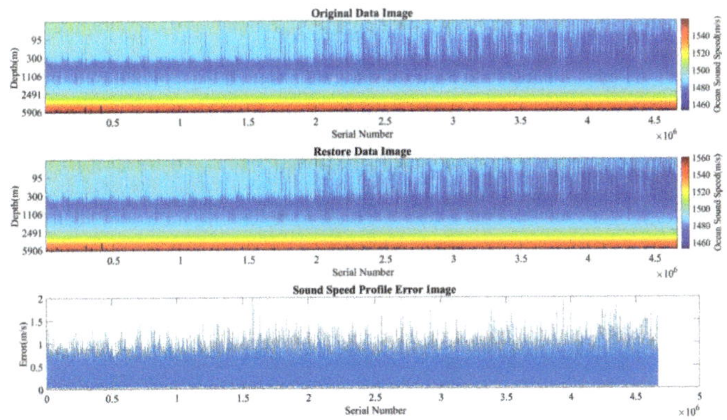

Figure 8. Performance of dictionary learning compression under spatial discrete structure ($k = 100$, $T = 3$).

Table 2 shows seven sets of experimental data of dictionary learning compression performance parameter under the spatial discrete structure.

Table 2. Dictionary learning compression performance parameter under the spatial discrete structure.

Average Error (m/s)	Maximum Error (m/s)	Compression Rate (%)
0.3745	1.9545	16.28
0.3544	1.7253	15.11
0.3578	1.9877	15.87
0.3978	1.8878	15.23
0.3147	1.9652	14.78
0.3799	1.8532	15.26
0.3783	1.4558	13.97

Under n = 10,000, $k = 2000$, and $T = 3$, δ_{comp} of compressed storage is stable at approximately 15%, MAD is less than 0.4 m/s, the maximum error does not exceed 2 m/s, the relative error is approximately 0.13%, and the data recovery accuracy is high. This method has a smaller error than the spatio-temporal structure. However, each point has a corresponding dictionary matrix, the storage data are increased, and δ_{comp} is affected to a certain extent.

3.2.3. Performance Comparison

Comparing the two data matrix structures, spatio-temporal structure has the advantages of being large scale and having a high compression rate, while spatial discrete structure has high accuracy, small error, and simple operation.

Spatio-temporal structure, combining the SSP data of all points into a whole, can be trained to obtain a unique dictionary matrix that matches original data. This operation is more concise when recovering data, and the storage resources occupied by the dictionary are greatly reduced; however, because of the matrix, the scale is vast, and substantial memory resources and computing resources are consumed when calculating singular value decomposition. The requirements for computer system configuration are relatively high, and the calculation time is relatively long.

For spatial discrete structure, dictionary learning is performed on the SSP data of each latitude and longitude point, and the corresponding dictionary matrix is generated in turn. Because the size of a single SSP is small, the difference due to compression is not large, and it is easier to generate a dictionary matrix with better performance and reduced error. The error is greatly reduced, the computer configuration requirements are not high, and the calculation time is shorter; however, because the number of dictionary matrices is determined by the number of latitude and longitude points, the relative space of the dictionary matrix occupies more storage space, and the compression rate increases slightly.

4. Conclusions

The data storage and compression of global high-resolution ocean SSPs have high scientific research value and play an important role in exploring ocean acoustic information and developing acoustic applications. The dictionary learning-based high-resolution ocean SSP compression storage method realizes data dimensionality reduction through dictionary learning, transforms a large-scale dense data matrix into a combination of a small-scale matrix and a large-scale sparse matrix, retains the changing trend of the SSP information, ensures a higher compression rate, and achieves precision control, providing a more reliable data compression storage method for high-resolution SSPs.

In dictionary learning, reasonable control of sparsity and dictionary scale can effectively control the compression rate and recovery accuracy of data after dictionary learning. According to different application backgrounds and application requirements, different parameters are selected for data processing, which increases the controllability and flexibility of the method. When the accuracy requirements are not high, for smaller δ_{comp}, the spatio-temporal structure can be adopted. On the contrary, when the precision requirements are high, the spatial discrete structure should be chosen.

The compression of high-resolution ocean SSP data through dictionary learning is mainly based on the sparse coding ability of dictionary learning. Some characteristics and laws of the dictionary itself have not been deeply explored. Further in-depth study of dictionaries is required.

Author Contributions: Conceptualization, Y.W. and K.Y.; methodology, Y.W. and K.Y.; software, Y.W. and K.Y.; validation, Y.W. and W.X.; formal analysis, W.X.; investigation, K.Y.; resources, W.X.; data curation, K.Y.; writing—original draft preparation, Y.W. and K.Y.; writing—review and editing, Y.W. and K.Y.; visualization, W.X.; supervision, K.Y.; project administration, Y.W. and K.Y.; funding acquisition, Y.W. All authors have read and agreed to the published version of the manuscript.

Funding: This research was funded by the National Key Research and Development Program of China, Grant No. 2016YFC1401800, and the National Natural Science Foundation of China, Grant Nos. 61379056 and 61972406.

Institutional Review Board Statement: Not applicable.

Informed Consent Statement: Not applicable.

Data Availability Statement: Data available in a publicly accessible repository that does not issue DOIs. Publicly available datasets were analyzed in this study. This data can be found here: http://mds.nmdis.org.cn.

Conflicts of Interest: The authors declare no conflict of interest.

References

1. Tosic, I.; Frossard, P. Dictionary learning. *IEEE Signal Process. Mag.* **2011**, *28*, 27–38. [CrossRef]
2. Bianco, M.; Gerstoft, P. Dictionary learning of sound speed profiles. *J. Acoust. Soc. Am.* **2017**, *141*, 1749–1758. [CrossRef] [PubMed]
3. Kuo, Y.; Kiang, J. Estimation of range-dependent sound-speed profile with dictionary learning method. *IET Radar Sonar Navig.* **2020**, *14*, 194–199. [CrossRef]
4. Sijia, S.; Hangfang, Z. Sparse Representation of Sound Speed Profiles Based on Dictionary Learning. In Proceedings of the 2020 13th International Congress on Image and Signal Processing, BioMedical Engineering and Informatics, Chengdu, China, 17–19 October 2020.
5. Qianqian, L.; Khan, S.; Fanlin, Y. Compressive acoustic sound speed proile estimation in the Arabian Sea. *J. Acoust. Soc. Am.* **2020**, *43*, 603–620.
6. Pali, M.; Rurtz, S.; Schnass, K. Average performance of OMP and thresholding under dictionary mismatch. *IEEE Signal Process. Lett.* **2022**, *29*, 1077–1081. [CrossRef]
7. Aharon, M.; Elad, M.; Bruckstein, A. K-SVD: An algorithm for designing overcomplete dictionaries for sparse representation. *IEEE Trans. Signal Process.* **2006**, *54*, 4311–4322. [CrossRef]
8. Rubinstein, R.; Bruckstein, A.; Elad, M. Dictionaries for sparse representation modeling. *Proc. IEEE* **2010**, *98*, 1045–1057. [CrossRef]
9. Mun, S.; Yazid, H. Performance analysis on dictionary learning and sparse representation algorithms. *Multimed. Tools Appl.* **2022**, *81*, 16455–16476.
10. Pati, Y.C.; Rezaiifar, R.; Krishnaprasad, P.S. Orthogonal matching pursuit: Recursive function approximation with applications to wavelet decomposition. In Proceedings of the 27th Asilomar Conference on Signals, Systems and Computers, Pacific Grove, CA, USA, 1–3 November 1993; pp. 40–44.
11. Huang, K.; Aviyente, S. Sparse representation for signal classification. *Adv. Neural Inf. Process. Syst.* **2007**, *19*, 609.
12. Zhang, Q.; Li, B. Discriminative K-SVD for dictionary learning in face recognition. In Proceedings of the IEEE Conference on Computer Vision and Pattern Recognition, San Francisco, CA, USA, 13–18 June 2010; pp. 2691–2698.
13. Chen, S.; Donoho, D.; Saunders, M. Atomic decomposition by basis pursuit. *SIAM J. Sci. Comput.* **1999**, *20*, 33–61. [CrossRef]
14. Shafiee, S.; Kamangar, F. The Role of Dictionary Learning on Sparse Representation-Based Classification. In Proceedings of the PETRA '13: Proceedings of the 6th International Conference on PErvasive Technologies Related to Assistive Environments, Rhodes, Greece, 29–31 May 2013.
15. LeBlanc, L.R.; Middleton, F.H. An underwater acoustic sound velocity data model. *J. Acoust. Soc. Am.* **1980**, *67*, 2055–2062. [CrossRef]
16. Yang, M.; Zhang, L.; Feng, X.; Zhang, D. Fisher discrimination dictionary learning for sparse representation. In Proceedings of the 2011 International Conference on Computer Vision, Barcelona, Spain, 6–13 November 2011; pp. 543–550.

MDPI
St. Alban-Anlage 66
4052 Basel
Switzerland
www.mdpi.com

Journal of Marine Science and Engineering Editorial Office
E-mail: jmse@mdpi.com
www.mdpi.com/journal/jmse

www.ingramcontent.com/pod-product-compliance
Lightning Source LLC
Chambersburg PA
CBHW061458210926
51850CB00008B/1758